Zworykin, Pioneer of Television

Zworykin,
Pioneer of Television

ALBERT ABRAMSON

Foreword by Erik Barnouw

University of Illinois Press   Urbana and Chicago

© 1995 by the Board of Trustees of the University of Illinois
Manufactured in the United States of America
C 5 4 3 2 1

This book is printed on acid-free paper.

Library of Congress Cataloging-in-Publication Data

Abramson, Albert.
    Zworykin : pioneer of television / Albert Abramson ; foreword by
Erik Barnouw.
        p.   cm.
    Includes bibliographical references (p.   ) and index
    ISBN 0-252-02104-5
    1. Zworykin, V. K. (Vladimir Kosma), b. 1888. 2. Inventors—
United States—Biography. 3. Television—United States—Biography.
4. Television—History. I. Title.
TK6635.Z93A63   1995
621.388'0092—dc20
    [B]   94-7464
                                                             CIP

This book is dedicated in loving memory of my father,
Joseph D. Abramson

Contents

Foreword

Erik Barnouw

The process of inventing television, which began over a century ago, became a long-running serial drama full of twists and turns that often seemed to reach its climax only to confront us with the message "to be continued." The drama started in the pre-electronic 1880s with mechanical systems in which the scanning of images called for a rapidly revolving disc. For half a century this disc was the centerpiece of television technology. Meanwhile the appearance of the cathode-ray tube in 1897 inspired different visions, in which the scanning and shaping of images would be done by electrons in a vacuum tube. Experimenters in many countries began tackling this challenge, but the problems were so complex that it took decades before solutions appeared and electronic television displaced the mechanical systems.

Albert Abramson, in books and numerous articles, has made himself the tireless chronicler of this international technical saga. Combing through old patent files, studying laboratory exhibits, and interviewing pioneers, he has weighed claims and counterclaims. He assures us that no one individual was "the" inventor of television. He mentions a score of inventors, of diverse nationalities, who played a significant part. Though scattered, they were aware of each other's works, read the same technical journals, and studied each other's patents. The process was fiercely competitive but also collaborative. Abramson likens it to inventing "by committee." Inventors sometimes replicated each other's experiments and built on each other's achievements. In the experimental stage all this was feasible. Later, with commercial use, patent rights came into play.

Among the inventors two men, Vladimir Kosma Zworykin and Philo T. Farnsworth, have most often been cited as preeminent—on somewhat different grounds. Zworykin, the Russian-born scientist chosen in the late 1920s by David Sarnoff to create an RCA television system, piloted his staff steadily toward that goal. As RCA president, Sarnoff maintained the staff's budgets throughout the Depression, and sometimes bought the patents of other inventors—$50,000 here, $100,000 there—to ease the way for RCA television. Such sums helped many an inventor continue television experimentation while they ensured Sarnoff control over whatever system would emerge. The Sarnoff-Zworykin drive triumphed with demonstrations at the 1939 New York World's Fair, followed later by worldwide commercial successes.

The other inventor, Farnsworth, was a Mormon farm boy who first encountered electricity at the age of fourteen. He became an ardent reader of electrical journals, and one day astonished his high school physics teacher with a series of designs showing how he thought television could be achieved electronically. (The teacher's testimony would later play a part in patent hearings.) In 1926, at the age of twenty, Farnsworth found a backer and was set up in a modest California laboratory. In 1929 he gave the world's first demonstration of an all-electronic (that is, no moving parts) television system. He received a number of patents for his work, covering ideas that RCA wanted and needed, particularly relating to Farnsworth's "image dissector" tube. Sarnoff made a $100,000 offer, which Farnsworth declined, saying that RCA could have use of his ideas on a royalty basis only. This was against Sarnoff's policy: RCA didn't pay royalties, it collected them. In patent hearings, RCA's lawyers sought to demolish Farnsworth's claims, but he prevailed. When Sarnoff was ready to proceed with commercial deployment of RCA television, he could do so only by coming to terms with Farnsworth, which he did with an unprecedented cross-licensing agreement. However, in RCA public relations it remained a Sarnoff-Zworykin story, in which Farnsworth was forgotten. He had been intent on his patents, not on the razzle-dazzle of PR, but the industry he helped to create tended to elevate image-making over substance. Ill health and early death ended Farnsworth's extraordinary career.

Zworykin was unique too, and Abramson has solid reason to embark on a Zworykin biography. In telling Zworykin's story, he is giving us a thread that runs through the whole inventing saga. In the years before World War I Zworykin was a student at the St. Petersburg Technical Institute and assisted Boris L. Rozing, his physics teacher, in one of the earliest efforts to create television based on the cathode-ray tube. A two-year apprenticeship in Rozing's laboratory was followed by a year at the Collège de France, under Paul Langevin.

The outbreak of war brought Zworykin back to Russia for service in its signal corps and plunged him into military electronics. After the revolution, amid the chaos of civil war, he escaped from Russia and made his way to the United States. In Pittsburgh he found work at Westinghouse headquarters, where he was allowed to launch his own television experimentation. This in turn led to the summons from Sarnoff. As RCA's electronic research director, he stayed through black-and-white television, color television, and their applications in war, space, medicine, and other fields. His career bracketed sixty years of the serial drama.

No one will ever be able to solve all the mysteries and disputes in this epic story, but Abramson has tried to judge fairly between diverse claimants. The ramifications of some of the claims are discussed in his voluminous notes. Abramson is critical of RCA publicity and patent operations, which he feels often distorted the historical record.

In recounting his final meetings with Zworykin, who died in 1982, Abramson mentions an ironic detail. In the corner of Zworykin's living room stood a seldom-used twenty-one-inch RCA set. Zworykin was apparently dismayed over what poured from the instrument to which he had devoted his life—as were Farnsworth and even Sarnoff. They had all been sure that what they were creating would "change the world." It has, but to what end? That, too, is an unfinished story.

Preface

This book is the result of some fourteen years of intensive research into the history of television, probably the first major invention that was achieved "by committee," that is, by the efforts of hundreds of individuals, widely separated in time and space, who were prompted to produce a system of "seeing over the horizon." Everyone knew what the goal was; in my research I have concentrated on how the various pioneers of television tried to get there. For television, the great stumbling block was the absence of practical technology. It simply did not exist. The motion picture for example did not face that problem. Once perforated stripe film had been produced, the machinery to make it operate was well within the technological capabilities of the time, and in a few short years, motion pictures were in widespread use. This was not to be with television. Most of the ideas were some twenty-five to fifty years ahead of the means to fulfill them. This did not discourage the visionaries at all. There quickly arose two schools of thought: "mechanical" television based on the rotating Nipkow disc and "electric" television based on the Braun cathode-ray display tube. Interestingly enough, working examples of both systems were displayed quite early (1909) in a rather uneven contest. While the Nipkow disc produced the earliest viewable results, the Braun tube was steadily being improved, albeit for other uses than television.

While no one man can claim to be the "Father of Television" (whatever that means), there were many pioneers, each of whom made invaluable contributions to the total effort and deserve some share of that title. The development of television was a cumulative process, with varying degrees of success for the people involved. Most of them were just dreamers; they produced paper ideas, recorded in articles or patents, with no means of fulfillment. Many made one notable contribution and then faded from view. Others appeared repeatedly through television's long gestation period. In most cases, their ideas were ahead of the times, with no means to supplement the limited technology that prevented them from producing a practical device.

The joint effort that resulted in television was made by inventors all over the world. No one country enjoyed a monopoly on the process; in fact several countries had two, perhaps three, outstanding "favorite sons." England had Campbell Swinton, Baird, and Shoenberg; France had Belin and Holweck, Dauvillier and Rignoux; Germany had Nipkow, Dieckmann, and von Ardenne; Hungary had von Mihaly and Tihany; Russia had Rozing, Takaev, and Konstantinov; Japan had Takayanagi; and the United States had Zworykin, Jenkins, and Farnsworth.

Of the Americans, the two most important pioneers who shared (although not equally) the title of *inventor* of television were Vladimir Kosma Zworykin and Philo Taylor Farnsworth. Zworykin, who worked for both Westinghouse Electric and the Radio Corporation of America (RCA) has been given the glory and Farnsworth has largely been relegated to the dustbin of history. This was mainly due to the treatment that Farnsworth had received at the hands of the RCA patents and publicity departments, who claimed that RCA had virtually single-handedly invented television and refused (as I discovered during my research) to give any other inventor or company the slightest credit. Farnsworth has virtually become a nonperson because of their efforts. In this he was not alone, others such as Major Edwin Howard Armstrong also received the same harsh treatment. Yet some of Farnsworth's accomplishments were equal to and perhaps of greater consequence than those of Zworykin. Lately there has been a great effort by his wife, Elma Farnsworth, to remedy this situation. To a certain extent she has succeeded, and today there is more awareness of Farnsworth's contributions to television.

However, in spite of the significant work Farnsworth had done, the present-day system of television owes more to Zworykin. Because Farnsworth's camera tube (the image dissector) had no storage capability, it bore the seeds of its own destruction. His picture tube (the oscillite) never could compete with Zworykin's kinescope.

Farnsworth may have been first with an all-electric (no moving parts) system of television, yet neither his camera tube nor his original picture tube survived. Zworykin's camera tube (the iconoscope) and picture tube (the kinescope) are clearly the forerunners of modern camera and picture tube technology.

In spite of my sympathy for Farnsworth and appreciation of what he had accomplished, my interest in Zworykin was whetted. My research into the history of television convinced me that indeed Zworykin, with the full backing of David Sarnoff, had been the single most motivating force behind the system of television that is in universal use today.

Yet his efforts have never been fully documented. I was quite surprised to find that of all the major television pioneers, Zworykin had not had a book written about him. This bothered me considerably. Baird of England has had at least five books written about him; Farnsworth, Belin, Jenkins, von Ardenne at least one each. So, in the July 1981 issue of the *Journal of the Society of Motion Picture and Television Engineers,* I published a well-received biography of Zworykin that featured his technical achievements. Part of the material came from a manuscript that Zworykin himself had given me several years earlier.

In May 1976, I had written Zworykin requesting certain information about his work beyond what was recorded in the literature and with the Patent Office. He replied that he was coming to Los Angeles in July 1976 and promised to bring with him certain documentation that might be of interest to me.

We met on July 22, at a rented cottage in Naples, California, and had a pleasant talk. At the end of the interview, which he would not allow me to tape, he offered me a set of notes outlining details of his private life up to the end of World War II. I took it home with me and read it carefully. Unfortunately the notes were badly written and very poorly organized, rambling and very confusing about dates and events. In short, I found them a hodgepodge of facts both real and imagined, much of which just didn't agree with the public record, my interviews with his fellow workers, or Patent Office files.

As he was still in the Los Angeles area, I tried to return the manuscript to him a week later, but he would not consider it. He had given it to me unconditionally and he said he was sure that I would be the person who would see to it that it was finally published. I later found out that there were about five more copies around and that he had tried in vain to have someone else publish it. Those who had read it before me apparently felt that it needed a great deal of work before it could become a viable biography. For instance, it had been offered to another historian who found it to be totally lacking in any of Zworykin's "hopes, aspirations and personal feelings." (Unfortunately I found this to be true.) However, in putting together my initial biography of Zworykin and in my subsequent efforts, I have found that by combining its information with what I had received from other sources I have been able to put together a fairly accurate account of his life, his work, and his times.

This book is the result of that compilation. I hope that it will present objectively and accurately the major contributions made to the development of television by Vladimir Kosma Zworykin.

I take full responsibility for all statements made and any errors that may have occurred. I welcome all additional information.

Acknowledgments

This book would not have been possible without the help of numerous persons and organizations. They have aided me in my research on the preparation of this book, for which I am most grateful. I particularly wish to thank members of the original Zworykin Westinghouse/RCA team, especially Harley Iams, Leslie Flory, Arthur Vance, and Dr. D. W. Wright, and such great television pioneers as Raymond Kell, Alda Bedford, Dr. Albert Rose, Dr. Otto Shade, Loren Jones, Felix Cone, Dr. James Hillier, Archibald Brolly, Harry Lubcke, and Albert Murray. I have also been helped by the great English historian of television Tony Bridgewater, Douglas C. Birkenshaw, Thomas M. C. Lance, Ray Herbert, Dr. E. C. L. White, Dr. James D. McGee, Dr. Sidney Fox, and Dr. Hans G. Lubszynski, and by two special pioneers of television, Baron Manfred von Ardenne and Prof. Kenjiro Takayanagi. Special thanks go to Al Pinsky, formerly of the RCA publicity department, who gave me unlimited access to Zworykin's files; Phyllis Smith, of the David Sarnoff Research Center; Dr. Marci Goldstein, of the Bell Labs Archives; Anita Newell, of the Westinghouse Archives; Brian Samain, of the EMI Archives and the BBC Archives; Jerry Sears, of the United States Patent Office; Birdy Rogers, of the Los Angeles Public Library; and the staff of the Alexanderson Library of Sheaffer College, Schenectady. I also wish to thank my dear French correspondents, Jacques Poinsignon and Jean-Jacques Ledos, as well as Gerhardt Goebel in Darmstadt and the late George Shiers. Thanks also go to the following friends who worked hard at translating technical articles and patents into understandable English: Günter Schmitt (German), Micheline Barkeley (French), John Inoue (Japanese), and Sharon Behry (Russian). Finally, I wish to thank my editor at the University of Illinois Press, Patricia Hollahan, whose infinite patience and literary skills will always be remembered. And last but not least my dear wife Arlene, who always encouraged me when I balked at the size of the task ahead.

For their kind permission to reproduce photographs from their collections, I wish to thank the Bell Telephone Laboratories, the General Electric Company, the Westinghouse Electric Company, and the David Sarnoff Research Center. I also wish to thank the many individuals who gave me material from their private collections.

Finally I would like to thank the late Dr. Vladimir Zworykin for his kind cooperation in our few interviews and, of course, for the wealth of information he gave me in his notes. Without them, this book would never have been written.

Zworykin, Pioneer of Television

CHAPTER 1

"Father of Television"

The private hospital room at Princeton Medical Center was half-lit. The patient was receiving plasma from the usual bottle hung on a metal stand. He was also connected to a series of machines that recorded his breathing, heartbeat, and blood pressure. At the moment, he was resting comfortably, his eyes closed and his respiration almost regular. There was a nurse in attendance, but it was obvious that she could do little to help. She was alone, no one from his family was present. The patient's beloved wife Katherine, also ill, was too sick to come to the hospital.

The patient was old, he would have celebrated his ninety-third birthday the next day. He was a small man and quite wrinkled. He had been in the hospital for over a week. His name was Vladimir Kosma Zworykin. He had been world famous. For over fifty years his name had been in the forefront of every major development in the world of television, an electronic mechanism that could project moving images instantaneously over the horizon.

There was no denying that he had provided two of the most important pieces to the puzzle. The first was a practical camera tube, the iconoscope; and the second, of even greater importance, was the kinescope, the picture tube that was used in millions of television sets all over the world.

For these contributions, Zworykin was usually called "the Father of Television" by the press and by the industry that he had helped found. Publicly he professed to be bothered by the title, since he was only too painfully aware that television had come about through the efforts of dozens, perhaps hundreds, of individuals who had slowly and methodically put the pieces together to create the most powerful communications medium in the world. Secretly, however, he cherished that name. There were others, of course, who also had claim to it. First was Zworykin's younger contemporary, Philo Farnsworth. In poor health for several years, Farnsworth had died eleven years earlier (March 1971) of pneumonia, unheralded and forgotten.[1] Then there was Zworykin's beloved Russian physics teacher, Boris Rozing, who had introduced him to the very concept of television, and the Scotsman Alan Archibald Campbell Swinton, who had really unlocked the door with his 1908 letter and 1911 paper, both entitled "Distant Electric Vision." Other names also came to mind, Americans such as Charles Francis Jenkins, Dr. Ernst F. W. Alexanderson of General Electric, and Dr. Herbert E. Ives and Dr. Frank Gray of the Bell Telephone laboratories. Then there were the British—John Logie Baird (one could not ignore Baird), Sir Isaac Shoenberg, Allan Blumlein, W. F. Tedham, and Dr. J. D. McGee, all of Marconi-EMI, who had done such magnificent work—and the Germans, who included Paul Nipkow, who had started the whole process, the pioneer Dr. Max Dieckmann, August Karolus, and of course Baron Manfred von Ardenne.

How could anyone overlook the French contribution? Zworykin had learned valuable lessons from the French; Édouard Belin and Dr. Fernand Holweck, Georges Rignoux, Georges Valensi, and Dr. Alexandre Dauvillier all came to mind. The French were great at conjuring up the machines of the future, everything from the automobile and the airplane to devices such as the motion picture and even television itself. Somehow, they soon tired of each innovation and went looking for new fields to conquer. But two Frenchmen in particular must have stood out in Zworykin's mind. The first was Gregory Ogloblinsky, his chief engineer at Westinghouse and RCA, who had been his right arm. The other was Pierre Émile Louis Chevallier. Together, they had set Zworykin on the road to fulfilling his dream.[2]

Last but not least there was David Sarnoff, who might well be called the "godfather" of television. Not an inventor but a hard-driving entrepreneur, in total control at RCA, he steered his forces relentlessly toward the goal that dominated much of his life, the launching of electronic television. His chosen instrument for achieving the goal was Vladimir Kosma Zworykin. Even during the Great Depression, when entire departments at RCA Victor were being discharged wholesale, Zworykin's research group never had a cent cut from its budget. Sarnoff bought up television patents internationally to place all means at Zworykin's disposal. Holdouts like Farnsworth and Armstrong, who invented FM radio, were challenged via litigation. In the years after World War II, extraordinary success crowned the Sarnoff-Zworykin efforts.[3]

The past twenty-eight years had been quite routine for Zworykin. Almost every day he had left his home at 103 Battle Road Circle and driven his trusty 1966 Cadillac to his office at the RCA research laboratories in Princeton, New Jersey, on Highway 95. The office staff enjoyed watching him come to work every morning, and he greeted everyone by name. He was really a legend in his own time, like Benjamin Franklin, Alexander Graham Bell, or perhaps Thomas Alva Edison. But he was quite unaffected and looked forward to the visitors who came from all over the world to have a few words with him.

As soon as he entered his office each morning, he would check with RCA public relations manager Al Pinsky and ascertain who was visiting that day. He was a charming, gracious host, greeting his visitors with a warm smile, glistening blue eyes, a friendly handshake, and kind words. He still had a heavy Russian accent after all the years he had been in America, and his hearing was impaired, which made for slow conversation.

He would show his guests a picture of his first camera tube, answer the usual questions, and always, but always, say with a twinkle in his eye, "Don't call me the Father of Television." This usually provoked a startled response, "But Dr. Zworykin, you are the Father of Television," and of course he would not argue.

After the visitors were gone, he would pick up his notebook and continue work on one of his many projects. Medical electronics had become his life's work since he had retired from RCA in 1954. In that year, he became the director of the Rockefeller Institute for Medical Research in New York City, though he remained in his office in Princeton.

He really enjoyed being at the RCA research laboratories in Princeton. He would work on his latest project, and when he tired of that he would wander through the various laboratories to see how the other researchers were coming along. He always offered a helping hand, although it must be admitted that lately some of the newest "high tech" projects were beyond his comprehension. He would then go back into his office and look over his correspondence. He answered every letter, no matter how trivial.

However, he only worked half a day. Around noon or so, he would clear his desk and drive home for lunch and a short nap. The rest of his day was spent around the house doing what he liked best, relaxing over his memorabilia and looking back over a long and fruitful life. His devoted wife Katherine served him lunch and asked him how things went at the office. They would often discuss sports, the latest news, and sometimes even politics.[4]

There was no television set in his study. A twenty-one-inch color set given to him by RCA sat in one corner of the living room and was very seldom used. Zworykin could not get used to the inanities that were constantly pouring out of that instrument. He had often stated that "It is awful what they are doing, television as entertainment, I don't like." He was distressed by its depiction of violence and crime, and complained that it was no good for children. He was often quoted as saying that the best part of a modern television set was the "off" switch. Was this the result of a half-century of hard progress? What had they done to his progeny?

The scarcity of educational and cultural programs really bothered him. They were relegated to the handful of public broadcasting stations in the UHF bands, and they did not make money. Half-baked dramas, usually based on sex and violence, childish game shows, and the ever-present daytime soap operas were the bread and butter of the television industry. News, sports, and education, that is what television was supposed to be. Well, there was no use in getting angry over it, this is the way it was, and this is the way it would continue.

On the other hand, Zworykin was proud that the first pictures of space, and especially of Earth's moon, had come from an RCA camera system used for the Ranger's series of space shots in 1965. Also, orbiting satellites were just coming into being and live news pictures were being flashed instantaneously all over the world. The "global village" of Marshall McLuhan was a reality, and there was hardly anywhere in the world that television could not reach. Events were often seen instantaneously, or at least within minutes of their occurrence.[5]

However, for Zworykin time had run out. His mind had recently become less sharp and his body just did not respond as it used to do. He had often stated his desire to keep on working, not just for the sake of doing so but because of the many things he hadn't gotten around to doing. He often said that he would only retire when he died. Even ninety-two years of a full, eventful life was not enough time for a man like Zworykin. He had had no fear of death, and now he was dying, of nothing more than old age.

For a few hours, things seemed to go well. However, around 11:00 he was having some difficulty and simply stopped breathing. The nurse moved over to him, felt his pulse slowing down; she looked at the machine and watched the rhythms getting smaller and smaller. Within seconds, it was all over. The great man had passed on. It was Thursday, July 29, 1982.

With worldwide communications being what they were in 1982, it should have been expected that newspapers all over the world would

feature the headline "Father of Television Dies at 92." This was not
to be. Word of his demise was treated just like any other ordinary
obituary. The RCA publicity department gave the press a handout
with a short résumé of his life's work, including an old picture of him,
and this appeared in many of the large newspapers across the United States, on page 22 or so. Most of the editors either had never heard
of Vladimir Kosma Zworykin or had presumed that he had long since
passed on. So his death was given very little attention. Even television,
the very medium that he had helped create, only reported his death
much later, and did so with very little fanfare.[6]

CHAPTER 2

Growing Up in Tsarist Russia

Vladimir Kosma Zworykin was born in Murom, Russia, on July 30, 1889. He was the youngest of the seven surviving children, five daughters and two sons, of Kosma A. and Elana Zworykin.[1] His birthplace was one of the oldest settlements in Mother Russia, dating back as far as 864. Favorably situated on a bend of the river Oka, a tributary of the Volga, it attracted people from the neighboring woods. At that time rivers offered a simple, economical method of transportation and so quite early Murom became a prosperous center of commerce.

Murom, located in present-day Vladimir Oblast, had played an important role in early Russian history. During the Mongolian invasion of 1232 it was repeatedly besieged and burned by the Tatars. It remained nearly deserted until 1393, when it was revived, and by the sixteenth century flowered with cathedrals and monasteries. With the rise of Moscow as a major principality only 240 kilometers to the west, Murom gradually declined and by the eighteenth century was simply a typical provincial, prosperous small town.[2]

Kosma Zworykin was a member of a well-to-do merchant family who had received a commercial education and, for his time, was a man of progressive ideas. He was quite active in the affairs of Murom, such as the library committee, and served on the city council and for one term as mayor. He had inherited a wholesale grain business from the family, and later acquired a steamship line on the Oka River, to which he devoted most of his days. As a result, he had very little time for his large family. They only saw him mainly at mealtime or in church, which he demanded the family attend regularly. As a result, the household was run by Vladimir's mother, who referred to their father mostly in times of crisis.

Elana Zworykin had married quite young and was actually a distant cousin of her husband with the same family name. They had had twelve children, and Vladimir was the youngest of the seven that had survived. With such a large family, she seldom had time for all of the children, who were actually tended by a large staff of nurses and aides. But there was also a large extended family of cousins, aunts, and uncles, whose visits were great occasions to be looked forward to. It was fun meeting all of these wonderful people and young Vladimir anticipated with joy the gifts they always brought with them. His was a carefree and joyful childhood, with the privileges typical of his class.

Vladimir grew up in the mansion that had been the family home for several generations, a three-story stone structure which was too big even for a large family such as the Zworykins. The house was located in the center of the prosperous provincial town that Murom had become, facing two churches and a bazaar where the local farmers came to sell their produce. The rear of the home faced the Oka River, with a magnificent view of the river and the woods surrounding it. Between the house and the river was a big fruit and vegetable garden. Part of the garden was on high ground surrounded on three sides by a deep ravine, overgrown with all types of trees and bushes. This was Zworykin's most favorite place to play and hide. The woods surrounding the city and the Oka River were the source of many adventures. Like most of his young friends, Zworykin thought that chasing and shooting rabbits, foxes, and other small animals was great fun. In the wintertime, the river froze over and there were other grand adventures such as sledding or skating on the ice. He had many young friends, in particular his classmate Vasili and his cousin Ivan. He learned the joys of walking through the woods and the love of hunting, which remained with him for the rest of his life.

He remembered an incident in which he and Vasili had gone hunting for ducks. In trying to retrieve a duck he had shot, he fell into the icy water and had to wait for Vasili to form a rescue party to get him out. Strangely enough, even though he spent almost an hour in the freezing water, he did not even catch a cold, although at the time he was disposed to such illnesses and was suffering from what was diagnosed as asthma. Later this was identified as an allergy to cats and dogs, which never reappeared.

Apparently he was a precocious child who often indulged in practical jokes of dubious merit. One example he remembered was the time he and Vasili celebrated an aunt's birthday by setting off some homemade fireworks that misfired and burned down a barn.

Young Zworykin had an interest in everything happening in Murom, which by the time he was born had a population of twenty thousand and boasted a number of schools, several textile mills, a railroad repair works, a machine factory, and various small industries, as well as a fairly large library, which he visited often. He remembered, for example, seeing the first private telephone lines brought to the city. He watched with great curiosity as the wires were strung on the various poles. What struck him most was how the telephone was used as a source of news. Murom had no newspaper, so news got around very quickly for anyone who had access to this modern instrument. Gossip and especially rumors were spread over this new means of communication. His fascination with a machine that could talk stimulated his imagination. His first adventure with anything electrical was the installation and repair of the doorbells in his house and the family passenger steamboats.

Like most Russians of the time, Zworykin became aware of the importance of religion to the family life at a very young age. Murom boasted twenty-three Russian Orthodox churches and three monasteries. His parish church was next door to his home and the whole family usually attended Saturday evening mass and Sunday morning mass.

The most memorable holy event of the year was Easter, preceded by seven weeks of Lent. As a susceptible child, he was very impressed with the solemn ceremonies, which to him were full of joy. The other great holiday was Christmas, which always meant decorated Christmas trees, tables laden with food, distribution of gifts, and continuously changing groups of clergy, relatives, and friends. For him it was also a time for ice skating, sledding, and skiing—and often frozen ears and fingers. His interest in the holy days led to an early devotion to the Russian Orthodox Church that languished as he grew older.

Zworykin's education began long before he was formally sent to school. His sister Anna had taught him to read and write, and he was sent to a private tutor to be prepared for entrance to the *Realschule* (a secondary school that did not teach classics or prepare students for the university). He loved his tutor, Elizaveta Ivanovna, who treated him as one of her own children, and he felt that a great deal of his interest in learning came from her. But his entrance into the first grade in the *Realschule* marked a change in him. He became more independent, less under the supervision of his sister and nurses at home.

Zworykin had a great interest in school, and his appetite for learning continued throughout his life. He was a good scholar and learning came easily. He liked natural sciences, and, in the higher grades, physics as well as the basic courses offered in the local *gymnasium*. At one time he was put in charge of a small collection of scientific instruments used to demonstrate physical laws, and he was very often picked to help the teacher demonstrate them.

There were other demands on young Zworykin's time. Since his older brother Nicolai had shown no feeling for the family business, his father tried to interest young Vladimir in it. So he was frequently taken on trips between Murom and Nijni and learned to meet the arriving boats and make arrangements for their departure.

Here he had the opportunity to watch his father conduct the family business. His father would often offer visitors a cigar from one of several boxes. Watching them accept and smoke the cigars aroused young Zworykin's curiosity about their taste. So one day, when no one was around, he decided to try one himself. It made him violently sick, and a doctor was called. When the illness was explained, young Zworykin was severely punished. But he had learned his lesson and from that day on never smoked again. Another temptation was the abundance of alcohol available. However, the presence of his father's chief steward, who kept an eagle eye on him, seems to have prevented him from acquiring that bad habit.[3]

Times were changing in Russia at the beginning of the twentieth century. There was a general dissatisfaction with the tsarist government. There was corruption and oppression from all sides, and a complete lack of freedom. The ruling class owned all of the wealth, enjoyed all of the privileges, and monopolized all of the political power. It had no intention of giving up any of its prerogatives and absolutely no use for the average Russian, who lived most often under deplorable conditions.

It was only a matter of time before the old tsarist government would have to collapse. Two recent events helped spark the change; the disastrous war of 1904–5 with Japan and the revolution of 1905.

Russia and Japan were having territorial problems and on February 8, 1904, without a declaration of war (just as at Pearl Harbor some thirty-seven years later), Japan assaulted the Russian fleet at Port Arthur and Chemulpo. The Japanese navy attacked at close quarters, pulled alongside the Russian battleships, fired torpedoes through their hulls, and turned away at top speed. Russia suffered severe losses during the ensuing conflict and was finally forced to accept peace in the Portsmouth Treaty of September 5, 1905.

On January 9, 1905, a demonstration of St. Petersburg workers bearing a petition to the tsar was fired upon by Russian troops. During the chaos that followed, a general strike in October brought the country to a standstill. But the revolution was premature. Order was soon restored and by the end of 1905 the revolution had spent itself.[4]

There were mild demonstrations in Murom, and, while young Zworykin tended to be neutral in politics (even later he never indicated any burningly devout allegiance to the left or right), he could not ignore the marches and demonstrations around him. He claims that he often joined in organizing strikes, marches, and so on, even though such participation was quite dangerous. He remembered one march that started peacefully enough but soon ran into the police, who fired over the heads of the marchers and started a panic. Zworykin, who had escaped by climbing over a fence, was waiting to see what was going to happen. He helped a young girl over the fence and they returned home together. This was the start of a very beautiful, though brief, romance.

At about this time, Zworykin graduated with honors from the *Realschule* in Murom, and although he admitted that he had no particular preference for engineering, he decided to go to St. Petersburg to take the entrance examinations for the St. Petersburg Institute of Technology. His choice of engineering was influenced by the fact that several members of the Zworykin family were professors of engineering, and his brother and several cousins were attending engineering colleges.

The entrance examinations were extremely difficult, as there was a surplus of applicants for the few vacancies. Zworykin did not make the first acceptance list, but he was allowed to enroll in the physics department of the St. Petersburg Polytechnic Institute on the merit of his high school diploma. He indicates that he had actually started to take courses there. One of the first lectures he attended was by the noted Russian physicist Abram Fedorovich Ioffe. Ioffe was an excellent teacher and later established a school of physics from which came many great Soviet scientists.[5]

When his father heard that Zworykin had enrolled in the Polytechnic Institute, he came to St. Petersburg and, as Zworykin later related, settled the matter once and for all. It seems that polytechnics were considered extremely "liberal" while engineering schools were assumed to have a very rigorous curriculum that left no time for meetings or riots. Zworykin had qualified for the second acceptance list and was allowed into the St. Petersburg Institute of Technology. The ordering of a uniform by his father closed the case.

In St. Petersburg he made many fine friends, a number of whom had become involved in the revolutionary activities of the times. His father was wrong; the Institute of Technology happened to be at the very heart of the revolutionary movement. Zworykin and other intellectuals and students from such provincial towns as Murom were participating in these seditious demonstrations as a part of their student life.

Just after Zworykin arrived at the Institute of Technology a students' strike was called and the institute was closed. A host of students had barricaded themselves inside the institute and Zworykin found himself attending noisy and heated meetings. After several days some sort of agreement was reached between the students and the police and normal life resumed. However, Zworykin was very leery of what was going on. He became quite suspicious of those students who seemed to be making the most noise. He noticed how certain individuals were caught and arrested, which ruined their lives for no apparent reason. He also had the feeling that the police knew everything that was going on (and he was quite right about this). So he learned to play the game, to take part in the student uprisings, marches, and strikes and, in spite of several brushes with the police, managed to avoid any real trouble.

During his second year at the institute, his sister Maria was permitted by their parents to come to St. Petersburg and enroll in the Women's Polytechnic School. She moved into Zworykin's apartment and for two years they spent most of their free time together. They enjoyed the same interests and visited museums, exhibits, operas, and concerts. However, after one year, Maria lost interest in engineering and entered medical school, from which she eventually graduated.

During Zworykin's third year, the institute organized a trip abroad for a large group of students, to familiarize future Russian engineers with European industry by having them visit Germany, Belgium, France, and England. The expedition was organized by the International Chamber of Commerce, who hoped to improve the international situation. Thus wherever the Russian students went, they were treated to official banquets and greeted by the mayors of cities and other officials. To his surprise, Zworykin had been chosen chairman of the administrative committee and was put in charge of a group of fifty students.

Zworykin's group of engineers visited Berlin, London, and Manchester. This was his first experience as an administrator and it was not a very happy event. Being in charge of such a large group gave him nothing but headaches. In almost every foreign city they visited, he had to bail students out of jail, settle arguments between them, and of course keep track of their whereabouts. Most of them had never been out of Russia before and did not know how to behave, which he attributed to their naiveté, that is, their lack of experience and self-discipline. He did not consider the trip abroad a success.

During his time at the institute in St. Petersburg, Zworykin's summers were spent getting actual experience in industry. He enjoyed this very much. At first he worked as a stoker on a railway, until he passed certain tests and wound up as an engineer in a switching yard. The next summer he spent in a steel plant, where he was assigned to the drafting rooms for designing bridges and other kinds of metal construction. The third summer was spent in a power station and the fourth in an experimental motor-testing laboratory at the institute.

Students at the institute were given report books which outlined all of the subjects and experimental work that had to be completed before they could graduate. All subjects had to be graded by professors, but in all other respects control was by the honor system. Since Zworykin was particularly fascinated with physics, he spent considerable time in the physics laboratory, not only doing the regular assignments but trying his hand on all the equipment available.[6]

At this time, the professor in charge of the physics laboratory was a certain Boris L'Vovich Rozing. Professor Rozing, who had been born and raised in St. Petersburg, had graduated from the department of physics and mathematics of the University of St. Petersburg in 1891. He began teaching at the St. Petersburg Institute of Technology in 1894 and by 1897 was the director of the physics department. He had a reputation as an excellent teacher who had an extraordinary rapport with his students. He was to have a profound effect on Zworykin's future life.

Zworykin claimed that at this time in his education he had a keen interest in a "novel type of vacuum system." It appears that Professor Rozing had noticed Zworykin's interest and Zworykin claimed that Rozing had also caught him working on some of the other students' problems. After a period of weeks, he took young Zworykin aside and instead of reprimanding him, gradually extended both his friendship and guidance. Then on one "unforgettable day," Rozing told Zworykin that he was working on a project of his own and could use a bright student to help him. Zworykin of course knew of Rozing's fine reputation and seized the opportunity to work with him.

The next weekend, Zworykin reported to Rozing's private laboratory, located across the street from the institute in the building which housed the Bureau of Standards, where Professor Rozing was also on staff. Here Zworykin discovered that Rozing was working on the problem of "electrical telescopy," something Zworykin had never heard of before. This was early in 1911.[7]

At this time, "electrical telescopy" (or television, as it later was called) was just a wild, visionary dream. Zworykin did not know that dozens of schemers had been studying the idea since the 1880s, or that Professor Rozing had been working on it in secret since 1902 and had made excellent progress. Rozing had been fascinated with the problem of seeing at a distance since the first cathode-ray tubes devised and built by Karl Ferdinand Braun in Germany in 1897 had been described in the various scientific journals. Rozing had applied for his first television patent in July 1907, and had actually started to build apparatus in 1908–9.[8]

In Rozing's laboratory Zworykin saw his first television apparatus. It was an incredible sight. There were a couple of dozen voltaic cells connected to electric motors which generated the necessary high voltages, with coils and wires all hooked up to a high-vacuum Braun cathode-ray tube. There were many types of electrical meters, pumps, beakers, and jars of chemicals, as well as the usual glass distilling apparatus and devices on various walls and benches. Other batteries were being continuously charged and the laboratory smelled of sulfuric acid.

The Braun high-vacuum tube was the heart of Rozing's invention. It was a round glass tube with a long narrow neck. The large end was about five inches in diameter and about twenty-one inches long. Several electrical connections entered the neck of the tube, which sat above a wooden board and was supported by three straps, two around the face of the tube and another at the neck. The straps were about twelve inches long and connected to three electrical insulators similar to those used on high-powered electrical lines. As it was a cold cathode tube (meaning that it had no hot filament to generate the flow of electrons for the electron beam), it needed some 20,000 to 30,000 electrical volts to operate properly. A large sign beside the machine warned "High Voltage: Do Not Touch!"

At this time, Braun tubes were very unreliable. They were expensive, deadly, and worst of all quite erratic. One could not be sure how the tube would behave until the next time the high voltage was applied to it. Braun tubes needed very complicated, expensive high-vacuum pumps to be processed correctly. As a result, they were usually ordered from a scientific instrument maker in Berlin. The main use for the tube was as an oscilloscope, a device that would graphically display wave forms of electrical current accurately. The face of the tube was usually viewed with the aid of a rotating mirror which acted as a kind of linear time base.

However, Rozing's Braun tubes were quite different. Inside the neck of Rozing's tube were two parallel metal plates that were used to electrically shift the electron beam itself before it was scanned and reached the screen. These two plates were connected electrically to the photoelectric cell in the "camera." Depending on the output of the photoelectric cells, the beam would be deflected up or down before entering the concentrating plate. Since this movement increased or decreased the amount of electrons passing between the plates, it had the effect of varying the brightness of the electron beam. (At that time, no other means of modulating the electron beam was known.) This was Rozing's ingenious method of modulating the brightness of the beam to coincide with the signals received from the camera in a large wooden cabinet that stood nearby.

The high-voltage generators were connected by wires to an anode (a flat, circular metallic plate) inside the rear of the tube. A stream of electrons was caused by the action of the high voltage, which freed them from this anode. These electrons went through a small aperture in a plate where they were concentrated into a small electron beam. This beam struck the face of the tube with high velocity. Here a screen was coated with a fluorescent surface such as potassium tungstate. Around the tube were sets of magnetic coils of wound wires that were used to deflect the beam in two directions (horizontally and vertically). When the beam was moved it would "draw" lines on the face of the tube which could be observed.

The electrical signals came from a wooden cabinet that was nearby. This was the camera or image pickup device that translated the light from a subject into electrical signals. At the front of the cabinet was a large lens for gathering the light from the object to be transmitted. This light was scanned by the mirror drums and, after being dissected, was finally focused onto a photoelectric cell at the rear of the cabinet. A series of meters were placed nearby to measure the amount of current from the cell.

Inside the cabinet were two sets of mirrored drums set at an angle to each other. They were connected to a large electric motor by means of belts and gears. One set of drums ran at a high rate of speed and the other at a relatively slower speed. This enabled the image to be scanned (or dissected) sequentially many times a second. Each mirror drum had a set of magnetic coils near each mirror face. As the drum revolved, these coils were driven across large magnets which in turn generated the scanning currents for deflecting the beam in the Braun tube. This fulfilled the important function of synchronizing the camera movement with the receiver. Rozing's device was the first true cathode-ray tube system in the world.

Professor Rozing described each of these items to young Zworykin. He explained that his system of television was based on the scanning principle. That is, by sequentially dissecting an object in a systematic fashion and rebuilding the image at the receiver, one could build up a picture line by line sent through a single channel. Through persistence of vision the human eye would create a replica of the original object on the face of the cathode-ray tube.

For some time Rozing had been building his own photoelectric cells. He was using potassium as well as sodium as the light-transducing surface. Cells made using selenium were too slow in reacting to light and had limited sensitivity. Because of Zworykin's interest in physics, Rozing decided that his first laboratory exercises would be to work with him in building photoelectric cells that were more reliable and sensitive.

For the next year and a half, Zworykin spent most of his free time in Rozing's laboratory. His main project was to build better photoelectric cells. He had to master the techniques of making and blowing the glass bulbs, inserting the various anodes into the envelopes, and distilling the potassium or other photoelectric substance in just the exact amounts. Layers that were too thick or too thin produced inferior tubes. Finally, Zworykin had to attach the cells to the crude, awkward vacuum pumps. The whole process required hours of physical labor to get anywhere near a working vacuum.

Then came the moment of truth, when each new cell was attached to a meter and a measured amount of light was focused on it. Unfortunately almost 99 percent of the tubes were bad, though Rozing was able to get enough current out of some of them to be able to carry out successful experiments. He could transmit very crude images across the laboratory from time to time. He was even able to demonstrate his machine; one early success came on May 9, 1911, and was witnessed by a host of university professors. For this, Rozing received the gold medal of the Russian Technical Society and several other honors.[9]

However, Rozing was not working alone on the problem. In 1909, Dr. Max Dieckmann of Munich described his own television device, which also used a Braun cathode-ray tube for a receiver and was capable of displaying crude moving images. However, it had a rotating wheel at the transmitter instead of a Nipkow disc, with metal contacts that actually touched the objects to be transmitted. Since Dieckmann never used a photoelectric cell as a transducer, in concept his machine was half telegraph (either "on" or "off") at the transmitter and half television at the receiver.[10]

At about the same time, another German experimenter, Ernst Ruhmer of Berlin, had demonstrated a simultaneous (without scanning) viewing device. It had a bank of twenty-five selenium cells set in a mosaic. Each of the cells was connected by wire to an incandescent lamp that lit up in accordance to signals from the selenium cells. This system could transmit crude silhouettes of simple letters or figures. However, by definition, without scanning or synchronization it was not a true television device.[11]

Also in 1909, two Frenchmen, Georges Rignoux and Professor A. Fournier, had described a television device that used a mosaic of selenium cells to pick up an image. The signals from this device were sent in sequence by a rotating commutator to a receiver. Here the signals were sent through a light valve (Kerr cell) in response to the signals received from the cells. This would modulate the light from an arc lamp, which would then be directed to a rotating mirror drum (which had as many facets as the cells) and would be displayed on a glass screen. This was a true mechanical television system, the first to be built and described.[12]

All of these experiments had been described in the various scientific journals available throughout Europe and America at the time. However, Professor Rozing felt that the ultimate solution would include the Braun tube as a display device. Rozing's work was given worldwide publicity and appeared in such magazines as *Scientific American*.[13]

In 1911, another important event occurred. On November 7, a Scotsman, Alan Archibald Campbell Swinton, became president of the Röntgen Society of London. For his presentation speech he chose the subject of distant electric vision. In his speech he described in great detail what would be needed to build an all-electrical television system that used cathode rays at both the transmitter and the receiver. While many pioneers had discussed the use of the cathode-ray tube for a receiver, his concept of using the cathode-ray tube as a transmitter was truly revolutionary. He described in great detail what would be needed to build such a system. This speech was an elaboration of an earlier letter he had written to *Nature* in June 1908. So Campbell Swinton must be given credit for the original concept of the electronic television method that is used throughout the world today. But at the time, because the Röntgen Society was quite small and its journal had a limited circulation, his ideas were not widely known.

However, Campbell Swinton's speech was reported in *The Times* (London) for November 15, 1911. Some four years later, in August 1915, his ideas received worldwide attention in a magazine called the *Electrical Experimenter.*[14]

Zworykin was unaware of all of this at the time. In his manuscript, however, he said he felt that what Professor Rozing had done was to give proof that cathode-ray television was feasible. He also indicated that even in 1911 he knew that intensive work on the development of the many components involved would be needed before it could be used for practical purposes. Simply stated, the technology required to support such a system still was not available.

All too soon, it seemed, this episode in Zworykin's life came to an end. Graduation day came and Zworykin was prepared to get his certificate. He had written a paper on diesel-powered rural power stations that was accepted for graduation. However, the inspector of the faculty showed Zworykin his report book, which indicated that he never received a credit in theology and therefore could not graduate. Zworykin immediately went to the professor of theology and convinced him that it was an oversight but that he had followed the curriculum, read all of the required books, and would submit to an examination.

At first the professor was upset, but he finally agreed to give Zworykin a provisional score and allow him to graduate. This is an early example of how well Zworykin had learned the gentle art of persuasion that later made it possible for him to alleviate a host of problems.

Vladimir Zworykin graduated in 1912 from the St. Petersburg Institute of Technology with an engineering diploma. He wanted to stay at the institute and continue to work with Professor Rozing, but his father wanted him to return to Murom and help run the family steamship business. Finally a compromise was reached—Zworykin would continue his education by going abroad for a year and then return home. He was supposed to choose either a German or an English engineering school. However, on Professor Rozing's recommendation, Zworykin chose to go to the Collège de France in Paris to study theoretical physics under Professor Paul Langevin, whom Rozing admired and knew personally. After a quick visit with his parents in Murom, Zworykin packed and left for Paris.[15]

At the time, Paul Langevin was one of the most prominent French scientists in the field of physics. His research included the field of magnetic phenomena, ionization of gases, and dynamics of electrons, and he delved into various problems in the theory of relativity. Upon his arrival in Paris, Zworykin met Langevin, who accepted him into his laboratory and suggested several research problems for Zworykin to work on.[16]

One of Zworykin's first projects was to repeat an experiment done by Professor Max von Laue on the diffraction of X rays by crystals, for which Laue later (1914) won a Nobel Prize. Zworykin admitted that he knew nothing about X rays, and knew nothing of Laue's work or his effect. However, though he had very little guidance, he later claimed that with his innate electrical as well as mechanical ability he was able to assemble a good set of equipment and reproduce the pictures of X-ray diffraction from various crystals as sharply as anyone had at the time.

In conjunction with this experiment, Zworykin had wanted to build better and more advanced equipment for the school, but there were no funds for such a project and he had to content himself with writing a paper on it. This was a pity, because Zworykin's ability to translate abstract ideas into working equipment was rare for a man of science and was one of his greatest attributes. Even though Zworykin considered himself primarily an engineer, he repeatedly crossed the tenuous boundaries between pure and applied science.

In that year (1912), time signal broadcasting started from the Eiffel Tower. Zworykin put together a radio receiver in his laboratory and later in his room experimented with various radio detectors. Zworykin claims that this was his first contact with radio broadcasting.

At first, Zworykin felt that the time spent in Langevin's laboratories was well worthwhile. Each Wednesday, for example, Langevin had an informal tea for his students, at which he discussed the latest news in the rapidly advancing world of physics. However, Zworykin soon realized that he was quite weak in theoretical physics, and that he should do more systematic study, which was quite difficult in the free atmosphere of the Collège de France. He therefore decided to move to a more formal German university after summer vacation.

Zworykin received permission from his father to spend his summer vacation in France rather than return home. Professor Langevin agreed with this and instead of loading him down with books to read on theoretical physics, advised him to go to the south of France, to get acquainted with the French people and, more importantly, to improve his French language skills.

Zworykin first went to Biarritz, which even then was quite a fashionable resort, but he didn't enjoy the idea of just lying on the beach, drinking, and golfing. Instead he left for Spain, and vacationed there. He had met a charming Spanish nobleman in Biarritz, and they traveled throughout Spain together until it was time for Zworykin to return to Paris. They did speak French throughout their vacation, however, and Langevin complimented Zworykin on his improvement with the language.

Upon the advice of a good friend and classmate of his brother Nicolai, Zworykin then visited Charlottenburg Institute in Berlin, where he was accepted for enrollment. But this was the summer of 1914 and there was the smell of war in the air. This quickly put an end to Zworykin's formal education and he prepared to return to St. Petersburg.[17]

CHAPTER 3

World War I

The German declaration of war against Russia on August 1, 1914, made it impossible for visiting Russians to remain in Germany, but also made it very difficult for them to get home. However, Zworykin was quite clever at making unusual arrangements and was able to get back to St. Petersburg, now called Petrograd in the wake of anti-German sentiment, by way of Denmark and Finland.

There, because of the desperate need for recruits and in spite of his education, he was immediately inducted into the Russian army as a private. As a graduate engineer with foreign postgraduate work, he was assigned to radio communication school. After a brief training period he was sent, still as a private, with a small party of specialists to the front lines near the city of Grodno.[1]

The eastern front was in utter confusion due to the disastrous defeat of the Russian army. Zworykin and his companions were unable to find the unit they had been assigned to join, and they found themselves completely isolated. No other unit was willing to accept a group of about thirty specialists they had no need for and would have to feed.

Zworykin soon found himself in a very peculiar position. The sergeant who was officially in charge of the group had great difficulties in dealing with the officers of other units, who paid little attention to him. Although only a private, Zworykin had an engineering badge on his tunic, and since engineers were held in high esteem in Russia, he had a better chance dealing with army officialdom. Soon he had become the unofficial leader of the group, and because he had received a considerable sum of money from his father before leaving Petrograd, he was also able to buy food for the men in his unit.

Searching for the unit to which they were attached, the group finally entered Grodno. Here they were marching along in dilapidat-

ed formation, dirty and practically in rags, when they were stopped
by a colonel in the uniform of the army engineers. He asked what
they were doing and where they were going. The sergeant in charge
of the group immediately had Zworykin explain their dilemma. At
first the colonel resented talking to a private, but when he saw Zworykin's engineer's badge he became interested in the unit. It turned
out that the colonel had a radio station in Grodno but no specialists
to operate it. He told Zworykin where to find the necessary equipment, which was in a nearby railroad yard, and ordered him to have
the station working by the next afternoon. The unit worked all night
and secured the equipment. By the next day they had requisitioned
a house on the outskirts of town and had the radio station operating. This seemed to be a lucky break for Private Zworykin, as he was
able to remain in Grodno for over a year.

During this period Zworykin had many problems with the Russian
officers running his radio outfit. It seems that he could not tolerate
the red tape and the inefficiency of his superior officers. He ran afoul
of several officers in his attempts, for instance, to communicate with
other bases without the proper codes.

Zworykin spent most of his time ciphering and deciphering radio-
grams. He claimed in his manuscript that he had learned the code
so well that soon he did not need a code book at all. This upset his
commanding officer no end. It was proposed that in order to have
physical possession of the code book, as the rules required, Zworykin
be commissioned an officer on the spot. But red tape prevented this,
since it meant a transfer from the engineering branch to the Signal
Corps and his commanding officer, a Colonel G., was loathe to see
him go. Finally the problem was solved by having an officer who was
nominally in charge of the code book stationed nearby, so a captain
with a small patrol was assigned to guard the station.

Zworykin later found that he was able to intercept news bulletins
from the German general staff at Nauen. When the colonel was informed of this, he immediately became apprehensive of Zworykin,
since any information not received through official channels was
suspect. Eventually Zworykin convinced him of the value of this material and was authorized to collect this data, translate it, and reproduce it on a hectograph machine. Some time after this, the colonel
told Zworykin that because of his efforts to keep their own communications secure, he was to be transferred to another unit, which the
colonel understood to be part of the intelligence service. This did not
please Zworykin at all. There were also rumors that the front was
being pushed back and that Grodno was soon to be evacuated.

The long hours of listening to Morse code, the many responsibilities he had taken on, and the uncertainty of the war took their toll, however. Zworykin developed a very bad case of insomnia and hallucinations, which he claimed came from fatigue and overwork. He went to the company doctor, who was by now a good friend. He was very interested in Zworykin's radio work and they had common interests in the theater and literature. Upon hearing Zworykin's complaints, the doctor recommended that Zworykin be sent to a psychiatric hospital in St. Petersburg, which had now been renamed Petrograd by the Soviets.

This transfer took place immediately. Upon his arrival, Zworykin was informed that the front had receded badly, and that Grodno had been evacuated, his radio station blown up, and his outfit scattered throughout the area. He had gotten out just in time.

He first had to report to the Officers Radio School. There he met an old acquaintance, one Colonel Ilia Emmanuel Mouromtseff, second in command and in charge of educational personnel, who asked Zworykin what he was doing there. Zworykin frankly explained his predicament, that he had been sent to the psychiatric hospital, and the reasons that led to it. Colonel Mouromtseff was very sympathetic. He told Zworykin that he needed an educated man for the school and took it upon himself to present Zworykin's case to the authorities. The next day Zworykin was excused from going to the hospital and was instead assigned to the teaching staff of the Officers Radio School. Two weeks later Private Zworykin was commissioned as a lieutenant in the Russian Signal Corps.[2]

Not too much later, an important event occurred in Lieutenant Zworykin's life; he met and fell in love with a certain Tatiana Vasilieff, a student at a dental school. After a whirlwind courtship, they were married on April 17, 1916. He informed his family by telegram of the marriage and to his surprise found no objections or complaints that such a step had been taken without consulting or even notifying his parents. Even in Murom life had been changed by the war. In answer to his telegram, he received congratulations and presents for his new wife. They quickly leased an apartment and began the life of a married couple, albeit under wartime conditions.[3]

Zworykin's army career also changed at this time. It happened that a French communications mission, headed by General Gustav Ferrie, arrived in Petrograd. The mission brought information and samples of the newest radio equipment including improved high-vacuum amplifying tubes. Zworykin was assigned to get acquainted with the performance of these tubes at the Russian branch of the Marconi factory, the Russian Wireless Telegraph and Telephone Company in Petrograd, where the Russian version of this type of tube was to be produced. This gave Zworykin the opportunity of getting possession of a few of these new tubes. He had also received a new British book by Dr. W. H. Eccles on radio communication, *Wireless Telegraphy and Telephony,* and with its aid proceeded to build a radio telephone transmitter in his apartment. The result of his work with these new vacuum tubes influenced his future activities, as he was then assigned by the Officers Radio School to the Russian Wireless Telegraph and Telephone Company factory, which was building radio equipment for the Russian army on the outskirts of the city.

Here he met the director and technical manager of the factory, Dr. Serge M. Aisenstein, and two remarkable physicists, Dr. N. D. Papeleksi and Dr. L. I. Mandel'shtam. Papeleksi was in charge of the development and production of a vacuum tube similar to Harry J. Round's tube, produced by the English Marconi Wireless Telegraph Company. Later both Mandel'shtam and Papeleksi wrote fundamental works on the nonlinear theory of oscillations.[4]

Zworykin was attached to the Marconi factory as an inspector of the radio equipment which it was supplying to the Russian army. It seems that this factory was doing outstanding work and was doing its best to meet the needs of the army as quickly as possible.

Zworykin quickly made friends with Dr. Aisenstein, whom he found very imaginative, well-educated, and, more importantly, receptive to his ideas. He mentioned his work with Professor Rozing on television at the St. Petersburg Institute to Aisenstein and to Aisenstein's chief engineer, A. M. Cheftel. It was decided that as soon as the rush of military work had ceased, that is, when the war was over, Zworykin would join the Russian branch of the Marconi company and form a group for the development of electric television. This, as we shall see, never happened.[5]

Zworykin then interested Colonel Mouromtseff and the radio school staff in a radio transmitter that he had been working on at home and got permission to test it in an airplane. Like other experimental transmitters of the time, it had been loosely put together on a wooden board. Other parts were attached to it by wires. Unfortunately, once it was in the air, the vibration of the plane made the pieces start to come apart. The test was not a success—Zworykin ended the flight with the battery leaking acid all over his feet and the gear tangled around him. He did not get another chance to test it, as he was to be assigned to other duties.

A few days later, he received a telephone call from Colonel

Mouromtseff, saying that the Officers Radio School was in the process of forming a special unit to be sent to Turgai, a small desert town near the border of Chinese Turkestan. The town was besieged by rebellious Turkomans, who had cut all of the lines of communications. A military expedition, complete with enough equipment to build three radio stations, was being formed to relieve the town and to reestablish communications.

The colonel needed someone who was well acquainted with the newest equipment, which was supplied by the Russian Wireless and Telegraph Company. Zworykin seized on the opportunity for two reasons. First, the thought of visiting a faraway place and having new responsibilities appealed to him. Second, sadly enough, he was having serious marital troubles. The relationship with Tatiana had been marred by friction and quarrels, and he thought that a temporary separation would help relieve the situation.

After several chaotic weeks of training the men and gathering of the materials, Zworykin was ready for this new adventure. One of the new officers turned out to be a brother-in-law of his who had been assigned to manage of one of the radio stations. Lieutenant Zworykin left by train several days after his unit, and happened to change trains at Orenbourg. When he got off, there was tremendous excitement in the air. Zworykin thought that the war was over. However, it turned out to be news of the assassination of the mad monk Rasputin on December 17, 1916. A celebration was in progress, since many Russians thought that his death marked the end of most of their problems and that life would soon return to normal. This of course did not happen.[6]

At the point of departure, Zworykin found that the first radio station was already deployed and operational, so he prepared to move his unit to the next site. As befitted his rank, he was given a military horse to ride. Having been raised with horses, he thought this would be a lark. It turned out that the military saddle was too large for him and the horse very big and exceedingly rough. After three days, he was totally incapacitated and had to spend the next three days riding in a wagon, much to the merriment of the other officers and Cossacks accompanying him. However, some smaller horses were captured in a minor skirmish with mounted Turkomans, and he was given one that let him ride to his destination without difficulty.

As soon as the expedition arrived at Kirghiz, Zworykin proceeded to set up the second radio station. After three days' rest, the unit set out for Turgai. Here the third radio station was constructed and made the connection with Kirghiz. Zworykin and his men were treated very well by the local warlords, who wined and dined them. But with the assignment finished, Zworykin had no desire to linger, feeling, as he said later in his manuscript, that he would go crazy or become a drunk if he stayed.

Returning to Petrograd, however, was not an easy task. It seemed that no one in charge would issue the orders for him to return on his own, but he decided to go back anyway. With his usual skill and a lot of luck, Zworykin convinced a certain general that he had never been instructed to remain there and was to be returned to Petrograd to resume his work in the laboratories. After a little intrigue involving the aid of the wife of a government official, the general agreed to give him a letter of dismissal and an order to return to Petrograd. As soon as Zworykin got back to Petrograd, he reported to Colonel Mouromtseff. He remained at the officers' school for a while and spent most of his time at the Russian Marconi radio factory.

One day he received orders to proceed to Moscow and take with him some radio tubes for installation at the Moscow radio transmitting station. After installing the tubes, he remained in Moscow for a few days to visit his sister Maria, who was attached to one of the local hospitals.

While he was at the radio station, he received a telephone call that a big fire had started in the center of the city and a riot was in progress. This was quite common at the time. A party such as the Russian Peoples' party had started a "patriotic" demonstration with the full protection of the police. The demonstrators were soon joined by hooligans and criminals, whose activities usually ended up as a pogrom in which intellectuals, students, and especially anyone suspected of being Jewish were the victims.

At this time, anti-Jewish sentiment was rampant and the rioters were looking for anyone with a Jewish name or a name that even sounded Jewish. As a result, many stores were looted and burned and many people were killed. Troops were sent to help the police, which only made matters worse.

Zworykin tried to take pictures of the rioting from one of the masts above the radio station, but without much success because it was illegal to raise the masts without permission. He heard some shouting and realized that his support was beginning to collapse. He had to be lowered quickly to the ground, dazed and shaking, just as the commandant was returning to the station.[7]

4

Revolution and Escape

When Zworykin returned to Petrograd in January 1917, he found a completely different atmosphere. Everyone knew that some sort of revolution was about to take place, though no one seemed to know when or how it would happen. As Zworykin recorded later in his manuscript, there was no return to the old way of life, and the feeling seemed to be that no matter what happened it would be an improvement.[1]

When the big event finally took place, hardly anyone noticed it at first. It happened so gradually, so unpredictably that it turned out not to be what everyone expected. On the surface, everything was peaceful and normal. However serious disturbances broke out in Petrograd, especially in front of food shops. At first the police and troops had the situation well in hand, but the street demonstrations grew in violence and there were frequent clashes with the police at police stations and public buildings. Isolated defections among the troops occurred on February 26, and by the evening of February 27, the fate of the empire was sealed.[2]

The rest of course is history. In a few days, on March 2, 1917, Tsar Nicholas II abdicated and power shifted to the Duma, the Russian parliament. The general appearance of the city was that of a holiday. No one was working, everyday life came to a standstill, Although casualties were few and the newspapers rejoiced over the "great bloodless revolution," it was very dangerous for army officers to show themselves on the streets. This didn't faze Zworykin, who recorded later in his manuscript that he, like most of the other officers, simply removed his insignia and epaulettes and walked around with a little red ribbon on his sleeve.

On the second day of the revolution things were so peaceful that Zworykin found himself strolling through the corridors of the Taurida Palace, headquarters of the Duma. Here he ran into a friend of his father, Aleksandr Ivanovich Guchkov, a member of the Duma and the minister of war and the navy. Guchkov asked Zworykin if he was still in the Signal Corps, because someone was needed at once to organize communications from the Taurida Palace to the island of Kronstadt, at the mouth of the Neva River. There seemed to be great danger at the moment that the commander of the fortress might be executed by the sailors of the fleet.[3]

Zworykin, who had been spending most of his spare time at the Russian Wireless Telegraph and Telephone Company factory, told Guchkov that he could get the necessary equipment if he had the proper papers. He also needed transportation, since the streetcars were not operating and it was a long distance to the factory. With his usual luck, Zworykin spotted a line of automobiles with chauffeurs, and sure enough one of them, by the name of Loushin, was an old friend from an earlier expedition. Through Loushin, Zworykin was able to procure a motorcycle with a sidecar, which became his personal transportation for several months, and Loushin's services as a driver. Zworykin was driven directly to the Russian Wireless Telegraph and Telephone Company factory where he met Dr. Aisenstein, who on the strength of a personal letter from Guchkov agreed to release the necessary equipment.

The next problem was how to transport the equipment, mounted on a truck, back to the Taurida Palace. Here Loushin suggested going to the Officers Radio School and asking for volunteers. Zworykin found the school in turmoil. Because he was an officer, Zworykin asked Loushin to explain the situation to the soldiers and ask for volunteers to get the equipment to the palace. The response was immediate, and they drove off to the Russian radio factory where they picked up the radio truck and returned to the Taurida Palace. By the end of the day, Zworykin had set up the equipment and was in communication with Kronstadt. He stayed with the equipment for two days after organizing the operation of the station.

On the third day he was passing through the palace and came upon a section labeled "Radio-Communications." He spotted a captain and asked him what their mission was. He was told that they were trying to set up the big radio station in Petrograd, but that at the present time it was not available.

Naturally the captain was shocked when Zworykin told him that he had already had a radio station operating from the palace. In fact, at first he was quite indignant that no one had informed him of this. After Zworykin explained that this had been done on the explicit orders of Minister of War Guchkov and that the equipment had come from the Russian radio factory, the captain was very pleased and on the spot appointed Zworykin to be his third assistant.

Zworykin's new appointment consisted mainly of sitting behind a desk and pretending to be busy. However, this didn't last long. Quite soon after consulting with Colonel Mouromtseff, who was still head of the Officers Radio School, he was told that he would be permanently stationed at the Russian Wireless Telegraph and Telephone factory. This of course pleased Zworykin very much.[4]

However, the revolution was taking its toll on Petrograd. Wherever one went in the factories, there was very little work being done. There were, however, continuous meetings, endless resolutions, and a complete breakdown of discipline, as well as numerous demonstrations and parades and endless lines in front of bakeries and food stores. On every corner was an orator, usually a soldier just returned from the front, calling for new freedoms and (as Zworykin put it later) "down with everything." Meanwhile, the situation at the front continued to worsen, as hastily organized troops failed to stem the German army's advance on Petrograd.

To make matters worse, Zworykin received a notice to report to a revolutionary tribunal. This frightened him as he had heard that officers who were summoned before these tribunals on the complaints of former subordinates, or sometimes for no other reason than being career officers, seldom returned. Zworykin was not amused to find that charges of "inhumane treatment" had been brought against him by an orderly who had been part of some earlier wireless telephone experiments but who had turned out to be quite slovenly and lazy. The claim of abuse was soon exposed as false and, much to Zworykin's relief, the judge dismissed the charges against him.[5]

Sometime later, Zworykin met an officer who proposed that he join a contingent of large artillery pieces that the officer was forming. He badly needed an engineer who knew motor generator and radio equipment. The thought of getting out of Petrograd and of visiting a faraway place and of new responsibilities appealed to him once again.

The acronym of the unit was T.A.O.N., which stood for "heavy artillery of special purpose." The unit was equipped with new six-inch guns and was completely motorized. Zworykin was to be in charge of the mechanical part of the outfit, as well as training chauffeurs, mechanics, and radio operators. Everyone in the outfit, officers and enlisted men, were volunteers, and the unit operated outside the chaos of the regular army. Zworykin remarks in his manuscript that he remembers this period of his life with great pleasure.

Zworykin had to learn everything about the complicated new gun mechanisms, tractors, trucks, and automobiles, which were mostly of French and American make. Fortunately he had succeeded in having Loushin transferred with him, which was a great help. Adding to Zworykin's activities, the commander of the unit insisted that the troops have some combat training, so he took them on forced marches and simulated battle front conditions.

When all the training was completed, the unit left for the front, to the south, but they were stopped near Kiev. There really was no such thing as a "front," as the German army was in Ukraine and had installed a "Hetman" (a Cossack chief) in Kiev. Zworykin's unit remained opposite Kiev on the other side of the Dnieper River. They were still in a very precarious state, caught between troops of Cossacks who were independent of everyone and detachments of partially demobilized troops who were leaning toward the Bolsheviks and therefore tended not to recognize the authority of the Duma and its government.

Everyday the troops were visited by propagandists from all kinds of political denominations who held meetings to try to persuade them, each in a different direction. No one knew anything for sure and everyone awaited instructions that never arrived. To complicate matters, Zworykin's wife Tatiana arrived from Petrograd with no place to stay. Zworykin found a room for her and prepared to move her to Kiev as soon as possible.

Finally, an invitation came to send delegates to a meeting of the different army units stationed in the surrounding territory. Zworykin and one noncommissioned officer were elected to represent their battery. The meeting was large, noisy, and fruitless, and adjourned without any real decisions. The army was divided by a multitude of political creeds from monarchists to communists, and although most of the parties agreed to support the upcoming parliamentary elections, the communists had already launched their campaign for "all power to the Soviets."

Following the meeting, Zworykin and his fellow delegate decided to return to their unit by train. The station was overcrowded and Zworykin was separated from his comrade-in-arms. A small disturbance began as rioters started to disarm and abuse any officers they could find. When Zworykin saw a small party approaching his car, he jumped out a window before they could get to him. A few shots were fired in the darkness toward him but he was not hurt. A little shaken, he walked safely to his unit, some five miles distant.

Overnight, the situation worsened. Without warning, his unit was awakened by shooting and found themselves surrounded by a regiment of Ukrainian soldiers. They surrendered without resistance, giving up weapons and equipment on the condition that all personnel would be permitted to keep their belongings and to go anywhere they wished. To achieve this, they were permitted to keep two trucks to deliver the men and their baggage to the railway station. Thus, they were demobilized by the Ukrainians, since they were in Ukrainian territory.

Zworykin was out of uniform at last. He went with Tatiana to Kiev and settled there for a while, but life was still far from peaceful. There were rumors that the front was reversing and that the Germans were going to evacuate the city. Again Zworykin had to make a decision, and again he and Tatiana were not getting along too well. She wanted to evacuate with the Germans; he wanted to return to Petrograd. This led to another separation, which they both knew might be final. It was finally decided that Tatiana would go with the Germans to Berlin while Zworykin returned to Petrograd.

Tatiana departed with a family from Berlin whom they had known before. At the last moment, Zworykin changed his mind about returning to Petrograd and decided to go to Moscow instead. With his usual good luck, he found a seat on one of the rare trains going there.

When he arrived in Moscow after a very hard trip, he found the city to be quite calm. He soon located his sister Maria, who was still working in one of the city's hospitals. From her he learned that their father had been dead for almost a month. This was a great shock to him, and as soon as possible he departed for Murom.

There he found the situation worse than in the large cities. Power had passed abruptly from the old-guard police and local officials to the local soviets, which consisted mostly of workers from the local factories and demobilized soldiers. They made life miserable for the former ruling class and completely disrupted the life of the city.

For some time, the average citizen didn't realize what had happened and could hope that soon everything would return to normal. But this was not to be. Zworykin found, for example, that his family home had been requisitioned for a museum of natural sciences. His mother and his oldest sister, both widows, had been permitted to use a couple of rooms temporarily, but there was no hope of regaining the whole house. He tried to persuade the women to return with him to Moscow, which seemed to offer more security, but they wouldn't leave Murom. This turned out to be a fatal mistake.

All of his friends and relatives were similarly victims of the times. His aunt Maria had been murdered in her own home; a cousin of his father had shot himself when his horses had been requisitioned. Zworykin finally located his school friend Vasili and found him to be an invalid, severely crippled during the war. Most of his former classmates had been killed or scattered all over the country. Zworykin decided to return to Moscow by himself.

Here he found that the Russian Wireless Telegraph and Telephone Company factory had been evacuated from Petrograd to Moscow. He visited it and was warmly received by Aisenstein, who immediately engaged him to work in the laboratory. However, production was almost completely disrupted. Some of the equipment had not yet arrived, and what was there took time to be installed. Most of Zworykin's time was spent searching the railway yards for missing machinery or waiting in long lines for food and necessities.[6]

On October 27, 1917, the provisional government was overthrown by the Bolsheviks. Zworykin did not consider this an improvement. In fact, work at the factory came to a standstill. In March 1918, the new government started to move its offices from Petrograd to Moscow, completely overloading the railways. This of course made the transfer of factory equipment increasingly difficult. Even the local communist cell that was now in charge of the Russian Wireless and Telegraph Company factory was powerless to help locate necessary equipment. In desperation, Aisenstein asked Zworykin to help a representative of the laboratory find some important cases loaded with instruments and get proper passes for their release from the railroad.

In order to get a pass for anything, it was first necessary to get a pass to enter the office where the passes were issued. As usual, Zworykin solved the situation by cunning—he got the appropriate rubber stamp and simply applied it to all of his papers, thus rescuing the vital laboratory instruments with a minimum of delay.

Zworykin finally decided that it was impossible to expect conditions to return to normal anytime in the foreseeable future, especially in research. He was restless and wanted to get out of the chaos and find a place where he could work in a laboratory. In addition, his demobilized status began to worry him, as the new communist government began to issue strict orders that all former officers were to report and be enrolled in the Red army. Zworykin felt that a civil war was brewing, and he decided to stay out of it at any cost, something he could only do by leaving the country.

This momentous decision was made for a variety of reasons. First, Zworykin felt completely alienated from the new communist regime, whose support came mainly from factory workers who had been converted to socialism by the preaching of past generations of intellectuals. The greatest opposition was from the wealthy, who were in the process of losing their lands, industries, and power. They were joined by former government officials, career officers, and the Russian Orthodox clergy. The intellectuals, already a minority, were divided into dozens of political factions, each going in its own direction.

The Russian *muzhiks* (peasants) were, according to Zworykin, a bewildered mass, quite happy with the seizure of their former landowners' properties and the resulting looting it provided but unsure whether the land now belonged to them or to the new ruling party. The majority of Russians were more or less neutral, going with whatever party promised them the most advantages or was the strongest at the moment.

Furthermore, the efforts of the Allies to keep Russia in the war actually aroused a new nationalism and unintentionally strengthened the position of the communist government. Zworykin, with his upper-middle-class background, felt no allegiance to any of these groups. He made up his mind to leave Russia, and while he admitted that he would go to any country where he could find a laboratory to work in, the United States of America looked the most promising.[7]

Once this decision was made, he put his plan in action. Getting out of Russia at that time was quite difficult, as very strict precautions had been taken to prevent people from leaving. The best way of getting out of the country was to obtain proper identification papers and some sort of official order to go abroad, or at least to the nearest border. His old friend Colonel Mouromtseff had been sent to the United States in 1917 as a member of the supply division attached to the Russian embassy in Washington, but Zworykin did not have the colonel's official connections and reputation and was forced to make other arrangements.

What Zworykin did have were friends who managed a large cooperative organization with offices in America and in the Siberian city of Omsk, which by then had become the capital of the White Guard government of Admiral Aleksandr V. Kolchak. From there he hoped to go on an official mission abroad.

Zworykin tried to get as many official-looking documents as possible, knowing that an ornate seal and impressive letterhead would do wonders with minor officials, who were mostly semiliterate. He quickly put together the necessary documents and carried them with him. Even though he was leaving the country legally he made sure that he could depart on a moment's notice.

This preparation suddenly paid off. He was going home from the Russian Wireless factory one day when he was stopped by a car driven by his old comrade Loushin, who was now working for the military police. Loushin had learned quite by accident that an order for Zworykin's arrest had been issued because he was a former army officer and had not registered for service.

Zworykin decided not to return home and had Loushin drive him directly to the Moscow railway station. Here he obtained a ticket to Nijni (now Gorki), where his family had a steamship office. It still had the Zworykin name on it although it had been nationalized. There he persuaded the man now in charge of the office, a former clerk with whom he had been friends before the war, to furnish him with a ticket to Perm (now Molotov) and some money in exchange for jewelry that Zworykin had with him. He had a pleasant trip on the boat, which he found to be reasonably clean and comfortable, and recounted later that the further he traveled from Moscow, the more peaceful the voyage became.

In Perm he found that rail travel to Omsk had been interrupted by fighting, so instead he took a train through the northern Ural mining regions and from there went by boat to Omsk. Zworykin later claimed that when he reached a mining region at the end of the railway line the local authorities believed that he was an engineer from Moscow who had come to inspect their steel plant. He denied it, but he finally agreed to sign a statement concerning the condition of the blast furnaces and shops. He was then told to take the same train back and change at the town of Ekaterinburg (now Sverdlovski), but he had no luck doing so because all the seats were reserved for the local soviets. After loitering around the railroad station for a while, he finally spotted a former colleague from the St. Petersburg Institute of Technology. Zworykin explained his predicament and at last was able to get on the train back to Ekaterinburg.

Here he immediately ran into more trouble. The military claimed that the Czechs were moving toward the railroad and put the city on military alert. They also apparently were not satisfied with Zworykin's papers, and he was taken into custody at a hotel turned into a prison, where he was joined by other former officers and people suspected of being friends of the Czechs. On the night of July 16–17, 1918, he heard that Tsar Nicholas II and his entire family, who had been under house arrest in a different part of town, had been shot. This completely panicked Zworykin and his fellow prisoners, who knew that periodically individuals and even groups were taken out of the hotel and did not return. The assumption was that they had been executed.

Soon Zworykin was called in for interrogation. Since he was traveling with a military overcoat, without shoulder straps but with his engineering badge, naturally he was charged with being a tsarist officer who was trying to escape the revolutionary forces. This Zworykin vehemently denied. He showed them his papers to prove that he had been demobilized by the Ukranians during the war, had the lowest commissioned rank, and was being sent to Omsk by the cooperative as a radio specialist. This didn't impress the interrogator at all. He decided to wait for verification from Moscow of Zworykin's credentials before releasing him.

Before this could happen, however, the Czechs reached the town and the guards disappeared, leaving the prisoners unguarded. Zworykin escaped, leaving behind his suitcase and most of his money. He was soon picked up by a Czech patrol and convinced them that he had come from the political prison. He was then befriended by a Czech sergeant who spoke German and had worked at the Skoda Armament Works. Zworykin mentioned the name of one of the engineers who had studied there and the sergeant became his protector. Under his aegis, Zworykin was allowed to go by train to Omsk, which was still in the hands of the provisional Kolchak Siberian government, who opposed the communists and had an agreement with the Czechs that permitted them to go unmolested to Vladivostok.

As soon as he arrived in Omsk, Zworykin located the offices of the cooperative from which he had traveling orders. He met with a very cordial reception, and the cooperative agreed that they needed more engineering information from America and would help him get there. In addition, as a radio expert Zworykin was commissioned by the government in Omsk to approach the Russian embassies in Copenhagen and London, and if necessary in America, to help him purchase appropriate radio equipment and bring it to Omsk. The problem, however, was that Omsk was surrounded on all sides, except the north, by various fighting groups.

Zworykin then found out that I. P. Tolmachev, a geology professor from Petrograd, was trying to go abroad with a scientific assignment. Zworykin met with Professor Tolmachev, who told him that he planned to leave Russia by way of the northern route, via the Irtish and Ob rivers and the Arctic Ocean. This was a much shorter way of getting out of Russia than going across the entire country, through war zones, to Vladivostok. Also, the professor had many friends both in the government and in the cooperative who were helping him to organize the expedition. Zworykin was quite desperate and asked Tolmachev if he could participate in any way. The professor agreed, and Zworykin found himself part of an Arctic expedition.

Sometime at the end of July 1918, Zworykin departed by boat from Omsk for the Arctic Ocean. The group traveled the length of the Ob, making various stops and landing at Obdorsk (now Solehard) on September 1, 1918. There they awaited the arrival of an ice-breaker, which finally came with a group of French radio specialists and their equipment. Zworykin then boarded the SS *Salambola* and set sail for Archangel.[8]

There he found that the city had been occupied by French, English, and American troops. Zworykin claims in his memoir that all the foreign embassies had moved there from Petrograd after the Bolsheviks took power, but actually they had first moved to Vologda when the new Bolshevik government moved its capital to Moscow, before moving to Archangel in August 1918.

Since Zworykin's instructions were to go first to London, he approached the British embassy for a visa. He was refused on the grounds that they did not recognize the Siberian provisional government. He then addressed the American ambassador, Dr. David Rowland Francis, from St. Louis, Missouri, a conservative, dedicated anticommunist who supported intervention by the Allies in the Russian civil war. Francis seemed to have great sympathy for Zworykin's plight, especially since he was on a mission for the Kolchak White Guard government that was fighting the communists. In addition, Francis knew that Zworykin had refused to join the Red army and was obviously seeking a way out of the country. Zworykin had certainly used all of his charm and guile to persuade Francis to give him his papers. At any rate, Zworykin claims that after having many lively discussions with the American diplomat about Russia and its culture, he was granted a United States visa. Francis also requested a transit permit for Zworykin from the British embassy.[9]

Zworykin also had letters to the Russian embassy in Denmark, so he decided to go to Copenhagen, which the American visa permitted him to do. He left Archangel and visited both Christiana (Oslo) and Copenhagen. He was quite impressed with Copenhagen and soon located an old acquaintance from St. Petersburg, which was fortunate because he was obliged to wait there for several weeks for the British visa. During that time he worked on learning English from a Russian refugee.

By the time the armistice had been signed on November 11, 1918, ending the war in Europe, Zworykin had arrived in London. There it took him almost a month of formalities to secure passage to the United States. Finally on December 21, 1918, he left Liverpool on the SS *Carmania* for the United States.

This was his first trip on an ocean liner and he was quite impressed by the luxurious mode of travel. He traveled first class and had access to all of the facilities of the smoking, reading, and living rooms and the promenade decks. He was particularly impressed with the elaborate menus in the ship's dining rooms and the choice of exotic foods, in great contrast with his way of life during the war. He also felt a bit intimidated by the surroundings. He had bought a new dark suit in London for the voyage, but was chagrined to find himself dining in this suit when everyone else was formally dressed. Obviously the other passengers were used to changing their attire several times a day. The trip across the Atlantic Ocean was uneventful and they arrived in sight of the Statue of Liberty on January 1, 1919.[10]

Zworykin was quite impressed with New York City, especially its tall buildings. He stayed first at the Waldorf-Astoria, but he soon found that it was too expensive for him and moved to the Ansonia Hotel on Seventy-second Street. He delivered his letters of introduction from Russia and located the offices of the cooperative organization under whose credentials he had been traveling. He spent several months with them, during which time he established communications with Omsk.

In the spring of 1919 he received orders from the provisional Siberian government to return to Omsk, as they needed a radio specialist and wanted him to bring back certain parts for their radio equipment. Zworykin's position with the cooperative was quite uncertain. He had traveled to New York using his own funds, and the cooperative had only accepted him upon the request of their officials in Omsk. Furthermore, the situation in Siberia was very unsettled and additional funds were hard to get. However, the government did agree to pay his way back to Omsk.

Why Zworykin decided not to stay in the United States at this time is unknown. He certainly could have asked for and received asylum, which supposedly had been his main reason to travel to the United States. He may not have been quite ready to give up his homeland, and his lack of fluency in English may have been a factor in this decision. He knew that to get a position in his profession without knowing English would be practically impossible. Another factor may have been that he had left all of his money in Russia and thought he could go back and retrieve part or most of it. Finally, there was always the chance, as remote at it may have seemed, that the new communist regime would not survive and things would somehow return to normal.[11]

At any rate, Zworykin wired Omsk for the date of his return. He received additional requests, some official and some private, and prepared for his return trip. Before he left, however, he went to Washington, D.C., and met his old friend Colonel Mouromtseff, who was working at the Russian embassy. Sometime during the first week of April 1919, Zworykin left New York City and took a train to the West Coast.

The return trip lasted about six weeks and took Zworykin to Seattle and from there to the port of Yokohama. After a rather pleasant voyage, he landed in Vladivostok, which was still occupied by the Allied forces. There his papers were accepted by the local authorities, but he was warned that passage by the Trans-Siberian railroad through the cities of Haborovsk (now Chita) or Harbin was uncertain. These cities were occupied by independent atamans (Cossack chiefs with military forces), Harbin by Lieutenant General Horwath and Chita by Lieutenant General Gregori M. Semenov. Semenov was holding a strategic point on the Trans-Siberian railroad. Despite the warning, Zworykin proceeded by train through Harbin to Omsk.

He was repeatedly detained for days and much of his baggage searched, though he was rescued each time by telegrams from Omsk. When he finally reached Omsk, he found that Admiral Kolchak had been named head of the "All-Russian Govt., Supreme Ruler of Russia, fighting the Bolsheviks." He also found that a full-blown civil war was raging between the White Guards and the communists.

Although the eastern part of the country was nominally under the control of the Kolchak government, there were several independent atamans who didn't recognize any government's authority and did basically what they pleased. Whole provinces were shifting from one authority to another and it was difficult to know which part of the general population suffered the most. The chaos was spreading, and about this time Zworykin decided to return to the United States permanently.

It appears that the Russian Ministry of Transportation was concerned with the delivery of goods purchased abroad and particularly from the United States. It was looking for someone to start organizing shipments for the next summer's navigational season. With his usual good luck, Zworykin was asked to take part in this activity in America, which suited his plans perfectly. He agreed to accept the position for at least one year, possibly two.

Once again, Zworykin was headed for the United States as a government courier with letters of introduction to various Russian organizations in the United States. Since he was now a seasoned traveler with a reputation for successful travel to the United States, he was loaded with all kinds of commissions from cooperatives and other organizations, letters from private individuals searching for relatives, and even a vial of myrrh (a blessed oil used in church services).

Zworykin retraced his steps from Omsk to Vladivostok via the Trans-Siberian railroad. He took a small boat to Tsuruga and then traveled to Yokohama by Japanese railroad. This time he spent a few weeks in Tokyo, waiting for a visa and a new steamship ticket. He finally boarded a Japanese ship, which took him to Honolulu and then to San Francisco. There he boarded a train for New York City, where he finally arrived in August 1919, with the firm resolution to remain there. It had taken him eighteen months to make these two journeys from Moscow to his new home in the United States of America.[12]

Alan Archibald Campbell Swinton. Courtesy of the Royal Television Society.

Dr. Max Dieckmann. Courtesy of *Bosch Techn. Berichte* (6, no. 5/6 [1979]).

Boris Rozing. From P. K. Gorokhov, *B. L. Rozing: Founder of Electronic Television* (Moscow: Hayka, 1964).

Vladimir K. Zworykin, Leningrad, 1911. Courtesy of the David Sarnoff Research Center.

Charles Francis Jenkins, with early prismatic ring transmitter, 1922. From *Scientfic American*, Nov. 1922.

John Logie Baird and early television apparatus, 1924. From A. Dinsdale, *Television* (London: Television Press, 1928).

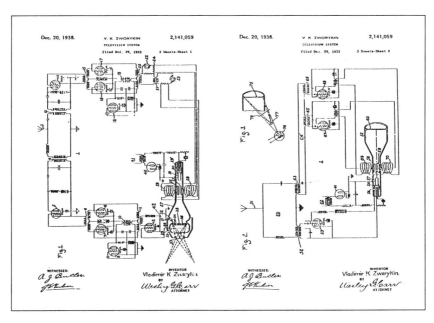

V. K. Zworykin, U.S. Pat. no. 2,141,059 (filed Dec. 29, 1923; issued Dec. 20, 1938), pp. 1 and 2.

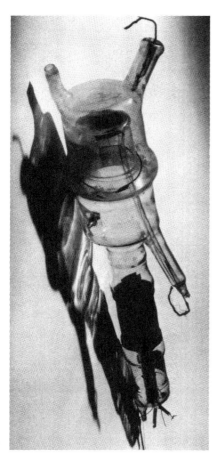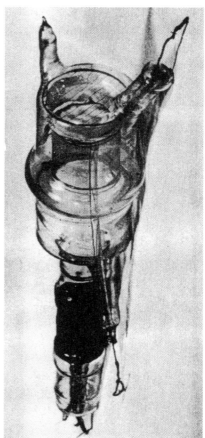

Photographs of Zworykin's 1925 camera tube. (Left) broken tube, three-quarters view, no grid in front of target; (right) three-quarters front view, grid in front of target. Courtesy of the David Sarnoff Research Center.

Zworykin with dual amplifying/photoelectric cell tube, Pittsburgh, Oct. 1925. Courtesy of the Westinghouse Electric Corporation.

Research Report R-429₄.

V. Zworykin,
June 25, 1926.

PROBLEMS OF TELEVISION.

Closing Report.

This order was used for experimental work on rapid picture transmission, using the modified Braun cathode ray tube both for transmitter and receiver. The receiving tube was developed using parts of a Western Electric Cathode Ray Oscillograph and gave quite satisfactory results. The transmitting part of the scheme caused more difficulties, particularly in the construction of the partition between the cathode ray tube and the photo-cell. This partition consists of a metal screen with an insulating layer and photo-sensitive substance on top of the insulation. Glass cloth for insulation has so far given the most promising results, except in the matter of electric leakage due to the difficulty of sealing it into the bulb. The cloth made of quartz or pyrex glass would be much more suitable, but unfortunately a firm could not be located which makes this kind of cloth.

This development has been temporarily discontinued in order to work with the mechanical method of picture transmission, which is still in progress.

The development of both problems will be continued on the new orders 6-4520 and 6-4522.

S.

CONFIDENTIAL

Research Report R-429 A, "Problems of Television," by V. K. Zworykin, June 25, 1926. Courtesy of the Westinghouse Electric Corporation.

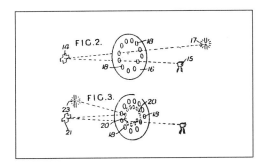

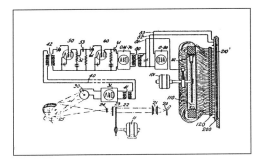

Flying spot patent of John Logie Baird (Br. Pat. no. 269,658; television apparatus; Baird, J. L. and Television, Ltd., Motograph House, Upper St. Martin's Lane, London; Jan. 20, 1926, no. 1696 [Class 40 (iii)]).

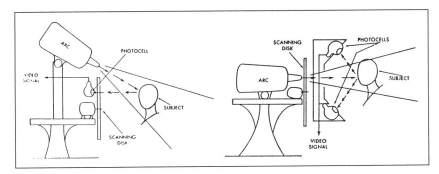

Flying spot patent of Frank Gray (U.S. Pat. no. 1,759,504; electro-optical transmission; Frank Gray, New York, N.Y., assignor to Bell Telephone Laboratories, Incorporated, New York, N.Y. a Corporation of New York; filed Apr. 6, 1927; serial no. 181,537; 20 claims [Cl. 178—6]).

Diagram showing the differences between direct and inverted flying spot scanning. Author's collection.

Herbert E. Ives and photocell, Washington, D.C., Apr. 1927. Courtesy of AT & T Archives.

Frank Gray (left) and John Hofele, with early television apparatus, Apr. 1927. Courtesy of AT & T Archives.

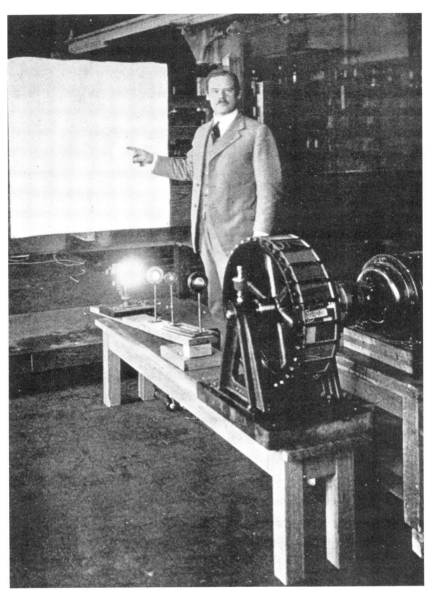

Ernst F. W. Alexanderson with his multispot projector. Schenectady, 1927.
Courtesy of the Hall of History Foundation, Schenectady, N.Y.

Alexanderson (left) and Ray Kell with 1928 General Electric television receiver. Courtesy of the Hall of History Foundation, Schenectady, N.Y.

RCA, GE, Westinghouse, and NBC executives present at the Aug. 8, 1928, Westinghouse television demonstration. Standing (left to right) Conrad, Goldsmith, Aylesworth, A. Taylor, Alexanderson, Chubb, Sarnoff, H. P. Davis, Bucher, C. R. Taylor; seated (left to right) Schairer, Eldridge, Baker, Ray, Horn, Kintner. *Wireless Age*, Sept. 1928. Courtesy of the David Sarnoff Research Center.

CHAPTER 5

An Invitation from Westinghouse

The first news Zworykin heard when he arrived in New York City was that the Kolchak government had collapsed. The Red army under Leon Trotsky was sweeping Kolchak from Siberia and driving him to the Far East and defeat. This of course meant that Zworykin's quasi-official status with the Russian Ministry of Transportation had ended, so he sought help from the embassy (which did not recognize the Communist government) in Washington, where his old friend Colonel Mouromtseff was the financial agent. Zworykin there presented himself to the Russian ambassador, Boris Bachmietieff.[1]

Bachmietieff offered Zworykin a position with the Russian Purchasing Commission, the financial agency of the embassy. This was headed by a Mr. Ugett and had offices in New York City. Completely out of funds at the time, Zworykin accepted this assignment. He was attached to the bookkeeping department as an adding machine operator, since they did not have a position for an engineer. He returned to New York and took lodgings in a boardinghouse in Brooklyn. There he continued to study English and to search for an engineering job in industry.

All of this time, Zworykin had been trying to locate his wife Tatiana. Several months after he arrived in New York, he found out through embassy channels that she was still living with friends in Berlin and was eager to join her husband in the United States. He was able to borrow some money and send for her, and in a few months they were reunited. They rented a house in Mt. Vernon, New York, and for a while life seemed to flow smoothly.

During his stay in New York, Zworykin never stopped looking for a job as an engineer. He finally received an invitation to come to East

Pittsburgh, Pennsylvania, to be interviewed for a position at the Westinghouse Electric research laboratories.[2]

At this time, the Westinghouse Electric and Manufacturing Company of East Pittsburgh was engaged in a life-and-death struggle with its archenemy, the General Electric Company of Schenectady, New York, the largest and most powerful electrical company in the world. During the war, GE had been furnishing certain electrical radio equipment, especially the Alexanderson alternator (a 20,000-cycle high-frequency radio transmitter which made long-distance communication possible), to the English Marconi Wireless Telegraph Co. Ltd. and the American Marconi Co. Ltd. to provide wireless communication between the United States and Great Britain.

At the end of the war, the English Marconi Wireless Telegraph Company had proposed to buy several more Alexanderson alternators, as well as patent rights, to expand its transatlantic services. This offer upset certain officials of the United States Navy who did not want wireless communications under the control of a foreign country. They suggested instead that GE suspend negotiations with American Marconi until after discussions with the Navy Department. These talks resulted in the formation of a new American-owned company, called the Radio Corporation of America (RCA), to take over the assets of the American Marconi Company. On October 17, 1919, RCA was incorporated, and on November 20 it took over the entire business of the American Marconi Company.

General Electric, which tried to establish control over the newly formed company from its inception, was the major shareholder and immediately set up a cross-licensing agreement with RCA in order to use the patents acquired from the American Marconi Company. Later, on July 1, 1920, a similar agreement was set up between GE and the American Telephone and Telegraph Company (AT & T) and the Western Electric Company, the manufacturing and research arm of AT & T.

Edward J. Nally of American Marconi became the first president of RCA. The chairman of the board was Owen D. Young of General Electric. Dr. Ernst F. W. Alexanderson of General Electric became RCA's chief engineer and David Sarnoff of American Marconi was named commercial manager. Sarnoff did not become general manager of RCA until April 29, 1921, and, despite the later claims, he played no role in the creation of RCA.[3]

Sarnoff, who had started as a telegraph operator with the American Marconi Company, had made excellent progress there, even before his promotions with RCA. Born in Uzlian, Russia, of poor and oppressed parents, Sarnoff had immigrated to the United States with his family when he was nine years old. Hard working and ambitious, he was determined to be a success in his adopted country.

Another myth about Sarnoff is the claim that during the tragic sinking of the ocean liner SS *Titanic* he was the sole communications link with the doomed liner through the SS *Olympic,* which assisted in the rescue operations. According to the best sources, this was not true; he was not alone. Wanamakers's radio station, where Sarnoff was manager, was closed when the first distress signal came from the SS *Titanic.* Only when Sarnoff heard of the sinking did he rush back to Wanamakers to become part of a vigil trying to establish communications with the ships in the area. This story, despite not being accurate, has become part of the legend of David Sarnoff.[4]

Certainly he was quite visionary and predicted the arrival of radio in the home. On September 30, 1915, Sarnoff sent his now-famous "Radio-Music Box" memorandum to his superiors at American Marconi. In it he said that "the idea is to bring music into the home by wireless. It was to come from a box which can be placed on a table in the parlor or living room, the switch set accordingly and the transmitted music received." Like most radical ideas, it was not well received by his superiors.[5]

The General Electric/American Telephone and Telegraph Company/Radio Corporation of America pact provided that the participants would make all their patents freely available to one another for ten years. GE was to have all the rights of wireless telegraphy and the manufacture of receiving apparatus. AT & T was to have exclusive rights in radio-telephony apparatus, plus the manufacture of transmitter apparatus. RCA had no manufacturing rights at all and was structured to operate the transatlantic service of the old American Marconi Company. All foreign patents were to be concluded through RCA, which was to act solely as a sales and service organization for the group.[6]

Westinghouse Electric and Manufacturing had also made plans to go into wireless communications after the war. For obvious reasons it had been kept out of the consortium, a situation which alarmed Westinghouse officials. Not wanting to be shut out of wireless communications, Westinghouse started to buy up all of the radio patents not yet owned by members of the consortium. Under the direction of Otto Sorge Schairer, manager of patent development at Westinghouse, the company acquired the patents of Michael Pupin, the powerful "feedback" patents of Major Edwin Howard Armstrong, and the radio patents of Reginald Fessenden.

A subsidiary, the International Radio Telegraph Company, from which many future Westinghouse managers came, was set up on May 22, 1920, to rival RCA. In addition, Westinghouse also started to hire many of the available scientists, including a cadre of recent Russian émigrés such as Stephen Timoshenko, J. Slepian, I. E. Mouromtseff, and others who had come to the United States after the Russian revolution.[7] Zworykin, of course, had no reputation at the time. However, the Russian émigrés at Westinghouse Electric were only too eager to help him land a position at the research laboratories.

When Zworykin visited the Westinghouse research laboratory at East Pittsburgh, he was so pleased with what he saw that he accepted a salary that was half of what he was receiving from the Russian Purchasing Commission. His friends in New York told him what a foolish thing he had done, but Zworykin was determined to go ahead. The only thing that held him up was that Tatiana was expecting their first child. He waited until his daughter Nina was born, on June 3, 1920, and then moved the family from Mt. Vernon to East Pittsburgh to start a new life in pleasant and less-expensive surroundings. Zworykin was able to rent nice rooms in the home of a telephone company engineer, and he was fascinated by his first assignment at Westinghouse as part of a group working on the development of a new amplifying tube (later known as the WD-11) for radio receivers of broadcasts from the newly constructed KDKA, the Westinghouse Electric radio station in East Pittsburgh.[8]

Radio station KDKA, which had been built and operated by Westinghouse engineer Frank Conrad, had started radio broadcasting in the United States as station 8XK with the presidential election returns in October 1920. It was so successful that other radio stations in the United States soon followed. It is ironic that Westinghouse Electric, as the traditional rival of General Electric, was the company that introduced commercial radio broadcasting, not RCA as so often has been claimed.

One result of Westinghouse's venture into radio broadcasting was that the company was finally allowed to join the consortium on June 30, 1921. Several other companies such as the United Fruit Company, the Tropical Radio Telegraph Company, and the Wireless Specialty Apparatus Company of Boston were also allowed to participate. This powerful group and its European counterparts now had such a stranglehold on radio patents and communications worldwide that it would be impossible for outsiders to build or operate a radio transmitter or radio receiver that was not covered by any of their patents.[9]

The Westinghouse research laboratory completely absorbed Zworykin. One of his early tasks was to prepare cathodes for radio tubes, which required coating a platinum filament with barium-strontium salts in order to make it a good electron emitter. At first the process was purely manual and the results were not uniform. Soon Zworykin proposed and built a semi-automatic coating machine that sped up the process and made the coating more uniform. In his manuscript, Zworykin also claims to have had other new ideas at the time, such as heating tube filaments by means of AC (alternating current) instead of by DC (direct current) using dry batteries, as was common at the time. Since these were new ideas, Zworykin was introduced to the preparation of patent disclosures. His biggest problem was in communicating, as he was having great difficulties learning to speak as well as read and write English.

A year went by quickly, and Zworykin had been promised a raise if his work was satisfactory. He knew that he had worked long and hard, so when he asked for the promised raise, he was shocked to learn that because of the recession that followed the war, he was expected to take a 10 percent cut in pay. He resigned on the spot and went looking for employment elsewhere.

The only work he could find was with a small company, C and C Developing Company of Kansas City, Missouri. He had obtained the position through correspondence and knew little of what was expected of him in the new job, but he took it on the basis of a salary that was double his Westinghouse pay.

Zworykin moved his family to Kansas City, where the director of the laboratory had rented him a large home in a pleasant part of the city. Once there, he found that his task was to build and equip a laboratory. C and C had a United States patent that claimed that high-frequency currents would speed the cracking process in oil refining. Zworykin's main task was to prove that this claim was valid.

Although Zworykin freely admitted that he knew nothing about oil refining, he had excellent knowledge of high-frequency currents. In a short while he was able to build the test equipment, complete the experiment, and submit a report on it to the board of directors. To his surprise and to the dismay of his employers, the experiment proved that the process did not work. The laboratory was immediately shut down and Zworykin was out of work again. He had been able to file for a few patents, in June and August of 1922, while he was with C and C, but they were of minor importance.

However, Zworykin liked Kansas City and found other things to do to make a living, starting by making custom-built radio receivers for a small radio company. They sold quite well over the Christmas holidays, but most came back for repairs because of faulty soldering. He then started testing a car radio that he claimed worked beautifully. However he was notified by the local police department that radios in cars could lead to accidents and he had to stop building and selling them.[10]

In February 1923, Zworykin learned that there had been major changes in the Westinghouse research laboratory, including the naming of a new manager, Samuel M. Kintner. Zworykin soon received an offer to return with a higher salary and a contract for several years. This was mainly because of the efforts of Otto Sorge Schairer, manager of the Westinghouse patent department, on whom Zworykin had made a great impression. Under the terms of the new contract, Zworykin retained rights to his prior inventions while Westinghouse acquired an exclusive option to purchase his patents at a later date. He accepted the offer by telegram, and a month later, on March 1, 1923, Zworykin returned to East Pittsburgh.[11]

At the laboratories, Kintner called Zworykin in and asked him what research project he would like to work on. Zworykin welcomed the opportunity to finally work on his long-held dream, an all-electrical (that is, without spinning discs or drums) cathode-ray television system. Zworykin drew up the plans for a completely electrical television system and submitted them to Kintner by April 4, 1923.

His system included cathode-ray tubes at both the receiver and transmitter. The receiving tube was quite conventional for 1923. It included a hot cathode, which was the normal means at the time for generating the electron beam, and used both electromagnetic (coils) and electrostatic (plates) deflection to move the beam on the usual fluorescent screen. There was no specific mention of the means for focusing the electron beam. However, as Zworykin mentions the use of argon in his camera tube, it may be presumed that he was using a similar gas for focusing his beam.

The scanning currents for deflecting the electron beam were to be generated primarily by rotating motor generators or triodes in oscillating circuits. As the camera and receiver deflection currents were driven by the same electrical signals, there were to be no problems of synchronization. The circuits for radio transmission were those available for the year 1923.

According to the patent application, the camera tube was of the most interest. The entire tube was filled with low-pressure gas such as argon. The photoelectric surface was deposited on a thin foil plate of aluminum oxide, which is an insulator. There was no connection between the grid, a simple metal screen with voltage on it, and the plate. However, wherever the beam struck the foil, it penetrated it.

This created a path (in the presence of the gas) to produce a connection between the foil and that particular section of the grid in the path of the beam. This caused conduction between the photoelectric layer and the grid that would be intensified by ionization of the argon vapor. In operation, the light from the image would cause electron emissions of varying intensity in accordance with the brightness of the subject. The degree of emission was dependent upon the amount of electrons freed from the surface covered by the cathode-ray beam. This current was then amplified and became the picture signal. This concept was very radical at the time and resulted in much controversy during the patent process.[12]

This was not the first patent application for a camera tube. A Frenchman, Edvard-Gustav Schoultz, had applied for the first camera tube patent in France on August 21, 1921. He was followed by a Russian, Boris Rtcheouloff, who applied for a camera tube patent on June 27, 1922. Even before them, though he had never applied for a patent on this device, Alan Archibald Campbell Swinton discussed an electric camera tube in his address before the Röntgen Society in 1911. Zworykin's tube was quite similar to Campbell Swinton's, with one major difference. Campbell Swinton proposed using a photoelectric target (the plate struck by the electron beam) that was composed of a mosaic of rubidium cubes, while Zworykin's target was to be made of thin aluminum foil covered with a layer of potassium hydride.[13]

It is interesting to note that in 1921, Marcus J. Martin published *The Electrical Transmission of Photographs,* which included a complete description and diagram of Campbell Swinton's 1911 system. It is almost certain that the Westinghouse patent department purchased a copy of this book for the research library, since they were committed to keeping up with the state of the art in phototelegraphy and television. Although he later denied it, this may have been where Zworykin first became aware of Campbell Swinton's work.[14]

As to the similarities between Zworykin's and Campbell Swinton's ideas, we must remember that the scientific world is full of simultaneous inventions, from widely separated parts of the globe. One example is the telephone, for which both Alexander Graham Bell and Elisha Gray filed for patents on the same day. Likewise, both Marconi in Italy and Popoff in Russia can legitimately claim to have invented radio in 1895. The same is true for the discovery of the aluminum extraction process by Charles Martin Hall in the United States and Pierre-Louis Touissant Heroult in France in 1886, or the creation of the electric light bulb in 1879 by Joseph W. Swan in England and Thomas A. Edison in the United States.[15]

Zworykin's plan was turned over to the Westinghouse patent department on October 8, 1923, and filed in the U.S. Patent Office on December 29, 1923. It immediately ran into opposition on several counts and took fifteen years to make its way through the Patent Office. There was no immediate effort in 1923 to reduce the patent to practice, and Zworykin continued to work on his many other projects.[16]

Meanwhile Zworykin was not the only one experimenting with television. The end of the war in Europe in 1918 had revived interest and stimulated new research. Édouard Belin, for example, who had been working on the transmission of still photographs by wire (facsimile) since 1906 and had sent the first still pictures over the Paris-Lyon telephone line in 1907 using a system called the Belinograph, gave a demonstration of "simulated" television at the Sorbonne in December 1922. Newspaper reports of this event prompted worldwide interest when they claimed that he was able to send "flashes" of light by wireless.

He was joined by other Frenchmen such as Georges Valensi and Dr. Alexandre Dauvillier, who had started their own independent research programs and made important contributions to the development of television. Valensi, chief engineer of the Postes Télégraphes et Téléphones Service, filed for his first television patent in 1922. So did Dauvillier, who was the chief of research physics of X rays at the Louis de Broglie laboratories. Interestingly enough, all of their patents had cathode-ray tubes as receivers. Somewhat further away, there were also reports of experimental work being done in Hungary by Dionys von Mihaly.[17]

In the United States, the idea of transmitting still pictures and television was taken over by an extraordinary inventor, Charles Francis Jenkins. In 1895, he had collaborated with Thomas Armat to invent the first motion picture projector, which was sold to Thomas A. Edison as the Vitascope. Sometime before 1922, he had turned his attention to television and was soon building working equipment. He applied for his first television patent on March 13, 1922, and by December 1923 was able to give the editors of two American radio magazines, Hugo Gernsback of *Radio News* and Watson Davis of *Popular Radio,* demonstrations of a very crude system using a unique scanning device, his "prismatic rings." These rings were a result of his attempt to devise a film projector that would use continuously running film. To do this, he designed a unique variable-area glass prism. As this prism rotated, it immobilized the film image as it sped through the projector. It did work after a fashion, but it created more problems than it solved and ultimately Jenkins abandoned it for the common Nipkow disc.[18] Still, the editors' reports were favorable. While they found the apparatus crude and cumbersome, they claimed that they could put their hands or other objects in the path of the light at the transmitter and see them on the receiver screen.

Jenkins's reputation was such that he was receiving valuable assistance from both General Electric and Westinghouse for his television experiments. He later thanked such high-ranking executives as L. C. Porter and Dr. W. R. Whitney of General Electric and both Harry P. Davis and S. M. Kintner of Westinghouse for their aid. For example, in his receiver he used the special neon glow lamp that had been developed for him by D. McFarlan Moore of General Electric, and in his transmitter he used the sensitive thalofide photoelectric cell developed by Theodore Case that was being used experimentally in GE's laboratories. These two devices were absolutely essential to any system of mechanical television.[19]

David Sarnoff, who had been promoted to vice president of RCA on September 18, 1922, while retaining the position of general manager, was advised of everything going on in the laboratories of General Electric and Westinghouse. He certainly knew of their efforts on behalf of Jenkins's television projects. It must be remembered that radio in all of its aspects, which would later include television, was the main concern of the radio group which included RCA, General Electric, and Westinghouse. Sarnoff would certainly have been informed of Jenkins's activities and he would have been most interested in any development that might hasten television's progress.

Surely these experiments were behind Sarnoff's comment in an address to the RCA board of directors on April 5, 1923, that for the first time "I believe that television, which is the technical name for seeing instead of hearing by radio, will come to pass in due course. The problem is technically similar to that of radio-telephony though of more complicated nature—but within the range of technical achievement. Therefore it may be that every broadcast receiver for home use in the future will be equipped with a television adjunct by which the instrument will make it possible to see as well as hear what is going on in the broadcast station." This prophetic speech by Sarnoff signaled the start of the long arduous path to develop the radio-television system that is in use today. Television had gained its most powerful supporter.[20]

Meanwhile in England, a Scotsman by the name of John Logie Baird decided to dedicate his life to perfecting the new science of television. Baird, who had failed in several earlier home enterprises, had decided that the Nipkow disc combined with modern photoelectric cells and electric amplifying tubes could produce a workable television system. With the help of Wilfred E. Day, a motion picture entrepreneur, he was set up in a laboratory at 22 Frith Street in London. He filed for his first television patent on July 26, 1923. He immediately started to build equipment that was admittedly crude. By early 1924, Baird was transmitting rough outlines of images using Nipkow discs with large glass lenses.[21]

For all of television research, the most significant event of 1924 was the publication of a paper, "The Possibilities of Television with Wire and Wireless," by A. A. Campbell Swinton in three April issues of the popular weekly radio magazine *Wireless World and Radio Review*. In this article, he described to a worldwide audience updated versions of his 1911 electric television scheme. After mentioning several ways of improving his original ideas, Campbell Swinton confessed that the task was hopeless unless taken over by "one of the big research laboratories like that of the G. E. C. or the Western Electric Co., one of those people who have a large staff and any amount of money to engage in the business, I believe that they would solve this thing in six months and make a reasonable job of it."[22]

This article stimulated a great deal of interest, and several companies, GE among them, started programs researching television technology. GE's program began in July 1924, under Dr. Ernst F. W. Alexanderson. By January 1925, Alexanderson gave a crude demonstration of television using a Benford crystal, a scanning device which consisted of a ten-sided glass prism that took the place of a rotating drum for film transmission. (Alexanderson wanted to avoid the use of Jenkins's prismatic rings and hoped that Benford's device would work instead.) The test was arranged by Dr. Benford, with the able assistance of his colleagues, Hoxie, Rockwood, Long, and others.[23]

Elsewhere, too, Campbell Swinton's article stimulated research into television. Manfred von Ardenne and Dr. August Karolus in Germany and Kenjiro Takayanagi in Japan had begun research programs of their own. Karolus started his experiments at the Physical Institute of Leipzig University in 1924, using the Nipkow disc as both a transmitter and a receiver. His main research was in the development of a Kerr cell to be used as a light valve (modulator) at the receiver.[24] Takayanagi later claimed that he had started his television experiments in 1924 at the Hamamatsu Technical Institute and that he actually built his first "camera tube" then. It failed, for which Takayanagi blamed inadequate technical know-how. He always used the cathode-ray tube as a receiver in his experiments.[25]

Nineteen twenty-four was a very important year for Vladimir Zworykin as well. On September 16 he received his naturalization papers and became a proud citizen of the United States. At the same time, he enrolled in the physics department of the University of Pittsburgh, which, by arrangement with Westinghouse Electric, offered him graduate credit for work performed in the laboratory. The university also gave him credit for his studies at the St. Petersburg Institute of Technology and his work with Professor Langevin at the Collège de France. In this way he was to receive a Ph.D. within two years, a goal which had become very important to him. In short, things were going well for Zworykin, and he was becoming very comfortable in his new life in the United States. Because his financial condition had improved, he was able to buy a small home in Wilkinsburg, a suburb of Pittsburgh, and a new car. He could even afford a maid to help his wife. At this time, too, his second daughter was born, a source of great joy. He had a many friends, most of them former Russian engineers and scientists who had fled the revolution and were now working at the Westinghouse research laboratories.[26]

Zworykin certainly learned of the Campbell Swinton magazine article from the Westinghouse patent department library, and he was able to use it to persuade Westinghouse to reduce his patent application to practice as soon as possible. His efforts to build and operate a cathode-ray television system started sometime in June 1924.[27]

This project became an officially sanctioned Westinghouse Electric laboratory experiment, and Zworykin was issued a work order authorizing him to obtain materials and manpower. His most important aide was Chris, the Westinghouse glassblower who made all the new camera tubes. The receiving tubes were to be converted Western Electric gas-filled oscilloscope-type tubes, which were available commercially and could be altered for other uses.

Zworykin often claimed that these early camera tubes closely followed his 1923 patent application. However, he complained of the problems of such a radical design at a time when sealing and vacuum techniques were in their early stages of development. Getting a proper layer of photoelectric material on the thin signal plate was more of an art than a science. The construction of the partition, consisting of a metal screen with an insulating layer and a photosensitive substance on top of the insulation, between the cathode-ray section and the photoelectric cell was particularly troublesome. Zworykin also claimed in a confidential report on the problems of television that he had a difficult time getting a suitable material for his signal plate and that a cloth of quartz or Pyrex glass would be more suitable but he had not been able to find a supplier for this.[28]

Not only was it difficult to construct the tube according to the specifications in his patent application, it was equally hard to get the new tube to operate in the way he had envisioned. Much later, in November 1962, he admitted that his camera tube worked in a rather unorthodox manner, based on what he called the principle of "bombardment-induced conductivity." There is no doubt, however, that these tubes actually did work, as several witnesses later testified.[29]

For his receiver, Zworykin used a conventional Western Electric oscillograph tube (224A) with a major modification. He claimed that he separated the filament from the cathode so that modulation of the beam was possible. To simplify matters in his patent specification drawings, Zworykin always specified alternating current generators (of the rotary type) to generate the currents for deflecting the electron beam. By gearing the two generators together he could produce one set of currents, one of, say, 1000 cycles for horizontal deflection and the other one of sixteen cycles for vertical deflection of the beam. This caused the beam to sweep back and forth both horizontally and vertically. This was known as sine-wave scanning. Since both camera and receiver were fed from the same signals, synchronization (keeping the camera and receiver in step with each other) was no problem.

Zworykin had to make or modify everything he used. Along with the special camera tubes and modified receiving tubes, he had to design and build amplifiers, power supplies, deflection scanning circuits, scanning coils, and so on. Another problem was that the stability of electrical circuits and their components such as vacuum tubes, resistors, and condensers was very poor. In general, the equipment was erratic and he had to repair and redesign many of the circuits to get them to work at all.

Zworykin claimed that in order to continue his work he needed a larger work space, more manpower, and a bigger budget. So sometime in late summer or early fall of 1925, he decided to give a demonstration to Harry P. Davis, the general manager and vice president of the Westinghouse Electric and Manufacturing Company.

While I have not been able to pin down the exact date of this demonstration, it seems to have occurred between August 16 and October 22, 1925, by which time a successful camera tube had been built and tested. In a letter to the commissioner of patents dated October 22, 1925, Zworykin's patent lawyer Wesley Carr stated that "the Nicolson [Western Electric's U.S. Pat. no. 1,470,696 of 1917] is a mere paper patent illustrating only an idea, while applicant's *device has been built and tested* and gives great promise of successfully solving the problem of transmission of motion pictures [emphasis added]."[30]

On the day of the demonstration, everything seemed to be in order, though Zworykin had had to work through the night to repair the damage to the circuits caused when he blew several condensers in a last-minute attempt to improve the system's performance. Present along with Davis were Otto S. Schairer, manager of the Westinghouse patent department and an old ally, and Samuel Kintner, manager of the Westinghouse research laboratory. It appears that Zworykin had several camera tubes available and picked out the most promising one for the demonstration. In order to simplify matters, he chose to project an X onto the target of the camera tube.

The camera tube was in fact working, but it sent out only a feeble signal and the image of the X that appeared on the face of the receiving tube was faint. Even this Zworykin took as a great achievement. To the managers present, however, it looked more like utter failure. The picture was dim, with low contrast; it had poor definition; and, worst of all, it was stationary. In addition, the use of motor generators for sine-wave scanning produced an erratic pattern on the screen.

Clearly Davis had expected something more worthy of his attention. After asking Zworykin a few polite questions about how much time and money he had spent on this project, Davis got up and walked out. Zworykin later claimed that he had naively "scotched his own case" by giving a vivid description of the technical difficulties which remained to be solved, how long it would take, and how much more it would cost. In any case, as he left, Davis whispered in a "off-stage" voice to Sam Kintner that perhaps it was better to get this "guy" to work on something more useful. This was quite a blow to Zworykin, who was then told that he was to do no more work on television but should stick to projects like photoelectric cells and phototelegraphy (the transmission of still pictures) which did have great commercial promise.

In spite of Davis's negative reaction, Zworykin's demonstration was important as the first of its kind in television history. No one had ever transmitted a picture from an electric camera tube to a receiver tube in this manner. Put simply, Zworykin had accomplished his goal of an operable camera tube.[31]

Despite Zworykin's later claims, the demonstration was not all-electric, since the deflection currents were created by motor generators instead of vacuum tubes. It would be some four more years before anyone could claim to have an all-electric television system and that someone would *not* be Zworykin.[32]

There was a bright side to this apparent disaster. All the information gathered from the laboratory experiments on the camera tubes that Zworykin had been building in 1924–25 was incorporated into a new patent application filed by Westinghouse on July 13, 1925. The new application was based on his December 1923 patent application, but it included many improvements such as the necessity for having a "mosaic" made up of individual cells in order for a camera tube to operate correctly. This patent also revealed the concept of "globules" (discrete elements) for the first time. Each globule was very active photoelectrically and for all practical purposes constituted a minute photoelectric cell. It also described a mesh target (with minute openings) which an electron beam could penetrate with ease. It is almost certain that these concepts were incorporated in some of the radical camera tubes actually built and demonstrated by Zworykin at this time.

Also included in the new 1925 patent was a remarkable color system based on a mosaic three-color (Paget) screen within a single camera tube envelope. At the receiving tube, a corresponding color screen reconstituted the color image in a similar single receiving enclosure. This patent ran into very few problems and was granted on March 31, 1927, in Great Britain and on November 13, 1928, in the United States.[33]

In addition to filing the new patent application, on October 2, 1925, the Westinghouse patent department decided to incorporate this information into Zworykin's pending December 1923 patent, adding that "the photoelectric material is potassium hydride, deposited in such a manner that it is in the form of small 'globules,' each separated from its neighbor and insulated from the aluminum·oxide." Both the Westinghouse patent department and Zworykin recognized that the idea of "globules" was imperative to his December 1923 patent application. The problem was that Zworykin had originally specified a "layer" of photoelectric material and it was now quite apparent that even if this layer was quite thin, the individual electrical charges stored in it would quickly disperse transversely across the plate and destroy any possibility of discrete information. This of course, meant no picture signal, only an average value of the illumination on the photoelectric plate. Unfortunately for Westinghouse and Zworykin, the United States Patent Office eventually ruled that this was "new material" and therefore not allowable.[34]

At any rate, the television project was taken away from Zworykin and turned over to Westinghouse engineer Frank Conrad. Conrad, who was highly favored by Davis for his work at KDKA, turned to more conventional methods such as mechanical scanning. At this time it was also disclosed that Westinghouse Electric had been using their own funds for this project. Although contrary to their agreement with RCA regarding radio research, this freed them of certain legal responsibilities to the patent pool. While Westinghouse and GE jointly owned and operated RCA, their mistrust of each other and of RCA had never diminished.

Under Zworykin's direction, the Westinghouse laboratories were also developing a new light valve for image display. This was Zworykin's mercury-arc device, which he had been working on with Dr. Dayton Ulrey since late 1923. Theoretically, if it could be rapidly modulated, it would provide a bright source of light to be used in a scanning disc receiver. The only light modulator available at that time was D. McFarlan Moore's neon glow lamp, which was quite limited in its brightness—and belonged to General Electric, which was also in the process of acquiring Karolus's patents on an improved Kerr cell (light modulator) to be used with an arc lamp.[35]

Prohibited from doing any more experimental work on television, though he was allowed to file for television patents, Zworykin continued building experimental photoelectric cells and vacuum tubes in the Westinghouse laboratories, being careful to direct his work along lines of interest to Westinghouse. This required adroit maneuvering. He had learned by this time that it was impossible to work on an idea in commercial research without camouflaging it, unless one could convince commercial people of its immediate profitability.

Up to 1925, the photoelectric cell was a separate part of any apparatus. Zworykin later claimed that to justify his work on a more sensitive photoelectric cell, he had decided to combine the vacuum tube amplifier and photocell in the same envelope. He produced a tube so sensitive that it could detect a whiff of smoke passed between the cell and the light source. It had commercial use as a fire alarm in unattended buildings, the turning on and off of lights, and even the matching of exact colors.

The Westinghouse publicity department was notified of this device and reported it to newspapers all over the country on October 21, 1925. They claimed, falsely, that this new cell was also the basis for the system of television being developed in the Westinghouse Electric laboratories by research engineer V. K. Zworykin. Zworykin was quoted as saying that "All the processes needed for television, or the projection of motion pictures by radio are already in existence." He continued, "The theory is all right, but at present the apparatus would have to be endless, cumbersome and uncertain. But it will be simplified. It will take some years; but we will have eventually the instantaneous or nearly instantaneous transmission of motion pictures."

From a scientific point of view, Zworykin's new tube was not of great value, and, ironically, it really had nothing to do with a television system. It is quite possible that the real reason for this media event was that Westinghouse had been prepared to disclose Zworykin's 1925 television system. Since it hadn't performed satisfactorily, the publicity department decided to promote his new tube instead.

Zworykin claimed that he was not quite pleased with this event and that he was embarrassed by the "undeserved publicity." However, he showed no reluctance to participate in such publicity events in the future and in fact seemed to relish the acclaim he received. This was his first exposure to the American public and from this time forward the name of Vladimir Kosma Zworykin was to become part of the American scene.[36]

Research into television accelerated worldwide during 1925. In January, the American Telephone and Telegraph Company also began a research and development program directed by Dr. Herbert E. Ives. The policy of the Bell System was "to develop all forms of communications which might be supplemental to telephony." They felt (and properly so) that they had a right to experiment with, build, and sell anything connected with the transmission and recording of sound and/or pictures. Their research laboratories at Western Electric and the newly opened (January 1, 1925) Bell Telephone laboratories in New York City were the finest in the world. Each had a superlative staff with an almost unlimited budget and the best scientific equipment available.[37]

In April 1925, Dr. Max Dieckmann and Rudolf Hell applied for a German patent on a unique camera tube which electrically converted the entire optical picture into an "electron image" instead of using a single electron beam to scan a target. It was a cold-cathode tube that used high voltage to free the electrons from the photoelectric plate. The electrical image thus produced would be moved sequentially by means of coils or plates across a single anode in an aperture. The current from this anode would be led to an amplifier and become the picture signal. This form of scanning of the image inspired the name "image dissector." While Dieckmann and Hell claimed to have built several tubes of this type, they had to admit later that they could never get any of them to work.[38]

On June 13, 1925, Charles Francis Jenkins gave his first public demonstration of radio-television in Washington, D.C. He was able to transmit the image of a small revolving windmill from the naval radio station NOF in Anacostia to his laboratories in Washington, some seven miles away. Jenkins's staff at the time consisted of Sybil

L. Almand, Florence M. Anthony, John N. Ogle, James W. Robinson, Stuart W. Jenks, and Thornton P. Dewhirst.[39]

In August of 1925, Dieckmann had a television system on display at the Munich Trades Exhibition. He was using a cathode-ray tube as a receiver, much like his 1909 apparatus. Although this was the first public display of a television system using a cathode-ray tube for a receiver, there is no record that it was actually operating at the time.[40]

Campbell Swinton's 1924 article had also been noted in the Soviet Union, where several patents based on his ideas were applied for in 1925–26. Among them were one by A. A. Tschernischeff and one by the team of B. P. Grabovsky, F. E. Popoff, and N. G. Piskounoff.[41]

Other forms of research prospered as well, as progress was made in the use of slides and motion picture film and sound recording. The Western Electric Company's research into sound recording was particularly successful. In 1925, Western Electric introduced an all-electrical method of sound recording that was quite superior to the mechanical-acoustical system then in use. This new system was offered to the Victor Talking Machine Company and Columbia Records and created a revolution in the recording industry.[42]

Also in 1925, using the same new recording principles, Western Electric gave a private demonstration of this new sound system to Samuel Warner of Warner Brothers Pictures. Using sixteen-inch discs running at 33⅓ rpm, they were able to synchronize the improved sound records with the film in a projector. This system, later known as the "Vitaphone," was used in *The Jazz Singer*, whose release by Warner Brothers on October 6, 1927, signaled the end of the silent picture.[43]

In addition to their new method of sound recording for records and films, AT & T and the Western Electric Company (now known as the telephone group) were very much involved in radio broadcasting. They had long ago divested themselves of their RCA radio stock, while still maintaining rights to the patent pool, and in 1923 they had started to build and operate radio stations. Now they were about to add toll broadcasting to their long-lines and local telephone service. This was considered a threat by the radio group (General Electric, Westinghouse, and RCA) and especially by David Sarnoff, vice president of RCA, who did not want the telephone group involved in radio broadcasting. Sarnoff's poor relationship with the telephone group exacerbated the continuing battle between the two giants of communications. Whenever a conflict of interest arose, Sarnoff was the focal point, and he was ultimately quite successful in keeping the telephone group at bay.

In 1926, after many difficult and secret negotiations, Sarnoff had finally come to an agreement with the telephone group that turned over radio station WEAF in New York City to RCA in return for about $1,000,000 and an agreement that the newly established National Broadcasting Company, originally Sarnoff's brainchild, would use AT & T's long-lines services. This agreement spelled the end of the telephone group's ventures into radio broadcasting.[44]

NBC was incorporated on September 9, 1926, and was jointly owned by RCA (50 percent), GE (30 percent), and Westinghouse (20 percent). The company officially began broadcasting on November 15, 1926, with two networks, the "Red," with radio station WEAF as its flagship, and the "Blue," with station WJZ. By the end of the year, twenty-five stations were operating under the NBC banner. Merlin H. Aylesworth was NBC's first president.[45]

In 1926, Zworykin was successful in developing a process to make a permanent photoelectric cell, that is, one that would not lose sensitivity, by electrolysis of potassium through a pure potash glass bulb with no soda. Using a bath of molten potassium nitrate (KNO_3), he made cells equal to the best sodium cells being manufactured at the time.

This was part of his studies at the University of Pittsburgh, and he was awarded the coveted Doctor of Philosophy degree in 1926. His thesis, appropriately, was entitled "Study of Photoelectric Cells and Their Improvement." Now it was Dr. Vladimir Kosma Zworykin, a title he was to wear well.[46]

Nineteen twenty-six was also an important year in television history, during which at least three demonstrations were given. The first and most important was given on January 27 by John L. Baird, who invited members of the British Royal Institution in London to a demonstration of mechanical television using Nipkow discs. He was able to show them flickering thirty-line images at about five frames per second.

Behind this successful demonstration was a process, known as "flying spot" or "spot-light" scanning, which Baird had discovered on October 2, 1925. This new way of scanning made it possible for the first time to televise faces with half-tones instead of just silhouettes or outlines of figures. Although it had already been patented by Georges Rignoux in France (1908) and A. Ekstrom in Sweden (1912), and a U.S. patent had been applied for by John Hayes Hammond, Jr. (August 1923), Baird applied for his patent on flying spot scanning on January 20, 1926, eight days before he gave this demonstration.[47]

At this time, due to the insensitivity and sluggishness of the photoelectric cells available, it was almost impossible to get any significant amount of electrical signal by directly illuminating the subject. The amount of light reflected back to the photoelectric cell was so minute as to be useless. However, by inverting the scanning process, that is, by projecting a powerful beam of light through the spinning disc directly onto the subject, it was possible to get enough light reflected back to the photoelectric cells to give a useful electrical signal.

Fear of industrial espionage led Baird to be extremely secretive about this new process, even to the point of misleading claims that its success was the result of a supersensitive photoelectric cell or some exotic circuit. Still, this public demonstration of half-tone television, crude as it was, was the first of its kind and is considered to be an important milestone in television history. It also gave Baird's backers much-desired publicity.[48]

In France, Édouard Belin had continued his work on cathode-ray television. He had been joined by Dr. Fernand Holweck, chief of staff of the Madame Curie Radium Institute. Together, they designed an operable television method, called the Belin and Holweck system, which they demonstrated on July 26, 1926, to General Gustave A. Ferri, head of the French military telegraph system, Professor Charles Fabrie of the Sorbonne, and René Mesny of the French Academy of Sciences. The demonstration was directed by Gregory N. Ogloblinsky, chief engineer of the Laboratoire Établissements Édouard Belin at Malmaison, who had constructed the apparatus. The receiver employed a special cathode-ray tube built by Holweck. It was metallic, continuously pumped (to maintain a vacuum), and used magnetic focus. It used negative grid modulation and operated with a high voltage of about 1000 volts. At the transmitter, two synchronous motors created currents for the magnetic coils and drove two small mirrors vibrating at right angles using the Rignoux flying spot system. These were used to pick up the images that were sent by wire to the receiver. The pictures were only thirty-three lines at ten frames per second and could only show silhouettes, not half-tones, but the process could show both moving objects and photographs.[49]

Just one week later, on August 2, 1926, the television apparatus of Dr. Alexandre Dauvillier, chief of physical research of the Louis de Broglie laboratories, was presented to the scientific community. It too used a cathode-ray tube as a receiver, though Dauvillier's was more modern than Holweck's because it was made of Pyrex glass and was sealed. Like Holweck's it used magnetic focus with a trace of gas. It had a willemite phosphor screen and used negative grid modulation of only one volt without displacement and deformation of the spot.

The beam was accelerated after modulation and before deflection. Dauvillier claimed that it operated at 300 volts to produce a very brilliant spot. He too was limited to the use of dual vibrating mirrors attached to tuning forks for picking up his images. Thus he could only produce forty-line pictures at ten frames per second. Dauvillier gave a public demonstration of his system at the Exposition de Physique in Paris on May 27, 1927, and both systems were widely reported in radio journals. Both were to have a great influence on Zworykin's future activities.[50]

On November 16, 1926, Frank Gray of the Bell Telephone laboratories reported to Dr. H. E. Ives that they had also been experimenting with cathode-ray tube reception. The standard Western Electric tube had been modified so that simple objects were reproduced, but without shades of tones. Gray noted that changes of intensity (modulation) of the electron beam caused changes in both the focus and the position on the screen, thus distorting the picture.[51]

A month later, on December 15, 1926, Dr. Ernst F. W. Alexanderson of GE revealed publicly for the first time that his company was developing a television system. He intimated that since it was so easy to transmit still pictures (the facsimile process), merely speeding up the process should ensure success. His preferred method was to use a multispot projector to get sufficient brightness on a screen. Yet in spite of all the publicity given the Alexanderson multispot projector, there is no evidence that it ever produced a recognizable picture. However, Alexanderson's remarks were enough to prompt the Bell Telephone laboratories to give a demonstration of their system as soon as possible.

Alexanderson's comments on cathode-ray television were also of interest. Obviously he was aware of the negative results of the Zworykin/Westinghouse demonstration, which led him to remark that "If we knew any way of sweeping a ray of light back and forth without the use of mechanical motion, the solution to the problem would be simplified. Perhaps some such way will be discovered, but we aren't willing to wait for a discovery that may never come. A cathode-ray can be deflected by purely magnetic means, and the use of the cathode-ray oscillograph has been suggested." However, he held out little hope for such a development.[52]

On January 31, 1927, David Sarnoff, now a lieutenant colonel in the United States Signal Corps Reserve, spoke to the Army War College on the same theme as Alexanderson's speech. He called attention to the multitude of uses of television in the war of the future. He said, among other things, that "It is conceivable that a radio-television transmitter installed in an airplane might be useful in transmitting a direct image of the enemy's terrain, thus enabling greater accuracy in gunfire."[53]

Just two months later, the first significant public demonstration of television was given by the Bell Telephone laboratories. On April 7, 1927, AT & T gave the finest exhibition of television yet, under the direction of Dr. Herbert E. Ives and Dr. Frank Gray. Pictures and sound were sent by wire from Washington, D.C., to New York City. There was also a wireless transmission from Whippany, New Jersey, to New York City, some thirty miles away. In addition to a speech given by Herbert Hoover, the secretary of commerce, there was a program of amateur vaudeville. Bell claimed that the pictures by wire were equal in quality to those sent by radio.

Bell's system produced fifty-line pictures at eighteen frames per second and used DC (direct current) restoration to maintain the black level. This was demonstrated on several small Nipkow disc receivers with two-inch screens that produced sharp, clear pictures with half-tones. There was also a two-foot bent glass tube screen that produced pictures of acceptable quality.[54]

The television project had started in January 1925 and had progressed rapidly. By July 1925, Gray and John R. Hofele, his chief laboratory assistant, were sending half-tone images across the laboratories from both slides and motion picture film. However, the concept of film transmission did not fit in with the proposed "person to person" service, similar to the telephone. Around May and June 1925, Gray also independently discovered the principle of the flying spot scanner. Although it worked well for film and slides, for some unknown reason he did not actually try it out on a live subject until February 10, 1926. At that time, Gray was able to televise live images with half-tones of good quality. By April 1927, when the impressive demonstration was held, Bell Telephone laboratories had spent countless thousands of dollars on research on the project and used up to one thousand men to complete it. It was the most consequential television demonstration given to that date.[55] But, since John Baird had still not revealed the secret of his earlier successful demonstration, it was Frank Gray's public disclosure of the flying spot scanner that made live half-tone television possible. For the time being, although the scanner could only be used in a darkened room, all successful demonstrations of television used this principle.

Among the many guests witnessing the demonstration in New York City was David Sarnoff. He was both astonished and dismayed to discover that the telephone group was light-years ahead in television research after he had maneuvered them out of radio broadcasting. In fact he now found that the telephone group had surpassed the radio group in the most important race of all, that for a workable television system. His immediate response was to order both General Electric in Schenectady and Westinghouse Electric in East Pittsburgh to speed up their research on television. Westinghouse was to concentrate on the transmission of 35mm motion picture film (a program which continued under the direction of Frank Conrad after Zworykin's removal), using their own funds.[56] At General Electric, where Alexanderson's group had been working on the transmission of motion picture film, a research program managed by William Arthur "Doc" Tolson, assisted by Raymond Davis Kell, now concentrated on live transmissions. Alexanderson abandoned his multispot approach, which had never really shown any promise, and adopted the flying spot scanner that Gray had recently revealed. Duplicating the telephone company's efforts, GE's researchers were soon transmitting twenty-four-line pictures at sixteen frames per second in their laboratories.[57]

In addition to the work being done at GE and Westinghouse, Sarnoff ordered a research program on television started within RCA itself. RCA had a technical and test department at 7 Van Cortlandt Park South in New York City, run by Dr. Alfred N. Goldsmith, chief engineer and a vice president of RCA. The research director was Julius Weinberger; the engineering director was Arthur F. Van Dyck; and Raymond Guy was in charge of broadcast station design. In addition to their main task of new product evaluation and planning, they were engaged in many kinds of original technical research. Late in 1927, as ordered by Sarnoff, they began an independent program of research and development into television headed by Theodore A. Smith, with eight laboratory assistants. RCA's program was separate from those of General Electric and Westinghouse, though all of RCA's electrical and mechanical components came from them.[58]

The Bell Telephone laboratories demonstration was soon followed by the successful efforts of Kenjiro Takayanagi of the Hamamatsu Higher Technical Institute of Japan, who claimed that he had been working with cathode-ray television since 1924 and had received his first image of the Japanese character イ on a cathode-ray tube on December 25, 1926. Takayanagi adapted Frank Gray's newly revealed flying spot scanner to his cathode-ray tube system and carried out successful experiments late in April 1927. He was using a Tokyo Denki gas-filled tube, similar to the Western Electric oscillograph, with a forty-hole Nipkow disc rotating at fourteen revolutions per second.

It appears that he was the first to use the cathode-ray tube with the flying spot scanner and get acceptable results. Part of his success came from the fact that his tubes were viewed from the phosphor side (through the tube) so as to get as much brightness as possible.[59]

In July 1927, the television apparatus of Georges Valensi was described. He too was using a cathode-ray tube as a receiver. It was gas-filled and used magnetic deflection. The transmitter was an arc projector that could only be used with opaques or drawings. Light from the object went through two special rotating discs that dissected the image before it reached the photoelectric cell. Synchronization was achieved by using two motors geared together so that only one sync signal had to be transmitted. There was no report on the quality of his images.[60]

In the midst of all of this renewed interest in television, Dr. Vladimir K. Zworykin at the Westinghouse research laboratories was perfecting a facsimile machine, a new radical photoelectric cell, and the recording of sound on motion picture film.

CHAPTER *6*

The Kinescope

The years 1928–29 were to be the most important in Zworykin's life. Early in 1928, he was involved in three different projects for Westinghouse, working on the recording of sound on motion picture film and on a new, highly sensitive photoelectric cell, as well as finishing up work on a new phototelegraphy (facsimile) transmitter and receiver. Assisting him on the facsimile project was a student engineer named Harley Iams, who had just graduated from Stanford University. Zworykin found him very compatible, and they were to work together for many years.[1]

The transmission of still pictures by wire or radio for news or business was quite important at the time. Most of the existing systems used special paper or needed to process the picture before transmission. The machine that Zworykin and Iams were working on was to be much simpler. The picture to be transmitted was wrapped around a cylinder and light from a standard source reflected back to a special magnesium-cesium photoelectric cell filled with argon gas. As the picture rotated, it was systematically scanned, turning the light into electrical signals. At the receiver, a similar rotating cylinder was covered with standard five-by-eight-inch bromide photographic paper. The incoming picture was scanned by means of a special helium glow tube that converted the electrical signals back into light fluctuations. Synchronization was assured by means of a seventy-cycle tuning fork used at the transmitter. Zworykin claimed that the machine could transmit a five-by-eight-inch picture in half-tones in forty-eight seconds or a written message at 630 words per minute.[2]

To make this machine practical, Zworykin and Dr. E. D. Wilson invented a new, highly sensitive photoelectric cell that combined magnesium and cesium. Cesium had two advantages—it is highly sensitive to light and its response to color is quite close to that of the human eye. The problem was that cesium is hard to handle; it is liquid at room temperature and in a vacuum is so volatile that it quickly distills and coats any insulating surface. Unlike earlier experimenters, who tried without success to incorporate cesium into a photoelectric cell, Zworykin and Wilson found that freshly distilled magnesium had the property of retaining cesium on its surface without altering appreciably the photoelectric character of the alkali metal.

Thus, they were able to produce a magnesium film in the normal fashion, by evaporation in a vacuum, which served not only to bind the cesium but also to act as a metallic conductor on the inner surface of the glass bulb. Zworykin and Wilson claimed that gas-filled cesium cells yielded currents of about twenty-five microamperes per lumen, many times the sensitivity of the common potassium hydride photoelectric cell, while high-vacuum cells yielded currents of about two microamperes per lumen. Zworykin filed for a patent on this revolutionary cell on March 3, 1928.[3]

Zworykin's other project, developing a special Kerr cell for the recording of sound on film pictures, reflected the fact that at this time the motion picture industry was in the throes of the conversion to sound pictures and almost every large electrical laboratory was trying to perfect a system that would compete with Western Electric's Vitaphone. Talking pictures had been around for many years, even back to the early work of Thomas A. Edison. Lee de Forest had been producing sound on film as early as 1923. By 1928 Theodore Case, who had invented the sensitive Thalofide photoelectric cell that had earlier been used by Jenkins and others for television, had produced a new cell called the Aeo-Light, which produced quite clear sound on film. Aeo-Light was purchased by William Fox and became known as the Fox-Case Movietone sound system. The German Tobis (Ton-Bild-Syndicate) sound system seemed to enjoy a monopoly in the European film recording industry, especially UFA and Klangfilm.[4]

General Electric had long had in its laboratory a sound on film recording system based on the 1919 Pallotophone (Pallophotophone) of GE engineer Charles H. Hoxie. It had been developed during World War I to record incoming radio signals at high speed by photography. GE's research showed some promise, but it had largely remained undeveloped. However, when Western Electric revealed its new Vitaphone process in 1926, David Sarnoff and GE decided to try to make their system competitive. As with radio broadcasting and television, Sarnoff would not allow the telephone group to have a monopoly on sound pictures. GE's sound on film system was renamed the Photophone process and laboratory work at GE's radio department proceeded under the direction of Elmer Bucher, executive vice president of sales. Dr. Alfred N. Goldsmith was vice president in charge of engineering and Elmer Engstrom was chief engineer for this project. In March 1928 a separate corporation, RCA Photophone, was set up with David Sarnoff as president.[5]

Westinghouse Electric was also engaged in research into talking pictures. Zworykin was working with L. B. Lynn and C. R. Hanna to develop a special Kerr cell that could be used as a light valve (modulator) to photograph the sound track on the film. The main advantage of the Kerr cell was its rapid response, with practically no frequency limitations, but it also had problems, such as rather poor optical efficiency. Zworykin, Lynn, and Hanna, however, claimed that they had overcome these problems and were able to produce a practical cell, similar to the Kerr cell designed by Dr. August Karolus, which was now owned by General Electric.[6]

Zworykin later claimed that as a result of his work on the Kerr cell he received a lucrative offer from a large motion picture company in Hollywood to continue his work there. Although two of his colleagues accepted the offer, Zworykin decided to stay and continue his research.[7] Most of this loyalty was due to the Westinghouse patent department under Otto Schairer, which was successfully processing Zworykin's many patent applications. Zworykin later told an interviewer that "I had considerable difficulty in securing patent protection. I didn't write English well and I was working in a new art. My experience in this regard had impressed me tremendously with the importance of a good patent lawyer in the process of invention." Of equal importance, the Westinghouse patent department was defending Zworykin from interference actions. For instance, in June and July of 1928 the Westinghouse patent lawyers had won two patent interference suits brought by a newcomer to the television field, Philo Taylor Farnsworth of San Francisco.[8]

Philo Farnsworth had been born near Beaver City, Utah, and raised on a small farm in Rigby, Idaho. He had very little formal education, but claimed that he had become interested in television as early as 1922 as a result of a magazine article. He arrived at some original ideas about a television system and later discussed them with his high school science teacher, Justin Tolman. By 1926 he had interested a small group of bankers in his ideas and with their help started serious work in Los Angeles. They soon moved him to a larger laboratory in San Francisco at 202 Green Street. With a very small but efficient staff he started to build all of his own equipment.[9]

Farnsworth applied for his first patent on what became known as the "image dissector" tube on January 7, 1927. He was quite unaware of Dr. Max Dieckmann's prior patent application as it had not yet been issued in Germany. The patent was for an electric television system using a unique cathode-ray tube for the transmitter. The camera tube was of unusual design and differed from the Campbell Swinton scheme in that it did not use a single electron beam to scan a photoelectric target. It was essentially a cold cathode (no hot filament) tube in which the electrons from the target were attracted to an anode by a high voltage. As in the Dieckmann patent, the entire optical image was converted into an electrical "image" that was deflected (horizontally and vertically) en masse toward a single electrical anode. Here the liberated electrons were sent to an amplifier that sequentially produced a picture signal. The release of electrons from the target was instantaneous, with no possibility of storing them.

Farnsworth's original patent did not specify a cathode-ray receiving tube as he was convinced that he could not patent such a basic device. However, at first he designed and built gas-filled tubes similar to those used in the Western Electric oscillograph. Then he converted to high-vacuum tubes with "long" focus coils (the full length of the tube) using magnetic focus and deflection. He also used rotating motor generators and tuning forks in his early efforts at scanning and synchronization. Farnsworth never resorted to any form of rotating wheel, disc, or mirror, and cathode-ray tubes were always used as receivers.[10]

By September 7, 1927, he was able to transmit a single horizontal line of light (no vertical scan) on the face of the tube, which proved that his principles were sound. Soon he was able to transmit the outlines of images from slides and motion pictures. On March 1, 1928, in an effort to sell his system, Farnsworth gave a private demonstration to James Cranston, vice president of the General Electric laboratories, and Dr. Leonard F. Fuller, of the University of California. It was not a success; the coils of the dissector tube heated up badly and the demonstration was finally ruined when the potassium became hot and distilled off the cathode. Cranston and Fuller did not think much of the system, since Farnsworth's picture tube (like most magnetically focused devices at the time) could at best produce small, dim images.[11]

All of this information about Farnsworth's progress was passed on to both the General Electric and Westinghouse laboratories. Because of these reports and the patent interferences, Zworykin was very much aware of what Farnsworth was doing. However, except for Farnsworth's camera tube, Zworykin was about four years ahead of him and was able to get most of his patent claims allowed. And even though the Dieckmann and Hell dissector patent was issued in Germany on October 3, 1927, it seemed to be of little concern to Farnsworth, General Electric, or Westinghouse.

Farnsworth's backers did not give up, though they were getting anxious about the mounting costs of development and the slow progress being made. On May 22, 1928, there was another effort to interest General Electric in buying up Farnsworth's television scheme. Cranston, who had seen the system, contacted General Electric patent attorney C. E. Tuller about Farnsworth's idea. Tuller referred the matter to Albert G. Davis, vice president of General Electric, who wrote him, "Our people do not think that Mr. Farnsworth's television scheme amounts to very much, but that he has done some pretty scientific work with very limited facilities." He continued, "if he wants a job with us, and is willing to recognize that the particular scheme which he has worked out is of very little if any value, as we now see it, there should be no difficulty in making arrangements with him." Finally, he stated, "if he should come to work for us, we should have to buy whatever he has invented to date and whatever he invents while in our employ comes to us under the regular engineering contract." This, of course, Farnsworth was not willing to do.[12]

Meanwhile, General Electric had been perfecting a mechanical television system since May 1927. On January 13, 1928, they demonstrated television with accompanying sound from their research laboratories in Schenectady. The telecast was under the supervision of Dr. E. F. Alexanderson, chief engineer. As usual, whenever a television demonstration was given, David Sarnoff, as vice president and general manager of RCA, was present to view the results. The system used the "flying spot" scanner which Dr. Frank Gray of the Bell laboratories had introduced as a pickup device. The picture was announced as being forty-eight lines at twenty frames per second, though it was actually twenty-four lines at sixteen frames per second, and it was about one and a half inches long and one inch wide. A special magnifying glass was used to enlarge the image up to three inches. The scanning disc receiver used the special neon glow lamp invented by D. McFarlan Moore. As part of the demonstration, receivers had been put in the homes of E. W. Rice, Jr., president of General Electric, E. W. Allen, GE's vice president in charge of engineering, and Alexanderson. The demonstrators claimed that the home receivers were getting pictures as good as those received at the GE laboratory.[13]

Later in 1928, on May 5, Charles Francis Jenkins gave a demonstration of motion pictures by radio to members of the Federal Radio Commission (FRC). For his transmitter he used an ordinary 35mm motion picture projector with its intermittent removed, and a forty-eight-hole disc running at fifteen frames per second. Jenkins showed his ingenuity by having his pictures received on a special "drum receiver" of his own design, which used translucent rods to transmit the images to a glass screen. For reasons that were never made clear, Jenkins was only able to transmit silhouettes of figures with no half-tones. He informed the FRC that he intended to sell his "Radio-Movies" instrument as an accessory to the regular radio receiver.[14]

Also in May of 1928, Kenjiro Takayanagi of the Hamamatsu Higher Technical Institute gave a demonstration of television to the Japanese Society of Electricity. He was using a custom-made cathode-ray tube for a receiver with Frank Gray's "flying spot" scanner for a transmitter, and a special phosphor screen built at an angle into the tube so that it was viewed from the phosphor side. He was able to show faces with half-tones on a screen of forty lines at fourteen frames per second. He later claimed that "images were recognizable and identification somehow possible."[15] Representing European research efforts, Dionys von Mihaly gave a demonstration of mechanical (Nipkow disc) television in Berlin.

In addition, the first demonstration of television at a radio show was given at a convention in Chicago in June 1928, by one Ulysis Sanabria.[16] And on July 12, 1928, the Bell laboratories in New York City gave the first demonstration of outdoor television using sunlight with a scanning disc, something which no other researchers had been able to do. Bell's success, another historical first for the telephone group, was mostly due to the use of a special Case Thalofide photo-electric cell. (This was not revealed by the Bell labs at the time.) It used a fifty-hole, three-foot disc with a new optical system using a six-inch lens.[17]

It was now time for Westinghouse Electric to reveal what it had accomplished in its laboratories. It gave a demonstration of "Radio-Movies" on August 8, 1928, from radio station KDKA in East Pittsburgh. Present was David Sarnoff, now executive vice president of RCA, who had just returned from Europe. This demonstration was so important that he was joined by the top RCA/Westinghouse/General Electric management. In addition to David Sarnoff (RCA), it included Frank Conrad (WE), Dr. A. N. Goldsmith (RCA), M. H. Aylesworth (NBC), Dr. E. Alexanderson (GE), L. W. Chubb (WE), H. P. Davis (WE), E. Bucher (RCA), E. R. Taylor (NBC), O. S. Schairer (WE), W. R. G. Baker (GE), and Sam Kintner (WE).

The demonstration of 35mm motion picture film was of sixty lines at sixteen frames per second. Pictures were sent two miles by land-line to the broadcasting station and transmitted by radio two miles back to the receiver. The film transmitter used a sixty-hole disc with a cesium photoelectric cell. The receivers used a similar sixty-hole disc with a special mercury arc device that provided the modulated light to the ground-glass screen. A 5000-cycle signal from a tuning fork was transmitted on a designated carrier wave to synchronize the synchronous motors at the receiver. Westinghouse was said to be contemplating the manufacture of commercial "radio-movies" receivers that would be sold through RCA.

Conspicuously missing from this demonstration was any mention of Dr. Vladimir K. Zworykin. True to its word, Westinghouse Electric had given him no active role to play in this presentation. However, Alexanderson confirmed that the cesium photoelectric cells used in the transmitter and the mercury-arc light modulator at the receiver were Zworykin's inventions.[18]

On August 11, 1928, General Electric attempted the first "outdoor remote broadcast" using a television camera to cover the acceptance speech of Governor Al Smith, Democratic candidate for president, at the capitol in Albany, New York. While the Bell labs outdoor demonstration had actually been accomplished using daylight, General Electric decided to use a different method. Since their optical system and cesium photoelectric cells were not as efficient as the Case Thalofide Cell, they used their indoor flying spot camera. This camera was actually a box with a spinning disc and a 1000-watt bulb inside. This was accompanied by a portable photoelectric cell unit that was used to pick up the picture signals. The camera was set up by Ray Kell, who planned to send the television signal some eighteen miles away to the General Electric laboratories. Here, another GE engineer, Alda Bedford, was to monitor the picture and rebroadcast it on General Electric station WGY. Though GE claimed that all went well at rehearsals, when the event started the newsreel cameras turned on their powerful lights and wiped out the picture. So while the demonstration made for good publicity, it was not a success.[19]

On August 24, 1928, Philo Farnsworth of the Crocker-Bishop Research Laboratory gave a private demonstration of his system to two members of the Pacific Telephone Company, J. E. Heller and L. A. Gary. They reported that the picture was of rather low intensity on a screen one and one-quarter by one and one-half inches. Images were hard to identify, but motion was easy to follow. Heller and Gary were impressed by the photoelectric transmitter (Farnsworth's dissector tube) and by the fact that it was scanned electrically.[20]

Ten days later, on September 3, 1928, Farnsworth gave his first demonstration to the press in San Francisco, who reported that his system was capable of transmitting some twenty pictures per second. The article in the *San Francisco Chronicle* showed Farnsworth holding his image dissector tube in one hand and a small cathode-ray tube in the other. The reporter noted that "The image was only 1.25 inches square with a queer looking little image in bluish light now, that frequently smudges and blurs." However, the article went on to assert that the basic principle had been proven and that perfection was just a matter of engineering.[21]

On September 11, 1928, the General Electric station WGY presented the first live television drama, "The Queen's Messenger" by J. H. Manners. It was telecast with three flying spot cameras on wheels to follow the action. An audience of scientists and reporters viewed the program from a nearby building and claimed, according to the *New York Times* reporter, that "The pictures were small, sometimes blurred, not always in the center of the screen and hard on the eyes because of the flicker."[22]

At this time Ray Kell at General Electric was also doing research into cathode-ray television. Although Alexanderson was more favorably inclined toward mechanical television, he allowed Kell to pursue experiments with cathode-ray tubes. On October 12, 1928, Alexanderson reported to C. J. Young of the radio engineering department that he believed that although there were many difficulties to be overcome, such as a very dim picture (which is exactly what Frank Gray of the Bell Telephone laboratories, using identical equipment, had also recently discovered), that without question this would be "a step toward the television receiver of the future."

On December 14, 1928, Kell reported to J. Huff of the patent department that he was transmitting images with the standard spiral disc of twenty-four holes at twenty frames per second to a standard Western Electric 224 oscillograph. He had devised a circuit for linear scanning of the beam to coincide with the motion of the scanning disc. With this device he was able to produce pictures of fair quality. This was the only cathode-ray tube research being carried out by the radio group at this time.[23]

On September 24, 1928, Zworykin was present, along with Charles Francis Jenkins, at a meeting of the Society of Motion Picture Engineers (SMPE) at Lake Placid, New York. He was there for the presentation of Lewis Koller's important paper on cesium-on-silver-oxide photoelectric cells. Here he also met with François Charles Pierre Henroteau from Ottawa, who spoke with both Jenkins and Zworykin about details for a patent on a camera tube with an electric mosaic.

He had filed for this patent on August 10, 1928. At this time Zworykin was in the process of finishing up his paper on the Westinghouse facsimile transmitter and receiver.[24]

On November 13, 1928, Zworykin received an American patent for his improved all cathode-ray television system. A British patent had been issued earlier, on March 31, 1927, and through this Zworykin's new system had come to the attention of A. A. Campbell Swinton. Campbell Swinton was so pleased with it that on June 1, 1927, he had written a letter about it to Alexanderson of General Electric. Campbell Swinton's enthusiasm may have been a factor in changing Alexanderson's attitude toward cathode-ray television. Zworykin's patent showed excellent promise, and its publication confirmed for the first time Zworykin's role in cathode-ray television.[25]

On November 18, 1928, the Sunday *New York Times* published an article by David Sarnoff entitled "Forging an Electric Eye to Scan the World." In it he predicted that "Within three to five years I believe that we shall be well launched in the dawning age of sight transmission by radio." Sarnoff divided television into two major categories: (1) radio motion pictures, which would transmit film directly to the home, and (2) radio television, which would instantaneously project light images either directly from the object in the studio or indirectly from the scene through the broadcasting station by way of remote control.

Sarnoff's reference to an "electric eye" was particularly interesting. While the article itself gave no details, its timing indicates that he had seen or heard something while he was in Paris that summer that sparked his interest in cathode-ray television.[26]

At any rate, the day before this article was published, Zworykin set sail for Europe aboard the SS *George Washington*. He claimed that he had been ordered by Westinghouse Electric to go to Europe to visit some laboratories connected by agreement with their organization. While it may have been Westinghouse, as his immediate employer, that sent him to Europe, the trip was certainly on the instructions of David Sarnoff because all radio and television research conducted by General Electric and Westinghouse was on behalf of RCA. Zworykin was the one person in the radio group who had interest in and understanding of cathode-ray television. In addition, because he was fluent in both French and German, Zworykin would be at ease in Europe. It has even been suggested that Sarnoff met personally with Zworykin beforehand and made arrangements with Westinghouse for his trip. In any event, the trip was hastily arranged. Zworykin applied for a passport on November 1, and was issued one on November 3.

This speedy approval indicates that perhaps the State Department took a personal interest in his application. He was scheduled to go to England, Germany, and France.[27]

According to Harley Iams, "Zworykin went to Europe to investigate the status of ideas applicable to cathode-ray television there." This would limit the field to Belin and Holweck, Dauvillier, and Valensi, all in France. Iams reported that he received postcards saying that Zworykin's trip was successful. In fact, this journey turned out to be the most rewarding trip of Zworykin's life.[28]

There is no record of where or whom Zworykin visited when he arrived in England. The Westinghouse licensee in London was Metropolitan Vickers; outside of receiving patent applications, they were not actively working with television at the time. Neither was the General Electric licensee, the British Thomson-Houston Company, nor the RCA licensee, the Marconi Wireless Telegraph Company.

From England, Zworykin went on to Berlin, where he visited the laboratories of Telefunken A. G., a company that was working on mechanical television under the direction of Dr. Fritz Shröter. Telefunken was the RCA licensee for all patents, including those of Westinghouse and General Electric, in Germany. Telefunken's forty-eight-line television system was using a flying spot scanner with the Nipkow disc and, like GE, used the very efficient Kerr cell of Dr. August Karolus as a light valve (modulator) for its receivers.[29]

From Germany, Zworykin went to Paris, where he first visited his old friend, Professor Paul Langevin, at the Collège de France. Although the French version of RCA was the Compagnie Générale de Télégraphie sans Fil there is no record of Zworykin visiting them. However, even though it had no business connections with Westinghouse, or with any member of the radio group, Zworykin went to visit the Laboratoire des Établissements Édouard Belin at Malmaison.

Here he was introduced to the staff by Édouard Belin, its founder. He met Dr. Fernand Holweck, who was quite famous in Europe for his design of demountable (able to be dissembled) electron tubes and vacuum pumps and was also chief engineer of the Madame Curie Radium Institute. Zworykin also met Gregory Nathan Ogloblinsky, a trained physicist who was Belin's chief engineer, and an engineering consultant, Pierre Émile Louis Chevallier.

Here too Zworykin was shown the Belin facsimile phototransmitter, which was quite similar to the device he was working on for Westinghouse. Both used a photoelectric cell for pickup, but Belin's used a Blondel (vibrating mirror) oscillograph as a light modulator at the receiver while Westinghouse's used a glow lamp.[30]

Belin, Holweck, and Chevallier must have felt secure in showing their revolutionary tube to Zworykin, since they had been notified by the French patent office on July 17, 1928, that their patent had been accepted and was to be issued on November 29, 1928.

Belin and Holweck then went on to demonstrate their latest television development. Zworykin had certainly known of the earlier Belin and Holweck apparatus of 1926. However, this device was now obsolete. Clearly, Zworykin had specifically come to Paris to inspect and evaluate a different device.

Belin's latest version of this apparatus featured an electrostatically focused cathode-ray picture tube, all metal and made in sections. It was continuously pumped (a Holweck specialty) to maintain a high vacuum, and there was a glowing cathode encased in a metal structure that produced the electron beam. It included a modulating grid that was part of the concentration electrode. There was an assembly for mounting internal plates for deflecting the beam vertically, and a set of magnetic coils wrapped around the neck of the tube for horizontal deflection. The electron beam was modulated, and then deflected and accelerated simultaneously. A fluorescent screen was attached to the large end of the tube. The proper difference of electrical potential between the concentration electrode and the insulated metal envelope which included the screen and the large section of the metal envelope surrounding it caused the beam to be sharply focused.

The pickup device was still the pair of vibrating mirrors geared together using the Rignoux flying spot system that Belin and Holweck had used since 1926. The pictures had no half-tones, and were formed in a zigzag pattern both horizontally and vertically. The screen showed a thirty-three-line picture with very little detail and medium brightness. However, Zworykin must have been impressed with the sharpness of the images.[31]

Of the three main methods for focusing an electron beam (gas, electromagnetic, and electrostatic), electrostatic focus had the worst track record. As L. T. Holmes and H. G. Tasker had found in experiments in 1924, its poor performance in the laboratory gave it very little chance for success, although part of their problem had been that their tube was filled with mercury vapor, which spoiled its performance.[32]

Gas focus, as used in the Western Electric oscillograph tube, had found the widest commercial acceptance. Its use made for a simple and relatively inexpensive tube. The beam was kept in focus by the presence of ions from the gas which attract the electrons in the center of the beam. However, the gas filling limited the tube's life because it slowly destroyed the oxide of the electrode, and the destruction was worse when high anode potential was used. It also had severe problems when modulated. This caused a variation of brightness accompanied by a loss of focus if the potential swing was high. In all, the use of gas focus in a "picture" tube was quite limited.

Magnetic focus, the oldest of the three methods, just didn't work well at low voltages. It was cumbersome and suffered from severe problems due to the action of electromagnetic (and electrostatic) fields on the electron beam. These displaced the beam's spiral path with an additional component to its velocity that blurred the focus. Its beam charged up the screen and after a while turned itself off. When modulated, the strength of the beam altered both its velocity and its focus. To get sufficient brightness, magnetic focus was used mainly in very high voltage devices that were continuously pumped; and these were found only in specialized laboratories.[33]

However, two German scientists, Walter Rogowski and Walter Grosser, had a different approach to solving the problem of electrostatic focus. In December 1923, they had filed for the first patent covering a glowing cathode oscillograph with "electron focus." This was to be accomplished by suitably arranging and shaping the main and auxiliary electrodes. They described a tube built on these principles in an article published in December 1925. An early explanation of this phenomenon, which became known as "electron optics," was offered in October 1926 by another German scientist, Hans Busch.[34]

Meanwhile the first person to apply this solution to television was Dr. Alexandre Dauvillier. In early February 1925 he filed for a French patent that claimed that "an electrical field between (figures) 18 and 16 caused the electron beam to be focused on the screen." This patent was issued on July 20, 1926. He could not get the device to work, however, and the tube he demonstrated in May 1927 had a slight amount of inert gas in it and was focused by a large magnetic coil surrounding the neck of the tube.[35]

About this time Holweck and Chevallier also started to work on a similar tube. Établissements Édouard Belin applied for a patent for it on March 4, 1927. Holweck and Chevallier made their new tube work by eliminating all gas from the tube and using a pure electron discharge. They found that it was possible to focus the beam sharply by applying the correct ratio of voltages to two properly spaced circular discs so that it could be modulated without changing either its focus or electron speed. Finally, they built a tube incorporating these principles. It was a metallic, gas-free tube in which the beam was focused by applying a set ratio (one-fourth to one-half) of electrical potentials between two diaphragms. The first was a concentrating element operating at about 600 volts. A higher voltage (2000 volts) was fed to an insulated section of the large section of the tube. This comprised the metal cone, the glass screen, and the body of the tube. The beam was modulated and then deflected while being accelerated and focused simultaneously.

This was Zworykin's introduction to electrostatic focus, and we can assume he congratulated Belin, Holweck, and Chevallier for their efforts. He knew that electrostatic focus, while an old idea, had not previously shown any promise in a cathode-ray tube. Yet now it seemed to be the most fruitful approach for a practical television picture tube. It was a tremendous advance over the Western Electric gas-focused tube that had dominated the oscillograph field in the United States and Europe and inhibited further research worldwide.[36] In the new tube, Zworykin had found the display device that the nascent television industry had sought for so long. He immediately knew the tremendous potential of this cathode-ray tube, but he also knew it could not be used in its present form. Using electrostatic focus as the principal building block, he envisioned certain alterations in the tube's structure.

Zworykin planned three important changes to the tube. The first was to move the deflection units (coils/plates) back between the first and second anodes. This would allow the electron beam to be modulated and deflected at low voltages. Then it would be accelerated and focused by the second anode, which was part of the pear-shaped portion of the tube. This was to be done by metallizing the interior of that part of the glass bulb. This now became an accelerating/focusing anode. This would allow him to apply accelerating potentials of voltages of thousands of volts to attain the tremendous brightness desired. The beam would be totally unaffected by the control and accelerating potentials. Second was to ignore the fixed ratio of potentials as prescribed by Chevallier, but to use higher second anode voltages for greater brightness while still maintaining electrostatic focus. And third, he planned to eliminate the vacuum pump and use a glass bulb of simple construction that was quite inexpensive, mobile, gas free, and sealed permanently.

Metallizing or coating the interiors of glass tubes was by now an old art, and the Hungarian scientist Paul Selenyi had recently described a tube in which this was done in order to increase the electron beam intensity. However, no one had ever used it as part of an electrostatic focus scheme. As often happens with inventions, all of the elements existed separately; it would be Zworykin's great task to put them together.[37]

First, Zworykin had to come to terms with the Belin laboratories. It appears that he was authorized to make some kind of business arrangement with Holweck on behalf of Westinghouse to buy one of these tubes and bring it back with him. Licensing and/or patent rights were to be worked out. It also appears that he promised to have Westinghouse Electric hire both of Belin's engineers (Chevallier and Ogloblinsky) at a later date and perhaps even to employ Holweck as a consulting engineer.

Zworykin returned to the United States on December 24, 1928, with the latest metallic demountable Holweck/Chevallier cathode-ray tube and a Holweck high-speed rotary molecular vacuum pump. He immediately gave a report of his visit to Sam Kintner. For reasons certainly relating to the strained relations between Westinghouse and RCA, Kintner seemed to show little or no interest in Zworykin's find.

Since Zworykin had gone to Europe on behalf of RCA rather than Westinghouse, though this fact has often been overlooked, Kintner suggested that a meeting with David Sarnoff would be in order. Zworykin immediately prepared to go to New York City and speak with Sarnoff personally.[38]

As executive vice president of RCA, Sarnoff wielded enormous power at this point. He was preparing to go to Paris on February 1, 1929, to participate in the conference on German reparations, and he was in the midst of a plan to separate RCA from its dominant senior partners, General Electric and Westinghouse. As early as 1927, both General Electric and Westinghouse had tentatively agreed to a plan that would allow RCA to build and sell its own equipment and would transfer elements of GE and Westinghouse radio facilities and staffs to RCA. Sarnoff's dream was to have his own production complex, and he had the huge Victor Talking Machine Company factory in Camden, New Jersey, in mind. RCA had arranged to purchase the Victor Talking Machine Company in January 1929 for $154,000,000. The deal included Victor's manufacturing plants in Camden, New Jersey, Oakland, California, and Argentina, as well as all the stock of the Victor Talking Machine Company of Japan, the Victor Talking Machine Company of Canada, the Victor Talking Machine Company of Brazil, the Victor Talking Machine Company of Chile, and the Gramophone Company of Great Britain. With these details worked out, Sarnoff was about to go to Europe confident that the new plan would be implemented by the RCA Board of Directors.[39]

Sometime around January 2 or 3, 1929, Zworykin met with Sarnoff in New York City. At this time, Zworykin was in New York for a meeting of the Institute of Radio Engineers (IRE), at which he delivered a paper on facsimile picture transmission. According to the *New York Times* of January 3, 1929, "Mr. Zworykin had arrived here yesterday after spending two months in Europe studying radio development work." He was quoted as saying that "This country is far ahead of Europe in television and facsimile transmission."[40]

This was not their first meeting. Sarnoff claimed that he had first met Zworykin in either 1926 or 1927, possibly after the successful Bell Telephone labs demonstration, and that at this time the broader aspects of Zworykin's efforts to develop an electrical television system were made known to him. It is certain that the two met early in November 1928, when Zworykin was ordered to go to Europe later that month. Their relationship seems to have been purely business and there is no evidence of any personal relationship between the two men. However, they had found a perfect partnership, in which each served the other's purpose in life—power and status for Sarnoff, accomplishment and fame for Zworykin.

At their meeting, Zworykin related the results of his trip, telling Sarnoff that this new picture tube was the answer to the main problem of television because it made possible a simple receiver that could be installed in the average home, was completely electrical, needed no maintenance, and would provide a bright picture in a normally lit room.

Zworykin also indicated that he had the tube operating in the laboratory. This was quite true, since he had brought back a working, electrostatically focused, demountable Holweck/Chevallier tube, though it needed a major modification to improve its performance. This mainly involved repositioning the deflection section so that he had the benefit of modulation and deflection of the beam at low voltage and then acceleration and focusing the beam at high voltages. In addition, the tube needed special scanning coils and electrical circuits to operate properly.

Zworykin also needed a source of picture signals. Since he had no camera tube at the time, he had a choice of either the flying spot scanner from General Electric or the Westinghouse/Conrad 35mm motion picture film projector. He chose neither. Knowing that it would be some time before he could show actual pictures, he still told Sarnoff, according to Lyons, that "basically the instrument was already operative, but it still required intensive—and expensive—development."[41]

When Sarnoff asked how much a workable television system would cost, Zworykin told him that he needed a few additional engineers and facilities, but that he hoped to be able to complete the development in about two years, at an estimated cost of about a hundred thousand dollars. Sarnoff replied, "All right, it's worth it," and the conference was over. The estimate, of course, turned out to be a far cry from the exact amount needed to finish the system. Sarnoff then ordered Westinghouse to provide Zworykin with the personnel and equipment necessary for such a project. This was no problem, as Westinghouse simply billed RCA for all expenses incurred.

Zworykin always boasted about his great salesmanship in getting Sarnoff to give him the financial backing that was so desperately needed at the time. Clearly, Sarnoff had great faith in Zworykin and really did not need convincing. He was pleased to promote and finance any project that would put RCA ahead of every other company in the field of television, especially the telephone group.[42] In any event, after this meeting Zworykin was set up in his own laboratory at Westinghouse. Sarnoff visited Zworykin in East Pittsburgh on January 27, 1929, just before he left for Europe on the SS *Aquitania* on February 2, 1929.[43]

The first item of business was an order to the Corning Glass Company for thirteen glass bulbs to be used for future picture tubes. Zworykin later said that actual work started on February 1, 1929. Two projects were begun immediately. First, as Zworykin had no camera tube at the time, he decided to use motion picture film as a source of picture signals. A standard Simplex 35mm motion picture projector, different from the Westinghouse/Conrad film projector, was ordered, which would be modified for television transmission. Second, while waiting for the glass bulbs to arrive from Corning, Zworykin decided to build several handmade glass tubes to prove that his improvements were correct.

Three men were assigned to this project by Westinghouse. According to Harley Iams, who was working with Zworykin on the facsimile project, he was put to work designing deflection circuits and the new cathode-ray tube. Another engineer, Cecil Friedman, was to modify a standard Western Electric oscilloscope tube by adding a grid to modulate the beam current. An Englishman, Dr. W. D. Wright, had been assigned to the project as optical engineer.

The purchase of the Holweck molecular rotary vacuum pump made it possible to process (evacuate) their own tubes. According to Arthur Vance, Zworykin used a laboratory De War flask (rather than the modified Western Electric tube) to incorporate all of his advanced ideas.[44]

At any rate, with the principle of electrostatic focus in mind, Zworykin soon produced an experimental glass tube that was portable and gas-free, and had the electron beam modulated at low voltage before deflection and acceleration. It had a second anode in the form of a silver (alkaline silver nitrate) coating on the inside of the bulb. After deflection, this second anode accelerated and focused the beam by high voltage with high brightness. Finally, he had the fluorescent screen connected to the anode so that it drained off the surplus electric charges that had previously accumulated there. Whichever device was used until the actual glass bulbs arrived from Corning Glass, Vladimir K. Zworykin had fathered the predecessor of all future picture tubes.

With this picture tube, Zworykin completely changed the course of the history of television. Several other experimenters, among them Chevallier and Holweck and Rogowski and Grosser, were working on the basic problems of electrostatic focus. In fact, on December 12, 1928, three Germans, Max Knoll, Ludwig Schiff, and Carl Stoerk, applied for a German patent incorporating electrostatic focus, and on March 22, 1929, Professor Roscoe H. George of the Purdue University Engineering Experiment Station presented a paper to the American Institute of Electrical Engineers (AIEE) describing a new metallic oscillograph tube that used electrostatic focus. George's tube, however, which he claimed to have had working prior to February 7, 1928, was clearly for oscillographic purposes only; it was designed to detect and measure lightning surges, and the absence of a beam-modulating electrode made it useless for television reception.[45] Still, on May 7, 1929, Dr. George and Purdue University signed an agreement with Grigsby-Grunow, manufacturers of Majestic radio receivers, to develop a television system using George's cathode-ray tube as the picture-reproducing device. Their first picture tubes apparently were all metal, continuously pumped, using variable velocity modulation of the electron beam, which means that the beam sped up to decrease its brightness and slowed down to increase brightness, but had no beam-control electrode. Electrostatic focus was accomplished by means of two circular discs.

This was very different from the sophisticated glass tubes Zworykin and his staff were building and operating at Westinghouse, but the fact that George's tubes did use electrostatic focus was later to cause RCA many legal problems. George and Howard J. Heim went on to develop a complete television system around advanced versions of this tube, ultimately similar to the kinescope, until 1931. Then, because of the Depression, Grigsby-Grunow stopped financing the project.[46]

For use with Zworykin's revolutionary picture tube, a standard 35mm Simplex motion picture projector was obtained and altered to remove its film intermittent. Wright, the optical engineer, designed a system of scanning the film with a simple vibrating mirror that was moved by magnetic coils fed from a simple vacuum tube "sine wave" generator. (Remember that Belin and Holweck's system used two small mirrors that moved mechanically in phase.) This mirror would transmit the light from the projector through the film onto a photo-electric cell by means of a relay optical lens system. At the picture tube, the electron beam could easily follow the movement of the mirror as it was fed the same signals from this generator. Thus the horizontal scan was simple sine wave scanning. With the film moving continuously, vertical scan was linear (sawtooth) using a toothed gear connected to a commutator driven by the film to indicate the start of each frame. The modified film projector was completely assembled by the end of April 1929.

As promised by Sarnoff, several topflight radio engineers were assigned to the project. In addition to Harley Iams, Cecil Friedman, and Dr. W. D. Wright, John Batchelor arrived in April, Arthur Vance in May, Gregory N. Ogloblinsky from Paris in July, and Randall Ballard in September 1929. Of these, Ogloblinsky was in some ways the most talented. He had arrived from France alone, since Pierre Chevallier apparently had other plans than coming to work for Zworykin and Westinghouse. This was later to have dire consequences for Zworykin and Westinghouse.[47]

Ogloblinsky immediately became part of the Zworykin engineering group. There was no question about his competence; trained in both physics and mathematics, he brought with him over four years of active work on both facsimile and cathode-ray television. He does seem to have had mixed feelings about his new American co-workers, however. He felt that their strength was in their diversity of talents, since they were all electrical engineers with a variety of backgrounds. He thought that their main weakness was in what he called a certain naiveté, and he claimed that he took pleasure in showing them the error of their ways. He related to Zworykin, however, and complimented him for his excellent management.

Ogloblinsky arrived too late to help with the picture tube project, then in the hands of John Batchelor and J. D. Harcher. Video amplifiers, deflection circuits, and high-voltage supplies were designed and built by Arthur Vance, who was the expert in electric circuitry. The power supplies were built by an engineer named S. Skypzak. The radio receiver was designed by Randall Ballard, and the general layout of the receiver was by Harley Iams and a student engineer named Pepper.

Whether Zworykin had picked his small but superb engineering staff or Westinghouse assigned them to him is not known, but they served him admirably. The project ran into very few problems, and the first satisfactory picture tube was finished by the middle of April 1929. Zworykin called it the "Kinescope" (*kineo* from the Greek meaning "to move" and *scope* "to observe"). The next step was to develop a suitable electrical circuit, and on May 10, 1929, the first demonstration of movie film over three separate sets of wires was given. Finally a circuit for the transmission of moving pictures over a single radio channel was developed using one radio channel for picture signals and synchronization. On August 17, 1929, pictures transmitted by radio were demonstrated before a group of General Electric and RCA engineers including Dr. Alfred N. Goldsmith.[48]

David Sarnoff had come back triumphant from his trip to Paris in June 1929. He was given a private demonstration of Zworykin's new tube when he came back and was quite impressed when he saw the Westinghouse logo displayed on its face. His gamble on V. K. Zworykin had paid off! After the August 17 demonstration, top management was so pleased that they asked to have the parts packaged into six self-contained cabinets to be sold by Christmas of 1929. This was done by November.[49]

However, Sarnoff was very disappointed to discover that the restructuring of RCA still left General Electric and Westinghouse very much in control. He threatened to resign if his original intention was not carried out. To satisfy him, a new company, RCA Victor, was formed with RCA owning 50 percent of its stock, General Electric 30 percent, and Westinghouse Electric 20 percent. Sarnoff was to be chairman of the board. RCA Victor was formed on October 4, 1929, and incorporated on December 26, 1929.[50]

The RCA engineering group under Goldsmith at Van Cortlandt Park had also made excellent progress using the flying spot scanner as a transmitter and the Nipkow disc as a receiver. They had applied for a television license from the Federal Radio Commission in April 1928, and were now in the midst of equipping a television station at 411 Fifth Avenue, adjacent to the RCA Photophone recording studio.

They had recently decided to go with the Westinghouse standard of sixty lines at twenty frames per second, rather than GE's forty-eight lines, within a 100kc bandwidth. On May 14, 1929, Julius Weinberger, Theodore A. Smith, and George Rodwin of Goldsmith's group presented a paper before the Institute of Radio Engineers requesting consideration of a set of standards suitable for a commercial television system. This was the earliest attempt by RCA to set standards for television transmission and reception.[51]

Late in June 1929, the RCA television station W2XBS began telecasting some two hours daily, from 7 to 9 P.M. This included many hours of transmitting images of a Felix the Cat doll standing on a rotating turntable. *Radio News* reported not only that "conservative" RCA was broadcasting on a regular schedule but that there were rumors of "impending RCA television receivers as well."[52]

Ogloblinsky was present at the August 17, 1929, Westinghouse/GE/RCA demonstration. Having apparently decided to keep Chevallier informed of everything happening at Westinghouse, Ogloblinsky wrote to him on September 11, 1929, relating details of the August demonstration. He stated that the demonstration had indeed gone very well. Concerning the new glass kinescopes, he said that their focus was quite good with a spot size of 2mm or so. He also reported that they were using the Holweck/Chevallier metallic oscillograph, which he claimed had even smaller beam size and better picture quality.[53]

Since his arrival from Paris in July, Ogloblinsky had been working with Zworykin on a new camera tube. Zworykin knew that the only competition was from Philo Farnsworth. This had been disclosed privately to General Electric and publicly to the press in September 1928. As far as can be ascertained, in spite of conflicting reports, Zworykin had no plans to experiment with the dissector. He was determined to build his own camera tube using the principles he had proposed in 1923 and 1925 and taking advantage of the newer techniques used in developing his picture tube. For instance, by using electrostatic focus, he could scan a photoelectric target with precision. Without gas in the tube, he had no problems with contamination of the photoelectric surface or the oxide coating on the filament. Finally, the new, highly sensitive Koller cesium-on-silver-oxide photoelectric surface could be incorporated into the camera tube.

Zworykin and Ogloblinsky built their first camera tube using the Holweck/Chevallier demountable tube, which was now available for other uses since Corning glass tubes were being used for the new picture tubes. Since the Holweck/Chevallier tube was demountable, it could be taken apart and rebuilt to any desired configuration. Zworykin stated in his report to Westinghouse that "after a study of its performance and proper alterations, a transmitter was assembled for 12 line [*sic*, but Zworykin must have meant twelve frame] pictures."[54]

The fluorescent screen was replaced with a transparent glass plate and a crude photoelectric target consisting of several rows of plugs or rivets set into holes in a thin insulated plate was built. The side of the plugs facing the light was silver plated, then inserted into the tube.

Here, in an atmosphere of oxygen, the silver was oxidized. From a receptacle in the tube, a mixture of cesium dichromate and silicon was heated by a high-frequency coil and deposited on the oxidized silver plugs. This was the photoelectric surface. The tube was given a final exhaust and purged of all gases by the Holweck vacuum pump.

In front of the target was a mesh screen that furnished the "polarizing" voltage. This was connected to one side of a resistor through a battery. The light from the scene (motion picture film) charged up each plug, and the charge was stored electrically. The electron beam scanned the rear of the target to discharge each of the plugs. Here a common anode connected the plugs to ground through the other side of the same resistor. This produced the picture signal.

The first tube, though it was a giant step forward, probably consisted of only a few rows of rivets of sixty to eighty elements each. It was installed on the same Westinghouse film projector that had been used earlier. However, the vibrating mirror was gone and the new camera tube substituted. The electron beam was blanked out on its return trace as it swept back and forth across the back of the target.

This type of insulated target or special plate with a large number of independent conductors inserted through it had first been described in the patent literature by Harold J. McCreary of Associated Electric Laboratories in April 1924. McCreary never built such a camera tube, and he lost a patent interference (Pat. int. no. 54,922) with Zworykin in February 1932.

According to Zworykin's closing report, the results were quite promising: "A rough picture was actually transmitted across the room using cathode ray tubes for both transmitter and receiver." The report also mentioned that "an interesting principle was found and verified experimentally." This was Zworykin's first, indirect, reference to the so-called storage principle. It simply meant that each plug coated with photoelectric material was a single condenser that was able to store an electric charge during the scanning of the target until the beam returned the next time. As such, it would have higher sensitivity than those targets that did not have storage capability.

The report continued, "The solution of the direct vision problem is thus considerably advanced and may be the next point of attack in the practical development of television." Of course, "direct vision" was the ultimate goal; to be able to transmit pictures in direct sunlight or with sufficient artificial lighting. No more work was done on Zworykin's camera tube until he moved to the Victor plant in Camden, New Jersey, in January 1930.[55]

The six home receivers requested after Zworykin's successful demonstration on August 17 were built by December 1929. Three were finished in October, and one of them was installed in Zworykin's home at 7333 Whipple Street in Swissvale, Pennsylvania, four miles from station KDKA. Westinghouse reported that it had been broadcasting motion pictures by the Conrad system daily since August 25, 1929. Zworykin was allowed to use this radio transmitter late at night (between 2 and 3 A.M.) some three times a week to telecast quite satisfactory pictures. He merely sent the television signal (picture and synchronizing pulses) from his film projector at the Westinghouse laboratories by land-line to the KDKA shortwave transmitter. Here the picture was telecast, without sound, using his system.

Zworykin claimed that he was able to receive sixty-line pictures at twelve frames per second quite well at his home. This was the first time in history that television images were sent by radio to an all-electric television receiver—another important first for Zworykin.[56]

The next step was to build a film projector for transmitting both sound and pictures. A special film was prepared at the twelve pictures per second frequency, and Zworykin's report claimed that sound pictures were transmitted across a laboratory room.[57]

In his notes, Zworykin said that at about this time he received a long-distance phone call from a man in Washington, D.C., who said that he had just come from London on behalf of a very large foundation. He claimed to have heard that Zworykin had made a new discovery in television (perhaps his picture tube), for which he might be eligible for a large cash award. This intrigued Zworykin, but he informed the caller that his work was confidential. The caller, who identified himself as Mr. Russell, then said that the award could be given only after their representative had witnessed the true nature of the invention. He was very persistent, claiming that it was the chance of a lifetime for Zworykin. Zworykin was fascinated, so he approached Sam Kintner and told him what transpired. Kintner replied that someone was obviously trying to find out how far Zworykin had progressed, but he had no objection to a demonstration in Zworykin's home. Mr. Russell was then invited to Swissvale and given a demonstration of sixty-line pictures from the Westinghouse transmitter. He seemed quite impressed with the demonstration, a Mickey Mouse cartoon, and informed Zworykin that he was assured of winning the award. The sum was so large that Zworykin questioned him and asked him what his real purpose was. With that Mr. Russell left, and of course was never seen nor heard from again.

Zworykin was quite upset by this incident; the loss of a large sum of money and the kidding of his co-workers, who took this spying as a compliment, upset him for a while. No amount of investigation ever cleared up who was behind the hoax.[58]

During the period from February 1929, when Zworykin actually started work on his new system, through October 1929, he had only filed for two minor patents on it—one on March 26, covering the film projector, and another on July 5, for transmitting all three signals on a single wavelength. It wasn't until November 16, 1929, that the Westinghouse patent department finally filed for a patent on Zworykin's new picture tube. After nine years (February 22, 1938), he was finally granted six claims, of which only the fourth related to an electrostatic field effective to focus a ray onto a screen. This was quite a long time to wait between inception and filing for a patent in a field where time is of the essence, and no reason has ever been given for the delay. Zworykin's patent application included all the advanced features he had actually incorporated into his tube. Chevallier had applied for his French patent six weeks earlier (October 1929) and for an American patent almost a year later.[59]

Just two days after the patent application was filed, at a meeting of the Institute of Radio Engineers on Monday, November 18, 1929, at the Sagamore Hotel in Rochester, New York, Zworykin delivered a paper entitled "Television with Cathode-Ray Tube for Receiver." It was part of a program of seven technical papers, an exhibition of parts, and a banquet.

For the very first time, Zworykin publicly used the word "Kinescope" to describe his new tube. He spoke of its many advanced features including its use of electrostatic focus. At that time he was not prepared to explain exactly how the high positive potential applied to the metallic coating functions to assist in focusing. He described the tube as being capable of showing a five-inch picture on a seven-inch screen made of willemite and said that the pictures were greenish rather than red, visible to a large number of people at once, and brilliant enough to be seen in a moderately lighted room. He claimed that framing was automatic, and the tube could be operated with no great mechanical skill—in short, another giant step forward for television.[60] This was one of the most important papers ever presented before the IRE, yet it was not printed in their *Proceedings*. Obviously, something had happened.

When Zworykin had finished his lecture and was having a lively discussion and answering those questions that he was allowed to discuss, a message arrived telling him to call the patent department in East Pittsburgh. He was told that permission to present his paper had been withdrawn. Obviously, he had to tell the patent department that it was too late. Curiously, the paper was published in *Radio Engineering* for December 1929. Neither Westinghouse Electric nor Zworykin ever explained why it was published there rather than in the IRE's journal.[61]

For some reason, the *New York Times* index claims that on November 19, 1929, the paper reported that Zworykin had given a demonstration of his new tube. This was not true; he had simply delivered a paper. For that matter, Westinghouse Electric had not even displayed one of his home receivers, although there were six available. This is an example of how many so-called demonstrations were nothing more than announcements of some new machine (and it is interesting to note that no such article appears in any available edition of the paper for that date).[62]

At any rate, Zworykin's secret was out. His efforts were reported in magazines and newspapers all over the world. There was a revealing picture in the Sunday *New York Times* for November 24, 1929, showing him fondly displaying his new tube. His confidence in it was revealed in his statement that he was ready to discuss "the practical possibilities of flashing radio images on motion-picture screens so that large audiences can view the television broadcasts of important events as sent out from a central station." He continued, "Visual broadcasts will be synchronized with sound. The cathode ray receiver has no moving parts making it more easily usable by the rank and file of the radio audience."

Yet, since he had not given a public demonstration of this device, most of the scientists who had tried in vain to produce a similar cathode-ray tube simply did not believe that Zworykin had done what he claimed. A. Dinsdale noted in *Radio News* that "the disadvantages of the cathode-ray tube as at present manufactured, are that its first cost is high, it requires expensive auxiliary equipment, it is difficult to focus the pencil of ray to a sufficient fine point and make it 'stay put' at that during scanning and the degree of illumination is low."[63]

This disbelief was shared by most of the scientists who had worked on developing a practical picture tube. It was summed up quite well in a private note from Dr. Herbert Ives of the Bell Telephone laboratories to H. P. Charlesworth, in which Ives says that Zworykin's work was "chiefly talk." Ives continued, "This method of reception is old in the art and offers *very little promise*. The images are quite small and faint and all the talk about this development promising the display of television to large audiences is *quite wild*." The task had been so formidable and the answer had eluded so many scientists for so long that it was hard to believe that the solution had finally been found.[64]

However, Zworykin's kinescope did attract the interest of Frank Gray of the Bell Telephone laboratories. In a few months, he reported on work done on a "new" Bell oscillograph tube. He claimed it was of the high-vacuum (no gas) type, used magnetic rather than gas focus, and had a lead (grid) in front of the hot filament to control the beam. Gray reported that "this tube gives television images equal to the quality of those on the disk and of the same order of brightness."

The kinescope also interested Kenjiro Takayanagi in Japan. Early in 1930, he started to produce a high-vacuum, multi-electrode Braun tube. He claimed that these tubes were built by Dr. Asao at the Research Institute of Tokyo Shibura Electric Company.[65]

The Bell Telephone laboratories had earlier reported a demonstration of a television system capable of transmitting color pictures. It was a simultaneous system in that all three colors were transmitted at the same time, and was based on the mechanical disc system first demonstrated in April 1927. It had, in addition, three sets of photoelectric cells used with the flying spot scanner. At the receiver, three sets of filtered neon glow tubes were viewed through mirrors behind the rotating disc. Bell Telephone claimed that the results were good, with the appearance of a small colored movie screen.[66] The 1929 color demonstration was to be the last major effort by the telephone group to achieve supremacy in the field of television. Zworykin's success with the kinescope, coupled with the oncoming Depression, had derailed most of their plans for continued research. There would be a demonstration of a two-way television system early in 1930, but their main focus was in research and development of long-distance transmission of television signals. Their running battle with the radio group was over.[67]

With the major exceptions of Philo Farnsworth in the United States, John Baird in England, and Fritz Shröter in Germany, David Sarnoff was now in firm control of television's future. All future RCA television research was to be consolidated at the Victor plant in Camden, New Jersey, and Sarnoff requested that Zworykin be in charge of the project. For Zworykin, this meant breaking up his home again, selling the house in Swissvale, Pennsylvania, and starting life afresh in Camden.

At this point, Vladimir Kosma Zworykin owed Westinghouse Electric nothing. Everything he had accomplished in the past year had been done on behalf of RCA and Sarnoff. He relished the idea of a laboratory of his own where he could continue his dream of building an all-electric television system, and as he made plans for the move, his future never looked brighter.[68]

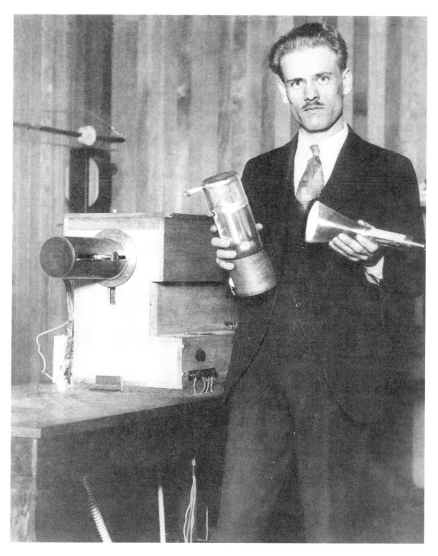

Philo Farnsworth and his 1928 dissector and picture tubes. Courtesy of the *San Francisco Chronicle.*

Zworykin with facsimile transmitter, 1929.

Harley Iams with Westinghouse facsimile receiver, 1929. Courtesy of the Westinghouse Electric Corporation.

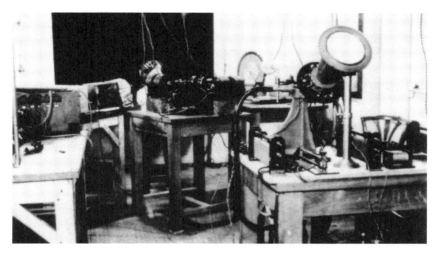

1928 Belin and Holweck cathode-ray apparatus using the Chevallier/Holweck electrostatically focused tube. Courtesy of Jacques Poinsignon.

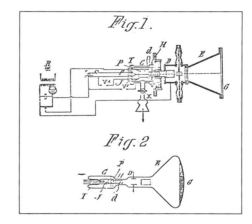

Chevallier's French patent (no. 699,478).

Chevallier's 1928 electrostatically focused tube. Courtesy of the Conservatoire Nationale des Arts et Mét-iers, Musée National des Techniques, Paris.

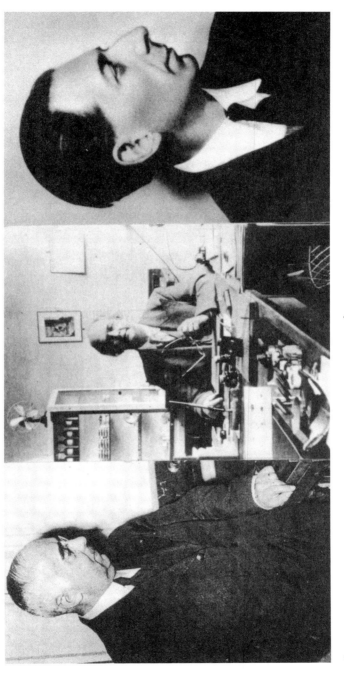

Édouard Belin (left), Fernand Holweck, 1940 (center), Pierre Émile Louis Chevallier, 1940 (right). Courtesy of Jacques Poinsignon.

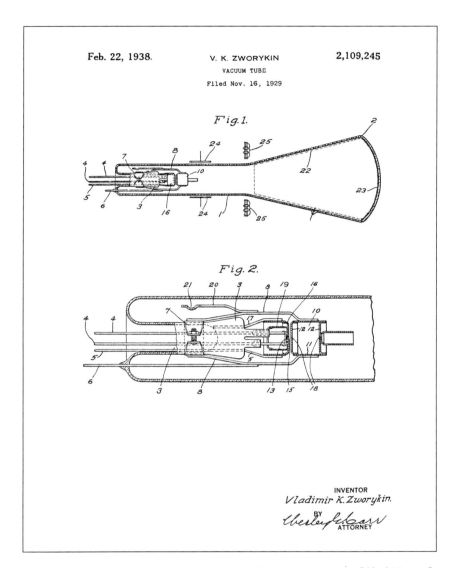

Feb. 22, 1938.   V. K. ZWORYKIN   2,109,245

VACUUM TUBE

Filed Nov. 16, 1929

Fig. 1.

Fig. 2.

INVENTOR

Vladimir K. Zworykin.

BY

ATTORNEY

Zworykin's 1929 picture tube patent (U.S. Pat. no. 2,109,245 [filed Nov. 16, 1929; issued Feb. 22, 1938]).

Zworykin with early kinescope picture tube, Pittsburgh, Dec. 1929. Courtesy of the David Sarnoff Research Center.

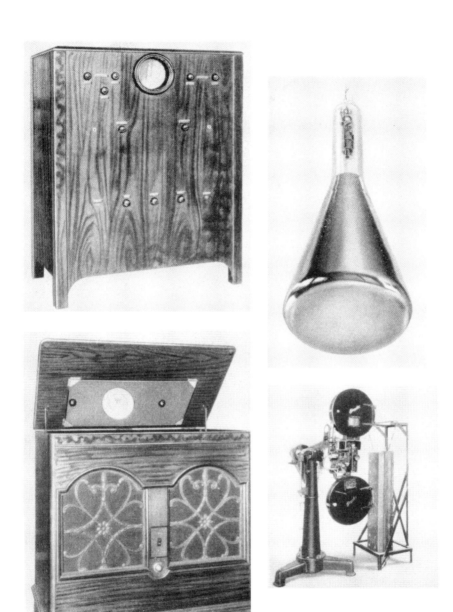

Westinghouse Electric's equipment (upper left) 1929 experimental laboratory receiver, (bottom left) home receiver, (upper right) early (1929) kinescope tube, and (lower right) Zworykin's flying spot film projector. Courtesy of the Westinghouse Electric Corporation.

Zworykin and unidentified woman with early all-electric television receiver,
Pittsburgh, Dec. 1929. Courtesy of the Westinghouse Electric Corporation.

Baron Manfred von Ardenne with 1930 picture tube. Courtesy of Manfred von Ardenne.

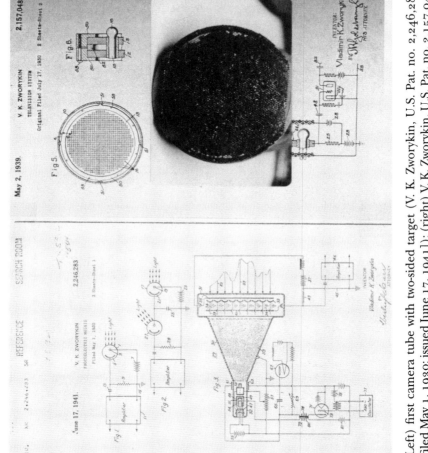

(Left) first camera tube with two-sided target (V. K. Zworykin, U.S. Pat. no. 2,246,283 [filed May 1, 1930; issued June 17, 1941]); (right) V. K. Zworykin, U.S. Pat. no. 2,157,048 (original filed July 17, 1930; renewed Jan. 30, 1937; issued May 2, 1939). (Inset, right) an actual mosaic from a tube of this type. Courtesy of Les Flory.

Randall Ballard (left) and Les Flory with experimental RCA dissector camera on the roof at Camden, 1931. Courtesy of Les Flory.

1930–31 RCA image dissector tube built by E. D. Wilson of Westinghouse. Author's collection.

First single-sided tube patent (Zworykin, U.S. Pat. no. 2,021,907 [filed Nov. 13, 1931; issued Nov. 26, 1935]).

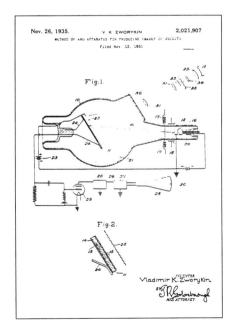

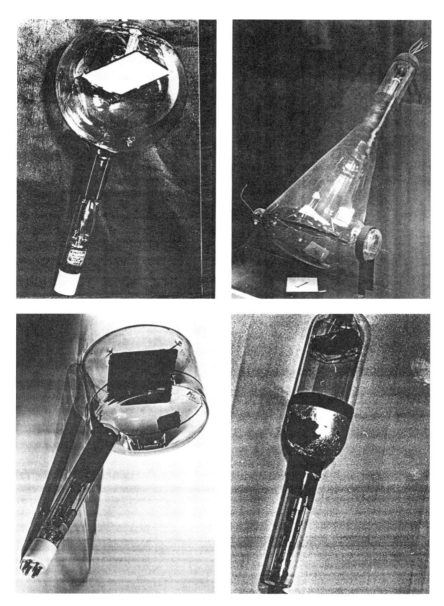

Four different types of experimental iconoscopes: (Top left) tube using circular bulb, 1931–32. Courtesy of the David Sarnoff Research Center. (Top right) early tube using kinescope bulb, 1931. Courtesy of the Franklin Institute. (Bottom left) "barrel" type tube made by RCA, 1933–34. Courtesy of the David Sarnoff Research Center. (Bottom right) very early iconoscope using sleeve type bulb, 1931. Photo in author's collection.

Gregory N. Ogloblinsky at RCA in Camden, 1931–32. Courtesy of Les Flory.

Ballard interlaced patent (R. C. Ballard, U.S. Pat. no. 2,152,234, television system [filed July 19, 1932; issued Mar. 28, 1939]).

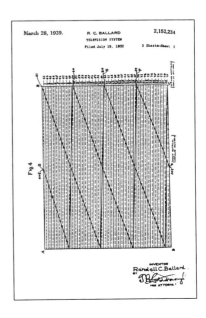

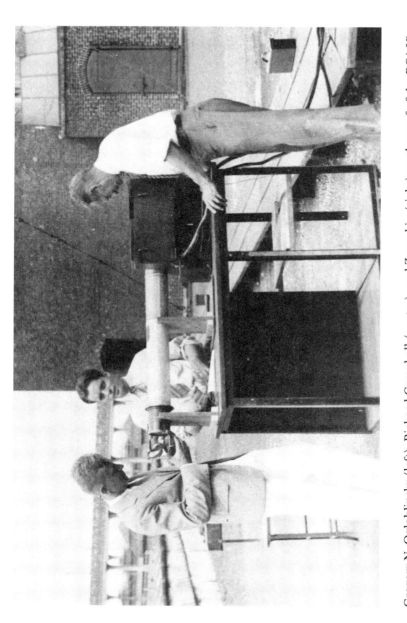

Gregory N. Ogloblinsky (left), Richard Campbell (center), and Zworykin (right) on the roof of the RCA Victor plant in Camden during a solar eclipse, Aug. 1932. Courtesy of Les Flory.

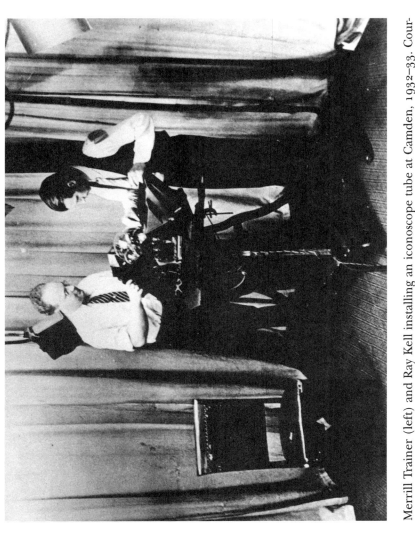

Merrill Trainer (left) and Ray Kell installing an iconoscope tube at Camden, 1932–33. Courtesy of the David Sarnoff Research Center.

CHAPTER 7

The Iconoscope

Moving Zworykin's laboratories from Westinghouse Electric at East Pittsburgh to the RCA Victor plant in Camden took several months. Zworykin tried to make sure that all of his engineers went with him, but he lost Harley Iams late in December 1929, when Iams became seriously ill and returned to San Diego to recover. He also lost Dr. W. D. Wright, who had to return to England due to family problems. The others, including John Batchelor, Arthur Vance, Randall Ballard, Gregory Ogloblinsky, S. Skypzak, R. Overholt, and the student engineer Pepper, were assured that they would be on the RCA payroll before April 10, 1930. At the same time, Zworykin prepared to move his family to 22 Merrick Villa, Collingswood, New Jersey, for a year.[1]

The stock market crash of October 1929, catastrophic though it was for the entire nation, had not affected RCA's long-range plans. On January 3, 1930, David Sarnoff became president of RCA, which acquired Westinghouse Electric's lamp works in Indianapolis, Indiana, and General Electric's Harrison, New Jersey, tube plant. Dr. W. R. G. Baker of General Electric became RCA Victor's vice president in charge of engineering, with L. Warrington Chubb of Westinghouse as his first assistant.

Two engineering groups were set up in Camden. The first was an RCA Victor research and development division headed by Albert F. Murray. Under Murray were radio receivers, directed by Dr. George Beers; acoustics, directed by Dr. Irving Wolff; and television, directed by Dr. Vladimir K. Zworykin. Of these, Murray was the only one who had not come directly from the radio group. Instead he had come from the Wireless Specialty Apparatus Company, owned by an RCA majority stockholder, John Hayes Hammond, Jr.

There was also a general research group that included Elmer Engstrom, division engineer, Photophone development and design; Dayton Ulrey, head, transmitting tubes; and Arthur F. Van Dyck, manager, engineering and test department.[2]

The official date of the transfer of all equipment and personnel from General Electric and Westinghouse Electric to RCA was to be April 1, 1930. General Electric made sure that the majority of key posts were filled by former GE engineers.

Setting up a new laboratory, which had to be equipped with a large variety of scientific equipment, was not an easy task. Some equipment was moved from Westinghouse but most was new, and it all had to be installed and put into operation. Zworykin complained that the new laboratory was first housed in an old factory building that was not suited for the purpose. Experimental work on the kinescope, receivers, and the new camera tube was held up for several months, but Zworykin's team did get started eventually.[3]

The first order of business at Camden was to set up a complete operating system. This was based on eighty-line scanning at twelve frames per second. A flying spot disc scanner and a mechanical pulse generator were installed, along with a different 35mm motion picture film projector (not the original one from East Pittsburgh) using an eighty hole Nipkow disc. With the use of the flying spot scanner for live pickup, Zworykin had to change over from sine-wave scanning to sawtooth scanning in order to use a Nipkow lens disc as a source of picture signals. This was true for both live and film pickups. This also produced better, more uniform pictures. The new nine-inch kinescope tube that was used for reception displayed excellent pictures. Now it was time to show off the system to visiting dignitaries.[4]

One of the first, coming from England, was A. W. Whitaker, who was in charge of advanced research development at the H.M.V. Gramophone Company. RCA had acquired certain financial interests in the company when it acquired the Victor Talking Machine Company in 1929, with the result that David Sarnoff now sat on its board of directors. The Gramophone Company had begun a limited research program into television in 1929. Whitaker's visit to the United States was to inspect the work being done by RCA, General Electric, Westinghouse, and the Bell Telephone laboratories. This was part of Sarnoff's policy to exchange information with companies allied with RCA, including licensee bulletins, patent applications, and just about everything of interest from the various RCA laboratories.[5]

On April 2, 1930, Whitaker witnessed Zworykin's television system for the first time. He reported that "his [Zworykin's] high-voltage cathode-ray tube presented a 5 inch image and was so bright that it could be viewed in a brightly lighted room." This was the first independent report of Zworykin's system. There was no doubt that Zworykin had accomplished his goal of creating a new, vastly superior means of television reception.[6]

About this time, Zworykin was informed that he was to go to San Francisco and visit the research laboratories of Philo Farnsworth, much as he had been sent to visit the Belin laboratories in Paris. The effects of the Depression were now being felt strongly and some of the stockholders, not Farnsworth or his immediate backers but the Carrol W. Knowles Company, were anxious to sell out their interest. They sent out invitations to RCA, among others, to inspect the laboratories and patents; this had been tried before with no success.[7]

Zworykin knew of Farnsworth's work from the reports and eyewitness accounts that had come to General Electric, as well as from the many patent interferences between Farnsworth and Westinghouse, and he welcomed the opportunity to see Farnsworth's system for himself. He did not relish being away from his new laboratory while it was still being equipped, but it was decided that he could afford to be away for the two or three weeks it took to get things organized.

Farnsworth, who did not know of these arrangements until the last moment, was at first surprised and then pleased when he found out that RCA was going to send Zworykin himself. It was assumed that he was representing Westinghouse Electric, his employer for the past seven years, since his move to RCA was not yet common knowledge. Farnsworth knew of Zworykin's work through his writings, patents, and patent interferences and had a very high opinion of him, but he knew very little of what Zworykin had recently accomplished in the Westinghouse laboratories in East Pittsburgh beyond the published reports. Still he welcomed Zworykin as a future ally, or at least as a source of much-needed financing, and felt that Zworykin would have a proper appreciation of what he had accomplished. Farnsworth knew the dangers involved in showing his system to a fellow researcher, but he felt that Zworykin already knew a lot about his system and that rather than stealing his ideas, Zworykin would be spurred to more intensive work on his own system. This of course is what happened.

Zworykin arrived in San Francisco on April 16, 1930, and spent three days at the Green Street laboratories. He was introduced to Farnsworth's financial backers and to his staff, which at that time consisted of Albert B. Mann, managing director; Harry Lubcke, associate director of research; Archibald Brolly, chief engineer; and Cliff Gardner, Robert Rutherford, and Harry J. Lyman. It appears that Zworykin and Farnsworth got along quite well. Zworykin was gracious in his praise of Farnsworth's results and seemed tremendously impressed with what he was shown. Farnsworth's system had been greatly improved since 1928.

Sitting at Farnsworth's desk on the first day after his arrival, in the presence of J. P. McGarger, president of Television Laboratories, Donald Lippincott, a patent attorney, and George Everson, one of the financial backers, Zworykin paid Farnsworth high praise by picking up a dissector tube and saying, "This is a beautiful instrument, I wish that I might have invented it." This of course pleased Farnsworth very much. Later Gardner, Farnsworth's tube builder and brother-in-law, showed Zworykin the success that he had had in sealing an optically clear disk of Pyrex glass onto the end of the dissector tube. Zworykin had been assured at both Westinghouse and General Electric that such a seal could not be made, so it was arranged to have Gardner show Zworykin how he had accomplished the sealing process.

Zworykin was given a demonstration of the entire system in operation. While he was pleased with the performance of the dissector tube, he was not all impressed with Farnsworth's magnetically focused receiving tube. Because it used a long magnetic coil (a "thick lens") wrapped along the entire length of the tube, it could display only small (two and a half inch), dim pictures. Farnsworth had had no better luck with it than any of the other top scientists anywhere in the world.[8]

The dissector camera tube was a different matter. With sufficient light, which was necessary because it had no storage capability, from either motion picture film or slides, it produced excellent pictures. It had no shading problems (had a flatness of field), and it delivered a clear, distinct picture with half-tones. It was inherently a directly coupled device with a constant reference level with all the background information.

Also, thanks to his work with Harry Lubcke, whom he had hired in February 1929 as a young electrical engineer from the University of California, Farnsworth had the world's first all-electric cathode-ray television system. It had an electric camera tube (the dissector) as a pickup device and an electric vacuum tube scanning and sync generator that produced sawtooth deflection currents and even blanked out the return scan lines. Finally, it used a cathode-ray tube as a display device.[9]

When Zworykin left, he thanked his hosts and promised them that he would report back to RCA what he had seen. From San Francisco, Zworykin went to Los Angeles for a few days and then went back to Swissvale, where his family was still living. Here he visited his former

co-worker Dr. E. D. Wilson of the Westinghouse laboratories and requested that he make up several dissector tubes for future experimentation. This, of course, is normal laboratory procedure, building a device and/or reproducing another's experiment to verify its
performance.[10]

When Zworykin arrived back in Camden, he wrote a report on the
Farnsworth television system which was turned over to Dr. Ernst F. W.
Alexanderson, chief engineer of General Electric, for evaluation.
Zworykin had praise for the image dissector tube, and while he commented that Farnsworth's picture tube was quite deficient, he praised
the use of sawtooth scanning and the vacuum tube electric scanning
generator.

Alexanderson's appraisal of this report indicated that he didn't
think too much of Farnsworth's efforts. He wrote that "Farnsworth
had evidently done some very clever work, but I do not think that
television is going to develop along these lines." He rejected the idea
of purchasing the system by stating that "I think that Farnsworth can
do greater service as a competitor to the Radio Corporation group
settling this provided he has the financial backing." He continued,
"If he should be right, the Radio Corporation can afford much more
for his patent than we can justify now, whereas, if we buy his patents
now, it involves a moral obligation to bring this situation to a conclusion by experimentation at a high rate of expenditure." This report
was based on two premises. First, Alexanderson had seen the Zworykin system in operation and knew of its excellent performance. Second, none of the General Electric people who had seen the Farnsworth system in operation were too pleased with it.[11]

At any rate, Sarnoff, knowing of Alexanderson's (and Zworykin's)
sentiments about Farnsworth's cathode-ray television system, was not
quite sure that either of them had been objective in their evaluations.
So, two months later, RCA sent both Albert Murray, the head of the
advanced development division, and Thomas Goldsborough, an RCA
patent attorney, to San Francisco to look over Farnsworth's system and
patent applications. They finally decided that RCA could do without
Farnsworth and/or his inventions.[12]

However, Farnsworth was quite right about Zworykin's visit spurring
him to put more effort into his own camera tube. On May 1, 1930,
Zworykin filed for a patent on a two-sided camera tube based on the
work that he had done with Ogloblinsky while still at Westinghouse.
It was his first application for a camera tube since July 1925, and was
completely different from the earlier proposal.

According to the patent, the proposed tube could be made in two versions, either "linear" or "planar." The linear version was a smaller tube for scanning motion picture film in which a single row of photoelectric cells could be used. The downward movement of the film through the projector would provide the vertical scanning component. The vertical scan could also be provided for by the use of a rotating polygonal mirror or similar device for live pickup. This meant that a large complicated mosaic was unnecessary, so the device was easier to construct.

The planar version was to be used to transmit the entire scene. This meant constructing a large mosaic of 6400 elements (eighty rows of eighty elements) of photoelectric cells to cover the whole image. It was to have a mosaic target built of rivets or plugs set into an insulated plate whereby each plug was separated from another. Each plug facing the scene was to be sensitized with cesium on a layer of silver oxide and the back side connected to a common plate, which in turn was connected through a resistor to an amplifier. In front of the target was a mesh screen that furnished the polarizing voltage and was connected to the other side of the resistor through a battery. This created the picture signal. The scanning was to be unilateral (sawtooth) in straight lines with a quick return. The patent application included means for a vacuum tube scanning generator so that sawtooth scanning and return beam blanking were accomplished both horizontally and vertically. The patent stressed that each plug would be continuously charged in accordance with the number of cells, thereby increasing the sensitivity of the tube. The larger the number of cells, the greater the electrical charge available. This was the heart of the patent.[13]

Zworykin and Ogloblinsky actually started work on tubes of this type at Camden beginning in May 1930. The first to be built were the linear versions, consisting of just a few rows of pins inserted into an insulated plate which were swept horizontally by the electron beam. The vertical movement was furnished by the moving film. This patent application was Zworykin's first to mention storage by individual condensers (not to be confused with his earlier claims for "globules"). Unfortunately, the application ran into a patent interference (no. 62,721) with Harry J. Round of the Marconi Wireless Telegraph Company and Charles Francis Jenkins and others (who were subsequently eliminated). They had all filed for patents that included elements of charge storage in order to make their devices more sensitive. Round had applied for a reissue of his patent, which had been issued May 20, 1930. Jenkins had been issued a patent on April 29, 1930, that included charge storage.[14] Jenkins's patent had enormous influence, since several inventors in other countries, among them Kenjiro Takayanagi in Japan and A. N. Konstantinov in the USSR, quickly filed for similar patents outside the United States.[15]

On November 27, 1931, three interferences were declared by the U.S. Patent Office between Round, Zworykin, and others. On October 1, 1932, the Westinghouse patent department moved to substitute the earlier (December 1923) Zworykin patent application, which was still pending, for the May 1930 patent application, in order to get an earlier priority date. Meanwhile the French patent office had issued Zworykin's patent on this tube on December 11, 1931.[16]

The Westinghouse patent department repeated the motion to substitute the December 1923 patent application for that of May 1930 in each of the three interferences. In each case this was rejected by the examiner of interferences on the grounds that Zworykin's right to the claims in his 1923 application depended on certain amendments which had been added to the specifications but which the examiner held to be departures therefrom.

In the matter of priority, Round relied upon the filing date of his British patent application, May 21, 1926. Zworykin stood on his December 29, 1923, filing date. The examiner of interferences awarded priority to Round, who was awarded the interferences on June 28, 1935. On February 26, 1936, the district court affirmed the examiner's actions.[17]

It is ironic that while Round had never built any device described in his patent, Jenkins had actually constructed a "sensitive plate transmitter" which consisted of a two-foot-square frame studded with half-inch-square light sensitive photoelectric cells, forty-eight rows of forty-eight cells each. Each cell was connected to a segment of a rotating switching device similar to the 1927 Bell labs two-foot neon glass large screen. Across each cell was a small condenser to store the accumulated electric charges, which represented the light from the scene, until discharged by the commutator every one-fifteenth of a second. Jenkins claimed that, although crude, "it performs." He had great hope for this principle, and of course he was quite right.[18]

However, while Jenkins had built an impractical "device" for demonstration purposes only, Zworykin and Ogloblinsky were actually building and operating camera tubes with storage in their laboratory in Camden. The first *linear* tubes consisted of a rectangular plate with only a few rows of sensitized pins inserted into it. Later, *planar* tubes were built which had up to 80 rows of pins or rivets. Unfortunately, there were enormous problems in building tubes with large two-sided targets of such complexity that were uniform and free of blemishes. However, Zworykin and Ogloblinsky did manage to build several that worked.[19]

By contrast, Wilson at Westinghouse had built several image dissectors that were working quite well. The dissector was a relatively simple tube to construct. The hardest task was to impart a proper layer of photoelectric substance on a target, which could be either a metal plate or a mesh screen. At the other end of the tube was a metal "finger-like" anode with a small aperture that allowed the electrons to strike a wire that led the electrical signal out to an amplifier. By now Westinghouse was using cesium on silver oxide (the Koller process), which made their tubes about a hundred times more sensitive than the ones that Farnsworth was building in his laboratory in San Francisco.

When used with the bright picture tube that Zworykin had developed, the Wilson dissector tube was able to present excellent eighty-line pictures. Of course, this was kept a secret. It would not do for either Zworykin or RCA to admit that a tube of the same design as Farnsworth's was far superior to the crude two-sided camera tubes being built in the RCA laboratories at this time.[20]

On June 20, 1930, Zworykin wrote to Harley Iams, who was recovering from his illness in San Diego, that "Ogloblinsky had gotten nice results with the transmitting tube (a modified Farnsworth type) which was about the same as our pictures of last summer." The sensitivity was quite high and Zworykin hoped that they would achieve direct vision soon. That was the ultimate goal, to be able to pick up an image by its own reflected light in either bright sunlight or minimal artificial lighting.[21]

About this time, Zworykin published *Photocells and Their Application,* an excellent survey of the art and science of photoelectric cells written in collaboration with E. D. Wilson of Westinghouse. This was Zworykin's first book, and the first of five major books that he would write with fellow laboratory workers.[22]

Meanwhile, Zworykin was having problems with the General Electric television research group. Although all television research had officially been transferred to RCA in Camden, Alexanderson's group felt that its mechanical system had merit, especially for large-screen television. While Zworykin had the support of Sarnoff for home television, Alexanderson also had Sarnoff's backing for large-screen television. Always looking for new fields to conquer, Sarnoff was chairman of the board of Radio-Keith-Orpheum (RKO), another acquisition by RCA to promote the new RCA Photophone sound film system and give RCA a foothold in sound pictures. Sarnoff was determined to break up the monopoly that his arch rival, the Western Electric Company, had with its Vitaphone motion picture sound system, so he wanted RKO to get involved in theater television in New York City. Sarnoff conferred with Alexanderson, who decided at first to transmit a television program under the direction of NBC from the roof garden of the New Amsterdam Theater in New York City to an RKO theater nearby. But Alexanderson reported that equipment had already been set up in Proctor's Theater in Schenectady and that "a preliminary test had indicated that the system would give entertainment results which might be considered sufficient for commercial purpose."[23]

So it was decided to use the television studio at the General Electric research laboratories in Schenectady. The theater would be the RKO Proctor's Theater. With Sarnoff's blessing, Alexanderson gave a large-screen television demonstration on May 22, 1930. The program consisted of an announcement by Merrill Trainer and a series of vaudeville acts. The finale was an orchestra number with the conductor directing from the television studio a mile away.

Alexanderson's system used the conventional forty-eight-hole flying spot scanner as a camera. A modified 175-ampere motion picture arc projector using Karolus's very efficient Kerr cell was installed at the RKO Proctor's theater. Ably assisted by Raymond Kell and Trainer, Alexanderson was able to project television images on a six-foot-square screen, accompanied by sound from a pair of full-range loudspeakers.

The picture signals picked up by the flying spot scanner and the accompanying sound were transmitted by radio to antennas on top of the theater. After being received and amplified, the picture signals were projected onto the large screen. The forty-eight-line picture displayed many defects, such as smearing, muddy halftones, streaks, and transmission noise. The pictures were only about half as bright as regular motion pictures, but with good halftones within the limits of illumination. However, the large size of the screen proved startling.

The size of the screen accompanied by the high-quality sound made for an excellent demonstration. In addition to getting a foothold in theater television for Sarnoff, Alexanderson had again proved the superiority of the radio group. And, of course, it was another television first for General Electric. On June 14, 1930, the *New York Times* reported that "RCA in affiliation with Radio-Keith-Orpheum was going to install television sets in theatres throughout the country."[24] However, despite the success of this demonstration, it was obvious that large-screen television was both premature and costly, and no more demonstrations were given by General Electric. Instead, RCA decided to concentrate on the home receiver, where the mass market would be.

Also, it was time to resolve the conflict between the General Electric/Alexanderson and the Westinghouse/Zworykin factions. General Electric was loath to cede anything to either Westinghouse or RCA. In order to resolve this issue, it was decided to evaluate the two systems at a test house in Collingswood, New Jersey. This took place on July 15, 1930.

Zworykin's group used their television system, based on eighty-line, twelve-frame scanning with the Zworykin kinescope for image reproduction. Picture signals were generated by the new unidirectional film projector. The entire system was furnished sync signals from a mechanical (Nipkow disc) rotating pulse generator. All pictures were to be displayed, of course, on the six Westinghouse experimental television receivers that were sent there. Elmer Bucher has even claimed that Zworykin's group had a "studio camera" available at the time, though there is no record of this device. While Bucher's report implies it was one of Zworykin's two-sided camera tubes, only the Wilson dissector tube was working quite well at that time, so that could have been secretly included in the demonstration. Arthur Vance was in charge of the Westinghouse/Zworykin demonstration.

The General Electric demonstration was supervised by Kell, who ironically had already acquired and been quite impressed by one of Zworykin's experimental kinescopes. According to all reports, General Electric simply relied on its mechanical forty-eight-hole, twenty-frame flying spot scanner and a standard mechanical General Electric home receiver.[25]

It was quite obvious from the start that the Westinghouse/Zworykin system was far superior to the General Electric/Alexanderson mechanical system. After a few days, a decision was reached to allot 90 percent of the research funds to the Zworykin system and 10 percent to the General Electric group. In addition, "it was decided to abandon mechanical scanning at the transmitter as soon as the Zworykin camera tube attained a higher degree of perfection. The results obtained with his system were so far superior to those of any other or prior system, that some enthusiastic observers declared the system ready for a public service: but the conclusion was overly optimistic, a large and more detailed picture was deemed essential for a telecasting service to the home."[26]

After this demonstration, Zworykin's group was reorganized, as five new engineers and a glassblowing group were added. This was now called the "Research Television" section, with Zworykin in charge.

Another important consequence of the demonstration was the decision to incorporate the three superb General Electric television engineers, Kell, Alda V. Bedford, and Trainer, into a special "Development Television Test" section at RCA Victor in Camden under the direction of William Arthur "Doc" Tolson, who had been the manager of the General Electric television program under Alexanderson. Tolson's group was to develop and integrate the various elements that were coming from Zworykin's laboratory and assemble them into a practical television system. When this arrangement began, in September 1930, it finally put an end to the rivalry between the General Electric and Westinghouse television engineers.[27]

The separation of RCA from General Electric and Westinghouse Electric was still underway. Since both General Electric and Westinghouse were still RCA's partners, having large stock ownership and powerful representation on the RCA board of directors, Sarnoff had not achieved real independence from them. However, this was about to happen, thanks to government intervention.

On May 31, 1930, the United States Department of Justice had filed an antitrust suit against RCA, the General Electric Company, Westinghouse Electric, and the American Telephone and Telegraph Company. They wanted the new agreement scrapped, the patent pool disassembled, and all exclusive contracts, both at home and abroad, made nonexclusive, so that the radio group would have to compete with each other in the production and sale of radio and other electrical equipment.

At this time, as the Depression was widening, the antitrust suit could have been fatal to RCA. Largely through Sarnoff's efforts, an agreement was reached with the Justice Department, and on November 21, 1932, a consent decree was signed. This was giant victory for RCA, as it kept everything it had gained through unification and it won complete freedom from the other corporations. The *New York Times* reported that "RCA is believed to be stronger because of separation from Westinghouse and General Electric and dissolution of patent agreement." Later, this was to cause much resentment on the part of both General Electric and Westinghouse, who felt that they had given up so much to RCA and received so little in return.[28]

The RCA technical and test department was disbanded on July 7, 1930, and the RCA experimental television station W2XBS was transferred to NBC for further development. This was the beginning of a plan for RCA and NBC to have a commercial television station on the air in New York City by the autumn of 1932. In September 1930, Theodore A. Smith, who had built the television station in 1928, left for another position with RCA Victor as a sales engineer. He was replaced in 1931 by Robert E. Shelby.[29]

On July 17, 1930, Zworykin applied for another patent on a camera tube. This was his first television patent for RCA. It was basically the same as his May 1, 1930, patent application (for Westinghouse) with a few minor changes in the construction of the target mosaic. He also applied for two more patents on improvements on his kinescope for RCA, the first on September 25 and the second, for improvements in a television system, on December 24, 1930.[30]

Zworykin wrote Harley Iams on September 12, 1930, that RCA had a successful receiver built around the new nine-inch kinescope. He stated that the picture was good enough for a commercial set, and noted that he was continuing his research to improve all the elements of his new system. He told Iams that he was working with Gregory Ogloblinsky on camera tubes and with John Batchelor on the kinescope. Arthur Vance and Randall Ballard were working on an advanced synchronizing and scanning system. In other laboratories at RCA work was continuing on home receivers and short-wave transmitters for sound and picture transmission.[31]

Ogloblinsky had turned out to be an invaluable asset. His work was brilliant and his relations with his American co-workers were now quite cordial. But he still had close ties to France and the Belin laboratories. He regularly corresponded with Pierre Chevallier, and as a result, Chevallier knew everything that was going on at Westinghouse and now the RCA Victor laboratories. Chevallier probably received Ogloblinsky's September 11, 1929, letter around the first week in October. Chevallier knew that the prior Belin patent of November 29, 1928 (Fr. Pat. no. 638,661), offered very little protection for him. It seems that he also had doubts about the seriousness of the many offers made to him by Westinghouse.

Chevallier, realizing that Zworykin had not yet filed for a patent (though it is certain that Zworykin had drawn up a patent application based on his improvements on the tube and was urging the Westinghouse patent department to submit it to the U.S. Patent Office), decided to act as quickly as possible. He evidently did this on his own as Belin's name does not appear on the patent.

On October 9, 1929, certainly with the knowledge of the Zworykin tube acquired from Ogloblinsky at hand, Chevallier submitted to the French patent office a simple patent with only one claim. This was for a smaller and more intense cathode-ray beam to be achieved by applying potential differences between successive parts of a metal or glass tube. There was no mention of the high-voltage capabilities of the new Westinghouse tube or of the fact that it could be modulated at low voltage and accelerated after deflection. These features were added later, in his American patent application. The French application was filed in the French patent office on October 25, 1929.[32]

Ogloblinsky wrote a letter on June 18, 1930, to Chevallier in which he bragged that the new picture tubes compared favorably with that of several movie projectors for domestic use. He also mentioned that the radio group would probably renew their offer to Holweck. Finally, he informed Chevallier that "the new emitting [camera] tube works but the pictures are imperfect. *(If you can, please don't advertise this fact.)*"[33]

On August 14, 1930, Ogloblinsky informed Chevallier that Zworykin was to be head of the research section, and that he had been told that it would become "a physics section and more." Chevallier was to be in charge of the new developments of cathode-ray tubes, emitting tubes, and projecting tubes. Ogloblinsky told Chevallier that he had been offered a salary of $150 per month, which Ogloblinsky assured him he could manage on quite well. He even invited Chevallier to live in the same boardinghouse he was living in.[34] Chevallier was to pay for his own trip, but he would be given an immediate raise for compensation.[35]

But Chevallier had decided not to come to RCA. None of the negotiations between Westinghouse and Holweck had been fruitful. Whatever the reasons, neither the oncoming Depression or the antitrust suit that was now about to tear the radio group apart helped the situation. Obviously, Westinghouse did not intend to do anything to help RCA when it too was fighting for survival.

He wrote Zworykin that he had a problem. Frankly, he said, he could not leave due to certain circumstances and really felt he would do better in France. He claimed that he had a good offer from Compagnie pour la Fabrication des Compteurs et Matériel d'Usines à Gaz of Montrouge with the help of René Barthélemy, a French television inventor.

He then explained to Zworykin that he had applied for an American patent within the one year allowed by the 1883 International Convention for the Protection of Industrial Property. The deadline he stated was October 20, 1930. This treaty gave any foreign patent application the benefit of the original filing date in the country of origin. The main provision was that the foreign application must be for the same invention as the original application. Chevallier's American, British, and German patent applications were basically the same as his original French patent application. However, they did include some subtle changes, in the drawings, for example, and claims that had been added after he had applied for the French patent. This letter seems to have been the first news Zworykin had received about Chevallier's American patent.[36]

On December 2, 1930, Chevallier sent Zworykin a copy of his U.S. Patent Office application. Zworykin promised to get in touch with Abraham Barnett, Chevallier's patent attorney in Washington. Zworykin claimed that "during the two years that have passed since I was in Paris, we have made much progress in the study and construction of the cathode-ray tube for oscillographic purposes. The one which we are now using and which gives us excellent results does not resemble at all the type which we discussed when I was in Paris. Therefore, very little in your patent will be of actual interest to our patent department, but I shall try my best to interest them and to help you sell your patent to the RCA providing you get sufficient claims in the United States." Zworkyin continued to say that he was sorry that Chevallier had decided not to join his laboratory. Finally, he stated that "we are having hard times in the United States. For that reason, all the appropriations are tied down and we are not free anymore to spend as much for our development as we were several months ago." That included his proposition to Holweck. He was afraid that it would be very difficult to renew the negotiations between Holweck and Westinghouse Company on the same basis as they had been a year earlier.[37]

On February 16, 1931, a French patent was issued solely to Pierre Émile Louis Chevallier for a cathode-ray tube incorporating "electrostatic focus." Neither Belin nor Holweck is mentioned as owner or assignee. The filing date was given as October 25, 1929, about three weeks before Westinghouse Electric had filed for the Zworykin kinescope patent. The Chevallier patent had only one claim: "The present invention has for its object the improvements brought to tubes, using either a fixed or modulated cathode ray beam, aiming toward making this beam more intense and of small section. These improvements consist of using gradual potential differences applied to successive metal parts or to conductors when the apparatus is made of glass, the relation between these various potentials being determined with great precision."[38]

On June 9, 1931, Zworykin was issued a French patent covering his picture tube. It, of course, disclosed all of the many Zworykin improvements. The RCA patent department contacted Abraham Barnett on June 16, 1931, and they were permitted to inspect Chevallier's pending U.S. patent file. They immediately decided to buy up the patent application in order to protect RCA's interest in Zworykin's tube.

On July 8, 1931, H. G. Grover, the RCA patent lawyer who processed all foreign patents assigned to RCA, became Chevallier's patent attorney. Chevallier then filed a second U.S. patent application on August 25, 1931. RCA allowed Chevallier's first American patent application to be divided and added several more of Zworykin's novel features to this second patent, to the dismay of both Zworykin and the Westinghouse Electric patent department.

The RCA patent department then went to great lengths to prove that indeed Chevallier was the first to originate the idea of electrostatic focus. (He wasn't, as we now know.) However, he had the three-week advantage of his filing date, which hurt Zworykin's claim and ironically inconvenienced both the RCA and Westinghouse patent departments. To make matters worse, on May 23, 1933, the U.S. Patent Office became aware of Belin's 1927 French patent (no. 638,661) and canceled twenty-two of Zworykin's claims to his picture tube (though, largely through the efforts of the RCA patent department, most of these claims were later given back to him).

However, the picture tube that Zworykin had so successfully developed and improved now belonged to Chevallier and was assigned to RCA. In addition, Chevallier was not only interested in money, he also sought revenge. Along with giving him a large cash settlement, RCA had the title of his American patent changed from "Methods for Operating Cathode-ray Tubes" to "Kinescope." However, Westinghouse still owned certain rights to the Zworykin patent and took control of it at a later date.[39]

There are many aspects of this affair that have never been resolved. For instance, why did the Westinghouse patent department, which was quite diligent under Otto Schairer, wait until November 16, 1929, to file for Zworykin's patent? Since he had brought the basic idea back with him from France in December 1928, he could (and probably did) have a patent application incorporating his many important improvements ready by April 1929, when the first successful tubes were made. Certainly he should have filed for a patent by the time of the successful demonstration (August 17). If Westinghouse had filed a patent application in either of those months, Zworykin's priority would have been assured. Clearly, something had stopped this process. It is possible that although Otto Schairer realized the value of this tube, he could not persuade top Westinghouse management, who had never been very interested in Zworykin's television work, of its worth.

Several other possible answers also suggest themselves. First, it is now known that Westinghouse was still negotiating with Holweck. They knew the Belin laboratories had already received a French patent on their basic device on November 29, 1928. So why should Belin file for another patent covering the same device?

Second, Westinghouse knew that Zworykin and all television research were soon to be transferred to RCA. Since RCA had paid for all the manpower and materials used in developing the kinescope, and Westinghouse had merely been acting as their agent, they may have waited for RCA to take over the negotiating process.

Third, and adding more complications to the problem, was the fact that Professor Roscoe H. George of Purdue's engineering experiment station had delivered a speech before the Institute of Electrical Engineers in Cincinnati on March 22, 1929, which was published in July 1929, describing a new cathode-ray tube using electrostatic focus.[40]

United States patent law states that prior publication or disclosure of an idea usually negates any chance of getting a patent on the device described. The Westinghouse patent department must have assumed that George had filed for a patent on this tube prior to his speech, though it turns out that he did not apply until September 14, 1929. George's patent was later bought up by RCA.[41]

Finally, since Zworykin was a Westinghouse Electric employee at the time, they may have insisted that because of their patent agreement, the original patent belonged to Westinghouse. This is what ultimately happened on July 10, 1932.

For whatever reasons, Zworykin's patent application was not submitted until November 16, 1929. The result of all of this was that Chevallier received a large sum of money and much-deserved credit for his original conception of the tube. Yet it is known that even though he had the tube working late in 1927, he had never developed it beyond the stage it was in when it was disclosed to Zworykin in December of 1928. In later years, while engaged in research in French television with René Barthélemy, Chevallier accomplished very little, and he certainly has not become famous. His contributions to television history are summed up by Marc Chauvierre: "The work of Holweck with the collaboration of M. Chevallier resulted in 1928 in a 33 line picture of remarkable quality."[42]

Zworykin certainly must have been resentful toward both Westinghouse and RCA for having failed to protect his remarkable achievement. He surely had no objection to Westinghouse paying Chevallier for the basic work he had done, but Zworykin's improvements had made the tube invaluable. David Sarnoff must also have been angry over this incident, which had been quite expensive for RCA. However, Sarnoff never lost faith in Zworykin, and RCA continued financing his work as if nothing had happened. By the time the Chevallier patents were issued in the United States in 1935, the topic had long been forgotten. Zworykin was solely responsible for the success of the tube that had transformed the newborn television industry. Finally, it seems that Ogloblinsky's duplicity in this matter was never known to either Zworykin or Sarnoff.

Zworykin had other problems as well. Due to the extreme secrecy with which RCA had surrounded all of its television research, there were many rumors (and accusations) about the integrity of Zworykin's research. The U.S. Patent Office became involved, and, as a result of the many interferences and lawsuits, it was alleged that Zworykin's work, his two-sided camera tube, was in fact an inoperative device.

In January 1931, Zworykin was requested to file affidavits showing that his device was operable. In a very unusual move, RCA invited the patent examiner, C. J. Spencer, to come to Camden and to witness for himself what they had accomplished. He arrived on February 20, 1931, and was given a demonstration of a two-sided tube based on Zworykin's July 1, 1930, application. The demonstration put an end to that allegation for all time. No court of law ever found that Zworykin's camera tube was "inoperative."[43]

Despite the successful demonstration a few months earlier, in May of 1931 Sarnoff was painfully aware of the difficulties that Zworykin was having with picture quality from his two-sided camera tubes at Camden. He received no word that they could soon (or ever for that matter) be made into a practical instrument, though his intense interest in the project and his management style ensured that nothing was kept from him.[44]

Even though Sarnoff was disappointed by the poor performance of Zworykin's crude two-sided camera tubes, he had supreme confidence in Zworykin's kinescope. According to a *New York Times* article published on May 6, 1931, "The use of this cathode ray tube had enabled him [Sarnoff] to clip four years off his estimate when Radio-Vision will be ready for the home." In fact, according to the article, Sarnoff publicly stated that "Several television transmitting stations next year will be installed in the New York area." For this, however, a viable camera tube comparable in quality to the kinescope was a necessity.[45]

In February 1931, RCA completed construction and installation of a 2.5kw VHF radio transmitter at the Camden plant and a receiving station at Collingswood, New Jersey. In July 1931, a similar transmitter was installed on the eighty-fifth floor of the Empire State Building, where RCA was in the process of building a complete television station. It went into operation on October 30, 1931. Elmer Engstrom was put in charge of this project.

The picture was to be transmitted on 44mH and the sound on 61mH. The radio transmitting equipment was designed and built by John Evans. The scanning rate had been raised to 120 lines at twenty-four frames per second. At this time, the transmission of motion picture film was considered to be a primary, if not the primary, goal of television. A flying spot scanner operating on this standard was available for live pickup. However, motion picture film gave the most satisfying results. All viewing was done on the Zworykin kinescope.[46]

Starting in November 1930 and continuing through March and April 1931, there was a great deal of publicity concerning Philo Farnsworth's all-electric television system in several of the popular radio magazines. The *Proceedings of the Institute of Radio Engineers* reported that on January 21, 1931, after Farnsworth addressed a meeting of the San Francisco I.R.E. section, seventy-one members adjourned to Farnsworth's laboratories and were treated to a demonstration film using a five- to seven-kilocycle bandwidth. Later, full details of his system were revealed. A. H. Halloran stated that "The picture quality was *supposed* to be better than the 72 line images shown by the Bell Labs!" This was quite a claim, as no one before had ever challenged the quality of the images produced by the telephone group. While Zworykin's laboratory pictures were far superior, they were wrapped in the secrecy that surrounded the Zworykin project.[47]

In May 1931, it was brought to RCA's and Sarnoff's attention that Farnsworth's television laboratories were again for sale. Since Farnsworth had the only alternate camera tube in existence and it was working quite as well, according to all magazine and journal reports, as in Zworykin's own laboratory, Sarnoff decided to visit Farnsworth in San Francisco and see what he had accomplished.

However, RCA was not the only company interested in the Farnsworth television system at this time. The Philadelphia Storage Battery Company (Philco), the largest manufacturer of radio sets in the United States, had warily been watching the evolving television situation. Philco resented being subservient to RCA in radio and decided that it would be free of RCA's dominance in television. Since the Farnsworth method was the only alternative system in electric television, Philco made a proposal to support his work. They offered to pay his research expenditures, which would be credited as prepaid royalties, leaving Farnsworth in control of his inventions. He was to move to Philadelphia, where they would set up a research laboratory for him on top of the Philco manufacturing plant.[48]

At the time, Sarnoff was unaware of any of this. In any event, he decided not to send an emissary to examine Farnsworth's work, but to go himself. It was of course quite unusual for the president of RCA to make such a personal visit. He had two reasons for doing so. First, he wanted to see what Farnsworth really had to offer, and second, in keeping with recent RCA policy, he was ready to buy up Farnsworth's enterprise or at least be sure no one else could do so.

Sarnoff went first to Hollywood and then to San Francisco, during the week of May 18, 1931. But he arrived too late to see Farnsworth personally, since he was already in Philadelphia with Jess McGarger negotiating a television agreement with Philco. Sarnoff was escorted through the Farnsworth laboratories by George Everson, who later claimed that Sarnoff had been impressed with what he saw. Sarnoff assured Everson that the Zworykin receiver avoided any of the Farnsworth patents, which was true. However, according to Everson, Sarnoff displayed an unusual interest in the image dissector tube. Sarnoff offered Everson $100,000 for the enterprise, including the services of Farnsworth himself, which was a generous offer in the midst of the Depression.

To his surprise, Sarnoff was told that such a deal was not possible. With that, Sarnoff left, stating that he had seen nothing that he could use. From here on, the intense rivalry between Farnsworth and RCA shifted to the U.S. Patent Office, where RCA would try everything, especially patent interferences, to get control of the dissector tube.[49]

Meanwhile, Zworykin's group was struggling with the coarse two-sided camera tubes made by Ogloblinsky. Though the tubes performed marginally, Zworykin and Ogloblinsky had never really resolved the myriad troubles of the two-sided targets; these included blemishes from pinholes, short circuits to the signal plate, and electrical leakage through the insulation. On the other hand, the image dissector tube, which was easy to construct, was working quite well and was being used in comparison tests. However, Zworykin, totally absorbed in the camera project, was determined to go ahead with a tube of his own design.[50]

Zworykin was always open to suggestions from his co-workers. As a result, he was soon able to produce his other great invention, a practical camera tube that was also to change the history of television. Since he and Ogloblinsky were having very little success with their two-sided camera tubes, it was decided to try another solution to the problem. Around May 14, 1931, Zworykin and his research group decided to experiment with a camera tube having a single-sided target. This would be done by simply turning the target around 180 degrees and having the electron beam scan (impinge on) the same side covered with the photoelectric surface. This was a bold step, which had never been tried before. At first, there was concern that the electron beam would physically damage the photoelectric surface. But Iams, who had come back to work in Camden on April 24, 1931, claims that it was he who convinced Zworykin that this would not be a problem. The other potential problem was the possibility that any residual gas from the hot cathode would destroy the photoelectric surface. In a two-sided camera tube, the mosaic plate formed a gas-free barrier, but this would not be the case in a single-sided tube. So this approach was quite a gamble.

By June 12, 1931, faint images could be detected, and this new approach showed so much promise that several tubes were built and research continued along these lines.[51]

The single-sided tube was built around a thin sheet of mica chosen for its uniformity and cleanliness. First, one side of the mica would be plated with a thin layer of platinum. Then a thin layer of silver would be deposited on the other side of the mica. This silver layer would be processed into a mosaic of separate elements, each element an insulated photoelectric cell. The researchers tried to produce such a mosaic by using a ruling machine or a fine-meshed screen, or by dusting the mica with a fine silver powder. All three processes were tried. Then this sheet of mica would be inserted into the glass bulb. The silver mosaic would be oxidized by letting in oxygen and ionized by a high-frequency current. Finally cesium was admitted on top of the silver and the tube was baked to get rid of all gases. The tube was then exhausted while the electron-gun assembly was heated to keep the cesium from coating it. This complicated procedure was only perfected after many months of trial and error. However, Zworykin's group was getting results; the few tubes that worked showed that they were on the right track.

A real breakthrough came when Essig accidentally left one of the silver mosaics in the oven too long. When he examined it, he found that the silver surface had broken up into a beautiful, uniform mosaic of minute, insulated silver globules. However, much experimentation was needed to perfect this process.[52]

By October 1931, Zworykin's group had achieved such good results with this new process that they decided to use a four-by-four-inch mosaic to get more picture signal from the tube. This required a larger glass bulb. With the larger bulb, the dogleg configuration of the tube was finalized. Zworykin, who was quite clever with names, called it the *iconoscope*, from the Greek *icon* meaning image and *scope* meaning observing. Many tubes were built and tested, but most performed only marginally, and it was not until November 9, 1931, that the first satisfactory tube was completed. According to Iams, tube number sixteen was the first iconoscope to give a reasonably good picture.[53]

In contrast to the kinescope affair, the RCA patent department quickly filed for a patent on this new tube on November 13, 1931, though the application later ran into the same sort of interference problems, mainly concerning the use of a single-sided target. First, there was a prior patent application by a prolific Hungarian inventor, Kolomon Tihany, who had applied for a British patent on June 11, 1928, covering the use of a single-sided target. This patent was never granted. English patent law is different from ours. First a "provisional specification" is filed. Then a "complete specification" has to be described so that any person skilled in the art could "carry it out" without further instructions. A single invalid claim makes the whole patent void, though it will still be published. RCA gave up four of Zworykin's patent claims to Tihanyi. Second, there was a prior patent application by François Charles Pierre Henrouteau, a Canadian who also held British citizenship, filed on August 10, 1928. This was for a camera tube with a single-sided target. Although this photoelectric target was scanned by a beam of light rather than an electron beam, it too anticipated the single-sided target of the new Zworykin camera tube. Just as they had done with Chevallier, RCA bought up both of these patent applications.

Further, the U.S. Patent Office had overlooked several prior British patents concerning the use of a single-sided target. The first was that of George J. Blake and Henry D. Spooner, who had applied for a camera tube using a single-sided target in England in February 1924. Another one was that of Riccardo Bruni, who had applied for an British patent on April 25, 1928. Interestingly enough, Bruni called his tube a "Photoscope."[54]

What was happening was that by the time Zworykin had reduced the iconoscope to practice, many of its important features had been conceived and patented by other inventors. For instance, the *Soviet Encyclopedia* claims that "S. I. Kataev invented the iconoscope in 1931." This claim was based on Russian Patent no. 29,865, filed on September 24, 1931, and issued on April 30, 1933. The patent includes two figures. Figure 1 shows a two-sided camera tube using a target of rivets or pins stuck through a plate. There is also a claim for "charge storage" of this "cellular" panel. Figure 2 shows a single-sided target in which the light goes through the target that was scanned by the electron beam. This was an impractical scheme at the time, as the light from the scene had to go through both the target plate and the photoelectric surface and charge the photoelectric elements from the rear, which would result in an enormous loss of sensitivity. However, with the advanced technology available in later years, it was made to work.

The other contender was Kenjiro Takayanagi, who claimed, "Before his [Zworykin's] acquisition of a U.S. Patent, I had applied for a Japanese patent of the same principle as his camera tube. I acquired a patent on this principle in 1933 although there was no physical iconoscope in operation." This patent covered an electron beam scanner. As it took time for Zworykin to build and operate his new tube, it was inevitable that he was going to be faced with a myriad of "paper" patents. So, of course, RCA did the only thing it could do—buy them up and put them in RCA's name. However the point to be made is that it was Zworykin who designed, built, and operated a working tube.[55]

In operation, the iconoscope had three great faults: First, it had very low electrical efficiency, only about 5 percent of its theoretical potential. Second, it produced large amounts of spurious signals that clouded the picture because the high-velocity scanning beam struck the photoelectric target with such force as to eject clouds of unwanted secondary electrons. This was due to the uneven distribution of secondary electrons over the surface of the mosaic. Third, it had no fixed DC (direct current) output, so that the electrical signal representing the picture level wandered up and down as the illumination on the mosaic varied.[56]

Zworykin's group began to investigate the electrical mechanism of the tube, trying to determine how the iconoscope actually worked. For instance its low sensitivity was contrary to expectations for a storage type tube; none of the prevailing theories explained its actions.

At first, Zworykin and his group assumed that a single-sided tube should work in the same manner as a double-sided tube. Later, when they found how the new tube actually functioned, they learned that the two operated quite differently. For instance, a two-sided target has more electrical capacity. It separates the primary electron beam from the photoelectric emission. So secondary electrons released from the point of impact with the anode are separated from the electrons released from the photosensitive surface. This prevents the two groups from intermingling. As a result, the mode of operation is different. A single-sided target operates in an unsaturated condition, thus lessening its sensitivity, while a two-sided target operates in a saturated condition.

It was also assumed that the electron beam acted as an inertialess commutator that discharged the elements on the mosaic as it swept across it. When struck by light, each of the tiny photosensitized silver elements stored a positive charge due to the emitted photoelectrons. The scanning beam returned the elements to their equilibrium potential with respect to a collector. Each illuminated element therefore went through a cycle of accumulating a positive charge during the entire picture period and releasing it each time the scanning beam swept over it. Since each element was coupled by capacity to the metallic base which formed the foundation of the mosaic, any change in charge on the former induced a corresponding charge in the latter. The change in charge produced a current in the lead connected to the coupling impedance (resistor) and amplifier that became the picture signal. This simple explanation enabled Zworykin and his group to continue their research on the tube.

But what was most important was that the iconoscope worked. It produced picture signals, though no one at the time knew exactly how. Every possible configuration was tried and tested. Zworykin's usual good luck had held; his tube produced usable electrical signals, which was all that mattered at the moment.[57]

By comparison, at this time the dissector tube was so efficient that they were able to take it out of doors and get live pictures in bright sunlight. This was because RCA was using the sensitive cesium on a silver oxide photoelectric surface (the Koller process) and also because of the low number of lines (eighty) then in use. Research went on to improve both tubes in Zworykin's laboratory, since Sarnoff was quite prepared to put the dissector tube into service if the iconoscope did not live up to its promise.[58]

At this time, with the success of Zworykin's new camera tube, RCA's patent department tried to strengthen his 1923 patent application. On October 8, 1931, they added still another claim, that "a photoelectric layer over said electron stream is deflected, is adapted to pass, and upon which the picture is adapted to be focused, and a circuit connecting the photo-electric layer and cathode." None of this really applied to the earlier patent application.[59]

While all of this activity was going on at the Zworykin laboratories in Camden, other important events were taking place in England. Due to the worldwide depression, the two major English record companies, H.M.V. (His Master's Voice) Gramophone Company and Columbia Graphaphone merged in April 1931. This was done to achieve certain economies of manufacturing, operation, and distribution of their products and services. This new company became known as Electric and Musical Industries (EMI). For several years, EMI existed alongside both H.M.V. Gramophone and Columbia Graphaphone. A third partner was RCA, which owned 27 percent of EMI's stock. As a result, Sarnoff, who had sat on the board of directors of H.M.V. Gramophone since March 15, 1929, when RCA completed its purchase of the Victor Talking Machine Company, now sat on the EMI board of directors along with John Broad, Marchese Guglielmo Marconi, Lord Marks, Edward de Stein, and E. Trevor Ll. Williams. Alfred Clark was chairman of EMI and Louis Sterling, managing director.[60]

The H.M.V. Gramophone Company had given a crude demonstration of mechanical television in London in January 1931, using a film projector and a mirror drum as a source of 150-line signals. Actually the image was divided into five sections of thirty lines each. This was sent to five photoelectric cells, each with its own amplifier. At the receiver, which consisted of a mirror drum with thirty mirrors, the picture was reconstructed from the modulated light of five Kerr cells onto a ground glass screen. It produced a picture some twenty by twenty-four inches wide. The demonstration was given by C. O. Browne, with the aid of J. Hardwick, W. D. Wright, and R. B. Morgan.[61]

One of Sarnoff's first actions on behalf of EMI was to have Zworykin send them several of his new kinescopes for experimental purposes. EMI immediately began a research project at the gramophone company's main plant at Hayes, Middlesex, under the direction of G. E. Condliff, C. O. Browne, and William F. Tedham. They were joined by engineers from Columbia Graphaphone—Allan Blumlein, P. W. Willans, E. C. Cork, and H. E. Holman, among others.[62]

Their first project was to perfect a 120-line twenty-four frame per second television system for the reception of motion picture film, with a receiver based on the Zworykin RCA kinescope. Tedham, an EMI chemist, was put in charge, with the assistance of H. Neal and J. W. Strange. H. G. M. Spratt was in charge of receiver design. Tedham's first project was to learn how to construct Zworykin's kinescope and if possible to improve its performance.

By May 1931, Browne, who was working with Tedham, M. B. Manifold, and Blumlein, reported that images as large as six inches had been formed by these revolutionary tubes and that good illumination had been obtained. In addition, W. D. Wright, who was now working for EMI, built a receiver around the kinescope and attempted to pick up the Baird television signals with it, with poor results.[63]

In August 1931, another television demonstration took place, this time at the Berlin Radio Show. It was the accomplishment of Manfred von Ardenne, a prolific television pioneer who had been working on high-brightness cathode-ray tubes for several years. He had finally developed an all-electric television system using cathode-ray tubes for both transmission and reception. This used the cathode-ray tube as a flying spot scanner. Von Ardenne never developed a camera tube; he left that task to Zworykin. He did claim that he was corresponding with RCA and with Zworykin personally, and was aware of Zworykin's efforts to produce an electric camera tube at Camden. He too acknowledged Zworykin's achievement, telling a German technical journal that "only Zworykin, using special tubes, seems to have met with any successful developments—at any rate on the transmitting side."

However, von Ardenne's cathode-ray tubes, which were a clever blend of both gas and electrostatic focus, were able to produce quite bright pictures. They had a Wehnelt cylinder surrounding the cathode, which, combined with a negative potential, gave a very convenient focus control. This helped reduce the destructive positive ion bombardment of the filament that had occurred in the Western Electric tube.

Because von Ardenne preferred to use a cathode-ray tube as a flying spot scanner, he was limited to motion picture films and slides. With the exception of a moving commutator to produce the frame pulses, his system was all-electric. Manfred von Ardenne was the first person to publicly exhibit a system using cathode-ray tubes for both transmission and reception.[64]

In September 1931, Captain A. G. D. West of H.M.V. Gramophone's research and design department visited Zworykin's laboratories in Camden and saw the group's latest results. He reported back that television was on the verge of being a commercial proposition. "They [RCA] intend to erect a transmitter on top of a New York skyscraper and if all is well, to market their receiving apparatus in the autumn of 1932." He went on to explain that the picture was six inches square and similar in quality to an ordinary cinema picture as seen from the back of a large theater. He also noted that RCA planned to sell receivers, self-contained in a single cabinet, for about £100 ($470).

As a result of his visit, RCA Victor offered EMI a complete television transmitter, similar to the one being installed at the Empire State Building, for £13,000 plus £2,000 for installation (about $70,000). Captain West recommended that EMI purchase this equipment, and this was authorized at first by the board of directors of EMI on October 28, 1931. However, Isaac Shoenberg, EMI's patent manager, turned this proposal down for a variety of reasons. First, he saw no reason to buy an American shortwave transmitter. He suggested that EMI see if his old employer, Marconi Wireless Telegraph, had a suitable transmitter available, and buy it instead. It happened that the Marconi Company did have one that could be modified for use on the six- to eight-meter band. This was a low-power SWB-Type Tx used for experimental transmissions between Rome and Sardinia on a ten-meter wavelength. It was leased to EMI and was delivered to Hayes on January 25, 1932.

Second, it was rumored at this time that the Marconi Company was eager to join forces with Baird Television, which might otherwise present problems in the future. This was because Marconi was the recipient of all the RCA television patents in England through its basic agreement with RCA. Thus Shoenberg saw a chance to get his old employer into the television picture as an ally. Captain West soon left H.M.V. Gramophone and was replaced by Shoenberg as head of research.[65]

Isaac Shoenberg had come from Columbia Graphaphone when it became part of EMI in April 1931. He had been chief engineer of the Russian Wireless and Telegraph Company from 1905 to 1914, at St. Petersburg, and had come to England in 1914, just before the war. He soon found employment with Marconi in London. Here, after many years, he became joint general manager and the head of the patent department of the Marconiphone Company, the Marconi Company's manufacturer in the field of home entertainment. He had resigned from this position in 1929 and joined Columbia Graphaphone before its merger with H.M.V. Gramophone.[66]

EMI received a steady stream of vital information in the form of licensee bulletins, patent applications, and exchange of ideas between the two companies. It was the beginning of a relationship between RCA and EMI that would give them almost complete domination of the new television industry.[67] In November 1931, RCA presented EMI with one of its most important patents, for the silvering and processing of Zworykin's kinescope. This became EMI's first television patent application.[68]

However, while EMI was free to produce its own television receivers, its agreement with RCA forbade it from working on the camera or transmission side in the field of communications. Sarnoff made sure that RCA had a monopoly on the building and operation of the new Zworykin camera tube.[69]

There were many inquiries as to the status of Zworykin's work at this time. In London, a meeting of the Television Society devoted to "Cathode Ray Television" was held at University College on November 11, 1931, chaired by Professor J. T. MacGregor Morris. Zworykin had promised to submit a paper on his new "cathode ray" system, and two other papers were to be presented. One was by Manfred von Ardenne and the other by C. E. C. Roberts, who had patented a form of image dissector tube in England in 1928. However, Zworykin canceled his paper and promised that he would resubmit it in January 1932. (This he did not do.) His cancellation only heightened curiosity about his work.[70]

By now, the worldwide depression had come home to RCA. Sales of radios, phonographs, and records were down tremendously, and radio broadcasting was the only part of RCA that was making money. As a result, the RCA Victor plant in Camden was closed for the last three months of 1930. Whole divisions were shut down; twelve thousand employees were laid off, and those who were lucky enough to keep their jobs had their salaries cut in half. Only a few departments remained open, and Zworykin's television laboratory was one of them. Sarnoff's faith in him had never faltered.[71]

CHAPTER *8*

"A Century of Progress"

At the bottom of the worst depression in modern times, RCA decided to lift some of the secrecy surrounding the Zworykin television system. The first demonstration, on May 16, 1932, was a closed affair, restricted to one hundred RCA licensees representing fifty radio set and tube manufacturers. The press was excluded, though the *New York Times* did report that the demonstration occurred.[1] The representatives met at the RCA Victor recording studios at 153 East Twenty-fourth Street and were addressed by David Sarnoff from the Empire State Building television transmitter, which had been in operation since March 1932, with pictures on W2XF and sound on W2XK.

The picture transmission was 120 lines at twenty-four frames per second, and there were both live and film pickups. The live pickup was accomplished by means of a flying spot scanner. The motion picture film was transmitted with the new 35mm film projector using a unidirectional film scanner. Sync pulses were furnished by a rotating mechanical impulse generator.

The pictures were received on six cathode-ray tube kinescope receivers with a picture about five inches square. The quality of the live pictures from the flying spot scanner was just fair. Observers claimed that "the images shifted frequently, lacked detail and the fluorescence was not brilliant." However, they were surprised at the clarity of the film transmission. "The images were well lit and the half-tones so good that the transmission of film would be quite satisfactory for the home."

Sarnoff told the audience that "We have arranged this demonstration in the spirit of the greatest possible cooperation with our licensees, so that you might be kept abreast with the progress of our laboratory research work and might judge for yourself the extent of the

technical progress we have made in television." However, he continued with the admonition that "There is a great deal of difference
between a demonstration (like this) and a practical, feasible commercial television service." He also stated that the system was not quite
ready to leave the laboratories and that much work remained to be
done. By this time he had decided to abandon his dream of launching television service in the autumn of 1932 and was thinking instead
in terms of winter of 1933, depending on how the economy was doing. At the bottom of the Depression, competition from television
could only have hurt the radio industry, which so far was weathering
the troubled times quite well.[2]

The demonstration was given to prove that the RCA television system was operational. It was an attempt to satisfy the RCA licensees
and the RCA stockholders by showing them something of what Zworykin and his engineers had accomplished in his laboratories at Camden. It was also the first presentation of Zworykin's revolutionary
picture tube. His iconoscope was neither mentioned nor used, because its performance was erratic and it was still considered a laboratory experiment. The iconoscope's very existence was top secret.[3]

It was to be the last demonstration of the Zworykin television system until April 24, 1936. The television system went back into the
laboratories for further development and the experimental radio
transmissions from the Empire State Building soon ceased.

However, Zworykin's research group was making excellent progress,
and the system was getting larger, better, and more sophisticated. As
it did so, Zworykin was hard pressed to continue his experiments and
supervise the laboratory at the same time. It appears that Zworykin had
never relished being a manager. His strengths lay in developing new
ideas and experimental designs, and in building and testing new equipment. It was obvious that the project needed someone who could oversee the entire television program, and that the running of the laboratory was better left in the hands of an experienced research engineer.
There would soon be one available.

By the summer of 1932, the television installation at the Empire
State Building was shut down. The RCA engineer in charge of its
development, Elmer Engstrom, returned to Camden, where he took
over the television test development group. He replaced William A.
Tolson, who was moved to other nonmanagerial duties. Engstrom was
also to coordinate the work being done in Camden by the Zworykin
group.

Engstrom brought to the Zworykin laboratory an intensified program of television research and development, similar to the highly

organized program at the Bell Telephone labs. He was among the first to institute the systems engineering procedures that have since become standard throughout the industry.

Engstrom's duties were to design and construct a complete experimental television system, and included a rigorous testing program to study television image characteristics such as (1) picture detail, (2) picture steadiness, (3) picture brightness, (4) picture contrast, and (5) frame repetition frequency. This study was to decide the number of scanning lines, picture size, and viewing distances for an optimum television system. Finally, Engstrom was also to oversee field testing and operation.

Zworykin's laboratory now became part of the larger research project run by Engstrom. Thus was born the Zworykin/Engstrom research group. Zworykin's group continued basic research into the operations of the iconoscope and kinescope and their circuitry, and Zworykin continued as director of television research for RCA Victor. It appears that the two men got along quite well.[4]

By this time, Zworykin and Ogloblinsky had improved the iconoscope's performance so much that it was producing better pictures than the image dissector. Its gray scale and definition were better, and by early in 1932 they had improved its efficiency and sensitivity so much that they were able to take many tubes and point them out of doors and pick up credible-looking images in bright sunlight. However, due to the use of the high-velocity electron beam, the tube still exhibited spurious signals (shading problems) with light and dark areas moving across the picture as lighting conditions changed.[5]

The Zworykin/Engstrom group began searching for solutions to this problem. A long-range project was set up to determine how the iconoscope actually functioned in order to improve its performance. The earliest explanation was that the scanning beam acted as an inertialess commutator that discharged each element. Since the charge on each element was proportionate to the light intensity, this discharge current was translated into the video signal.

However, this simple explanation really did not describe how the iconoscope worked. It turned out that the scanning beam sweeping over the mosaic acted like a resistance commutator. In a very complex process, two sources of signal appear, one the stored charge on the entire mosaic surface, the second the sensitive line ahead of the scanning beam. At low levels, the surface sensitivity contributes the major portion of the signal but, at high illumination intensities, line sensitivity can contribute a large fraction of the signal.[6]

Many different configurations of the single-sided tube were built and operated, and Zworykin continued to be certain that his tube would prove to be a more efficient instrument than Farnsworth's image dissector.[7]

The Zworykin/Engstrom group were also making improvements on the kinescope. The tube's brightness had increased several times (six to ten ft-lamberts). With higher brightness came another problem: it was now exhibiting pronounced flicker at the twenty-four-frame-per-second rate. (Flicker is directly related to brightness.) The twenty-four-frame rate had been chosen on the assumption that the showing of motion pictures would be a major objective of television. At low brightness this had been no problem, but at higher brightness it had become quite annoying.[8]

One solution would be to raise the number of frames per second and still be able to run motion picture film at twenty-four frames per second. In the motion picture theater this was accomplished by means of a rotary shutter in the projector that interrupted each single film frame either two or three times to give a forty-eight- or seventy-two-cycle frequency. This completely eliminated all flicker from the large screen. However, this was more difficult to accomplish in a television system. One obvious answer would be to raise the frame rate to forty-eight frames per second, which would allow it to run motion picture film by scanning each frame twice. However, this meant that the radio transmission channel would have to be twice as wide.

Another possible answer was to use an old television technique known as interlaced scanning. This was a method whereby the line scanning was not in sequence but skipped one or more lines, leaving a space or spaces that would be filled in later. This technique had been used by many experimenters, including John Baird and U. A. Sanabria. Sanabria had a rotating disc with three sets of holes set in a spiral path. This gave the same total number of lines (forty-five) while cutting the flicker rate by a third. This method had also been proposed by Dr. Fritz Shröter and Manfred von Ardenne for cathode-ray presentation, but they had never been able to demonstrate such a technique.[9]

The answer came from the Zworykin/Engstrom research laboratories. Randall C. Ballard, who had come with Zworykin from Westinghouse, devised the ingenious solution of electrical interlaced odd-line scanning that has since become part of every television system. At this time, late in 1931, Ballard was designing the first RCA electric (tube type) sync generator using blocking oscillators to keep a fixed relationship between the horizontal and vertical frequencies. No previous sync generator, including Farnsworth's, had been locked in this manner, and doing so was an important step in producing stabilized pictures.[10] He then turned to the problem of getting rid of the twenty-four-frame flicker by dividing each frame in two. Each frame was split into two "fields," each with one-half the number of scanning lines. As each field was one-forty-eighth of a second in duration, together they produced a twenty-four-frame picture. To make the system practical, Ballard chose an odd number of lines (eighty-one at the time). Each field of forty and a half lines was scanned in sequence. Ballard found that, because of the odd number of lines, as the scanning beam flew back at the end of the field, it would automatically fit (intermesh) between the lines of the previous field. The two fields would blend and, to the human eye, would present a single frame. As a result, the flicker rate had been doubled (forty-eight cycles) and flicker was almost entirely eliminated. In addition, doubling the field rate, which cut the number of lines in each field in half, afforded a considerable saving in channel bandwidth.

To produce the special synchronizing pulses needed, Ballard had to resort to a mechanical scanning disc with an auxiliary row of slits. It would be about two more years before an electric (tube type) interlaced sync generator would replace it in the RCA laboratories. Ballard's patent on this generator was one of the most important in the history of television, and his generator became one of the keystones of RCA's (and later on, EMI's) television system. RCA applied for a patent on this method on July 19, 1932.[11]

However, several other problems soon appeared. The first was that forty-eight frames per second was not high enough to completely eliminate the flicker. Second, the American electric power standard of sixty cycles was creating problems with twenty-four-frame interlaced pictures. Third, a new phenomenon, "interline flicker," arose because of the small number of lines then in use. Finally, though Ballard's solution was simple in theory, in practice it was hard to get it to work properly. Even though this process showed great promise, the Zworykin/Engstrom group continued to use sequential scanning for their laboratory experiments. By the middle of 1932, the RCA line standard was raised to 180-line sequential scan at twenty-four frames per second.[12]

Meanwhile the battle for control of the Farnsworth image dissector moved to the U.S. Patent Office. In March 1932, Zworykin and Ogloblinsky applied for a patent for an improved dissector tube. This patent was concerned with keeping the electron image in focus by applying corrective voltages to it. This application became the subject of two interferences by Philo Farnsworth, both of which he won.[13]

In addition, there was to be one other major attempt to wrest control of the dissector tube from Farnsworth. On May 28, 1932, a patent interference was declared between Zworykin's still-pending December 1923 patent application and Farnsworth's original January 1927 patent, which had been issued on August 16, 1930. This was the now-famous patent interference no. 64,027, Farnsworth v. Zworykin.

There was only one point in contention, Farnsworth's claim 15, concerning the forming of an "electrical image." The RCA patent department tried to prove that Zworykin's transmitting tube formed and used an electric image in a manner similar to that of the image dissector. The attorneys for Farnsworth convinced the examiner of interferences that only a tube of the dissector type could produce an electric image that was scanned to produce television signals. Farnsworth's attorneys also proved that the two types of tubes operated differently and that RCA had no right to make such a claim.

RCA brought in experts such as Colonel Ilia E. Mouromtseff, who testified that he was shown the Zworykin television idea in 1919; Professor Joseph Tykociner of the University of Illinois, who stated that the disclosure was made to him in 1921; and George F. Metcalf, expert on photocells and cathode-ray tubes, Dr. Dayton Ulrey, and Samuel Kintner, all of whom were told of the device in March or April 1923.

Farnsworth's lawyers brought in expert witnesses such as Dr. Leonard B. Loeb, professor of physics at the University of California, Leslie Gorrell, George Everson, Archibald H. Brolly, Cliff Gardner, Professor F. E. Terman, and Arthur Crawford to describe the operation of the Farnsworth system.[14]

Zworykin's testimony in this case did little to enhance his reputation. The attorneys for Farnsworth characterized Zworykin as follows, "Zworykin has shown throughout that he is perfectly ready to make any statement on the stand which he believes, at the moment, will be to his advantage. His memory is alternately super active and distressingly vague, and throughout his cross-examination he was so evasive as to make the elicitation of any fact whatsoever extremely difficult."

One of the key issues was the very existence of Zworykin's camera tube. Farnsworth's lawyers complained that "at no time during the proceedings had the party Zworykin produced and offered in evidence this tube which is alleged to have been constructed in accordance with the disclosure in the Zworykin application involved in this interference. Nor had the party Zworykin offered any explanation why this tube could not be produced in evidence." This charge was repeated at the final hearing on April 24, 1934. This portion of the motion involved certain testimony and sketches prepared on the witness stand by Tykociner and Metcalf respectively, which related to a device which the witnesses alleged they saw in operation. The motion was based on the grounds that this evidence was secondary because the device itself, "although available, was not introduced in evidence and was not produced during the taking of this testimony. This contention is clearly untenable." It continued, "The failure to introduce the device itself goes merely to the weight that shall be given the oral testimony." Further, it stated that "the device itself which was alleged to have operated is still in existence as shown by the fact that it was produced by Zworykin counsel, but that the device thus produced was not identified other than by a mere ex-parte statement and it was not even produced until the witnesses who had testified to seeing it in operation were no longer available for cross-examination!" Farnsworth's counsel objected repeatedly, in vain, and repeated demands that the best evidence be offered. That it was not cannot have been inadvertent.

Thus, while Zworykin's tubes existed, they were never presented in evidence, probably because they differed from those described in the 1923 patent application in construction and perhaps even in operation. Counsel for Farnsworth noted that "there is nothing in the record to show who built this particular apparatus, when it was built, under what circumstances it was operated and in accordance with what plans or specifications it was constructed."[15]

In October 1934, in a move to dispel doubts as to whether Zworykin's original 1923 device was operable, the RCA patent department instructed Harley Iams to build several models of the 1923 Zworykin camera tube. This was done, and in a sworn affidavit, Iams claimed that a tube built exactly to the patent specifications "produced an electro-optical image which was a reasonable likeness of the light image falling upon the transmitting unit which I built." At any rate, the original tubes (there were probably more than one) still existed in 1934, and while several witnesses claimed that they had produced an image, the RCA patent department was hesitant to show them to anyone.[16]

On June 22, 1935, the examiner of interferences awarded the junior party, Farnsworth, the priority. The case was closed on March 6, 1936. This was a great victory for Farnsworth and gave him complete control over the image dissector. This was the last attempt by the RCA patent department to secure rights to the dissector. Later, in 1939, RCA would have to come to terms with Farnsworth through contract negotiations.[17]

There still seemed to be much confusion about Zworykin's new picture tube in Europe, and many of the descriptions of it were inaccurate, even though Capt. A. G. D. West of Baird Television had seen it in operation and W. D. Wright and Arthur Whitaker of EMI were actually using the kinescope in the laboratory. However, Wright and Whitaker's reports to EMI were confidential, part of a campaign to conceal the intimate cooperation between the RCA and EMI laboratories.

This was very much in keeping with RCA's policy of maintaining the highest secrecy about Zworykin's work. In an answer to one letter about his progress, Zworykin wrote on September 20, 1932, "Our company does not consider it advisable to release for publication the results of laboratory research before this development had reached the commercial stage. The business depression is largely responsible for the delay in commercialization." In this he was only repeating Sarnoff's position that this was not the proper time to introduce television service in the United States.

Also in September 1932, in a reply to a query from the French journal *La Nature* on the future of television, Zworykin stated that he had high hopes for a cathode-ray transmitter which would allow for more detail and permit the transmission of outdoor scenes. But he gave no further hint as to what was going on in his laboratories at RCA.[18]

However, field testing under Engstrom continued at a great pace. Early in 1933 RCA had a complete television system working at Camden. For the first time, the Zworykin iconoscope was integrated into the system. The flying spot scanner and Wilson's image dissector had finally been discarded. With the introduction of Zworykin's iconoscope camera tube, the line standard was raised to 240 lines sequential scan at twenty-four frames per second. The only mechanical part remaining was the synchronizing pulse and scanning generator. There were two shortwave radio transmitters that sent out both picture and sound signals. All the receivers used advanced versions of Zworykin's kinescope.

Experiments were set up simulating the conditions that were likely to occur in a regular television service. Two transmitters were built with antennas mounted on masts on the roof. The studio and control equipment were located in Building 5, one thousand feet away. A remote pickup using an iconoscope camera was located one mile east. An outdoor program was televised, relayed one mile to the studio, and retransmitted to a television receiver four miles east in Collingswood, New Jersey. RCA had the most advanced television system in the world in their laboratories at Camden.[19]

Information about these developments was leaked to *New York Times* columnist Orrin E. Dunlap, Jr. On January 1, 1933, he wrote that Zworykin was on the track of a successful image receiver. He also mentioned that Zworykin "was an *expert* on cathode-ray tubes which scan the image electrically, thereby dispensing with motors and scanning disc that whirl in side the receiving cabinet as well as at the 'transmitting' end." This was the first publicity ever from RCA indicating that Zworykin had built a cathode-ray transmitter tube.[20]

With Engstrom effectively managing the television project, Zworykin was also afforded the opportunity of investigating more closely some phenomena he had observed in his early television work. He concentrated mainly on the new field of electron optics, the study of the behavior of an electron beam under the influence of electrostatic and electromagnetic fields. Electron optics disclosed that electrons traveling in a region containing such electrical fields experience deflections that are comparable with those of a ray of light in a region containing different transparent media. An electron beam can be focused by means of coils carrying direct current or by systems of charged conductors. In either case, the resultant magnetic or electric field can be regarded as a lens.

Zworykin wrote a paper "On Electron Optics," which he presented to the Optical Society of America on February 24, 1933. In it he discussed both electromagnetic and electrostatic focus of an electron beam in a high vacuum. He noted that "the idea of using the electrostatic field for both acceleration and focusing of the beam is very attractive and a great deal of effort has been concentrated in this direction." He took credit for applying this principle for the cathode-ray oscillograph that had been described at the November 1929 meeting of the Institute of Radio Engineers. He also mentioned the use of an "electron microscope" described by Bruch and Johanson.[21]

Another major advance, the use of a short coil (a thin lens) for magnetic focusing of an electron beam, was described by Zworykin on March 5, 1933. While this was an old idea, Zworykin was the first to use it in a two-anode tube. By using a thin coil of wire wrapped around the neck of the tube, he could focus, modulate, and deflect the electron beam at low velocity, as in his electrostatically focused tube. Then the second anode could accelerate the electron beam for high brightness. This meant that the sensitivity to deflection was vastly improved.

In addition, it was found that tubes using magnetic focus with a short coil could be sharper, have a smaller spot, and be more free of aberrations than tubes using electrostatic focus. This was another important step forward in cathode-ray technology. For the first time, a magnetically focused, two-lens, hard-vacuum tube with multi-anodes could be used with high brightness.[22]

At this time, it appears that all was not well at the RCA Victor plant in Camden. In May 1933, Albert Murray, head of the advanced development division, resigned to become chief television engineer for the Philco Corporation in Philadelphia. He took with him a half-dozen key RCA Victor personnel. He had, of course, complete knowledge of all the RCA developmental projects, including the entire Zworykin/Engstrom television program.[23]

It seems that the Philco Corporation, after less than two years of depending on Philo Farnsworth to give them an alternative to the Zworykin system, had decided to sever their relations with him. Farnsworth was uncomfortable in the atmosphere of the large, organized laboratory run by Philco, and it was obvious that they were not compatible. Farnsworth typified the "lone" inventor or innovator who was much more content away from the discipline of a systematized research group. During his second year, it became clear that Farnsworth's goal of obtaining a broad patent structure was not compatible with Philco's production program.

By now it had become apparent to Philco, especially after witnessing the RCA demonstration in May 1932, that the Farnsworth television system, as ingenious as it might be, could never compete with RCA's system built around Zworykin's bright picture tube. As a licensee of RCA, Philco knew most, if not all, of what was happening in the Zworykin/Engstrom laboratories. By hiring Albert Murray, an excellent television engineer, they were assured that at least they could produce a system comparable to the one that RCA had in its laboratories.[24]

This, of course, was quite a blow to RCA and Sarnoff, since it meant that their monopoly on cathode-ray television was now threatened. All the "secrets" of Camden were now in Philco's possession. Interestingly enough, because of the existing licensing agreements, Philco could develop any system of television in its laboratory as long as it did not market any of this equipment.

However, one result of this event was that RCA decided it was time to reveal most of the details of the Zworykin/Engstrom/RCA television system to the public. This meant disclosing the iconoscope camera tube for the first time, but it would assure RCA's priority for all the pioneering work done by Zworykin and his staff. It is also worth noting that this decision reflects the fact that RCA's patent department had won several important patent interferences (nos. 57,145,

62,025, 64,026, 64,035, and 68,337) on May 19, 1933, and RCA's patent situation seemed quite secure.

Sarnoff himself made the preparations for the disclosure of the iconoscope. In a May 1933 magazine article, he assured the public that "no other development threatens to supplant radio in our national life. What is promised is not a substitute for radio, but rather an extension and elaboration of radio services."[25]

The city of Chicago was planning to hold a World's Fair, called "A Century of Progress," in the summer of 1933 to celebrate its one-hundredth birthday (and incidentally to help Chicago recover financially from the Depression), and the Institute of Radio Engineers was to hold its yearly meeting there during the fair. RCA decided to have its most prominent inventor, Dr. Vladimir K. Zworykin, present a paper describing his new camera tube, the iconoscope, in detail. Again, as with the kinescope in November 1929, only a paper was to be read. There would be no demonstration of equipment to either the press or the public.[26]

The *New York Times* for Sunday, June 25, 1933, gave this anticipated paper great publicity. Orrin E. Dunlap, Jr., reported that Zworykin was to reveal his iconoscope or "electric eye" in a lecture before the Institute of Radio Engineers in Chicago the next day. Oddly enough, this article featured a picture of Zworykin with his once-famous 1925 combination amplifier/photoelectric cell.

Zworykin told Dunlap about his early work with Professor Boris Rozing in St. Petersburg, not knowing that Professor Rozing had recently died (April 20, 1933) of a cerebral hemorrhage while in exile in Archangel. Rozing had continued his research into television and other allied problems at the Leningrad Experimental Electro-technical Laboratory in Leningrad from 1924 to 1928. He continued his work at the Central Laboratory for Wire Communications from 1928 to 1931. In 1931 he was arrested and deported for three years to Archangel, when he became a victim of one of Stalin's purges of intellectuals and others dangerous to the state. Rozing had continued his research at the Physiks Laboratories of Archangel's Forestry Technological Institute until he died.

Dunlap also quoted someone who had seen Zworykin's pictures as saying that "this is not merely a step forward in television. It goes over the wall that had stopped all experimenters. Television is here." However, Zworykin went on record that "the fate of television now rests in the laps of the financial and merchandising experts." The article concluded with the news that "the inventor plans to sail for Europe in July and has invitations to lecture on television in London, Paris, Berlin and Moscow."[27]

On Monday, June 26, 1933, at the Eighth Annual Convention of the Institute of Radio Engineers in Chicago, Zworykin presented his historical paper, "The Iconoscope—A New Version of the Electric Eye." He began with a short history of television to this time, and reported how mechanical scanners could not get enough light to produce an outdoor picture. He also obliquely referred to other nonstorage devices such as the image dissector. Zworykin then described how his new camera tube worked. He noted that it was able to store energy between scans, each photocell storing its charge until discharged by the scanning beam, and claimed that it had the same light sensitivity as a high-vacuum cesium oxide photoelectric cell (about the same as photographic film) and the same spectral sensitivity. Zworykin claimed that certain tubes were capable of producing up to 500 lines of resolution, though he admitted that the tube had its faults. He then gave general details of the rest of the system, including the kinescope picture tube. The paper included pictures of the iconoscope tube, several diagrams of the system, and other illustrations. It was printed in the *Proceedings of the Institute of Radio Engineers* in the United States, the *Journal of the Institution of Electrical Engineers* in England, *L'Onde Électrique* in France, and *Hochfrequenztechnik und Elektroakustik* in Germany.[28]

On Tuesday, June 27, 1933, the *New York Times* made Zworykin's paper a page one story. An article by William L. Laurence, the science editor, gave details of the paper as presented by Zworykin. More information and the first published picture of the iconoscope and an RCA television camera appeared in the Sunday *New York Times* for July 2, 1933.[29]

For all intents and purposes, the disclosure of the iconoscope in June 1933 marks the beginning of the age of electronic television. The RCA system had a practical electric camera and a bright picture tube, as well as all the means for radio transmission and reception (receivers) of television images and sound available. While a mechanical sync generator was in general use, an experimental electronic interlaced pulse and scanning generator was in the laboratory. All future television developments would be based on this RCA system.

News of this disclosure was sent all over the world. To most television researchers, and the public in general, it came as a surprise to realize how far the Zworykin/Engstrom/RCA television system had progressed. For instance, at that very same time, U. A. Sanabria was still demonstrating a mechanical forty-five-line projection system at Macy's department store in New York City. And even though Philo Farnsworth had a basic electric pulse and scanning generator in his laboratories, his camera tube was quite insensitive, his picture tube did not have the quality or brightness of the Zworykin picture tube, and he lacked the means to eliminate flicker in his pictures. As the leader in television research, Zworykin had reached the pinnacle of his career.[30]

Researchers at the Bell Telephone laboratories, for a change, were very impressed with this paper. Unlike his reception of news of the kinescope, in November 1929, Dr. Herbert E. Ives was now convinced that Zworykin had made a real breakthrough. Incredibly, Ives finally admitted that mechanical television was now dead and that perhaps it was time for Bell Telephone to build an iconoscope of their own just to see what Zworykin had accomplished. The telephone company had long since ceased to be a major competitor with RCA in the field of television, and they had cut back their television research severely due to the Depression. Bell Telephone's main interest now was with the long-distance transmission of television signals by means of coaxial cables.[31]

At any rate, if the American public believed that "television was just around the corner," they were soon confronted with reality. The revelation of the Zworykin/Engstrom/RCA television system was but one of the highlights of the 1933 Chicago World's Fair. After a brief flurry of excitement, very little was heard of Zworykin's revolutionary new invention. The camera tube was neither shown publicly nor demonstrated, and it was quietly returned to the RCA laboratory for three more years of improvements. Interest in television soon waned, and Sarnoff's beloved NBC radio network was secure for the time being.

Now, however, the scene was to shift to Europe. It had been decided to send Zworykin overseas to deliver his paper on the iconoscope in person. On July 4, 1933, Zworykin set sail for England on the SS *Europa*. In London he presented his paper before the Institution of Electrical Engineers on July 17, 1933.

After the meeting, Zworykin visited the laboratories of EMI at Hayes, Middlesex, where he met with Isaac Shoenberg, who had become director of research on August 22, 1932.[32] Shoenberg had a superb staff working for him, including an advanced development division under G. E. Condliff with such outstanding engineers as Alan Blumlein, W. F. Tedham, C. O. Browne, J. Hardwick, P. W. Willans, Dr. J. D. McGee, Dr. E. C. L. White, and Dr. W. D. Wright. Later it included Dr. Leonard F. Broadway and Dr. L. Klatzow.[33]

As noted before, EMI had been getting all the licensee bulletins as well as patent applications from RCA since April 1931. Using this information, EMI was having considerable success with Zworykin's kinescope. Tedham had independently worked on it and made it into a practical device, which was claimed to have a life of over 1700 hours.[34]

On the other hand, EMI still had no electric camera tube and was using antiquated mechanical means for pickup of both live and film images. Since the agreement between EMI and RCA forbade EMI from being in the communications business, they could build cathode-ray tube receivers but not cameras. While Alfred Clark, EMI's chairman, and Sir Louis Sterling, managing director of EMI, may have agreed tacitly (they were no match for David Sarnoff), Isaac Shoenberg had other ideas in mind. He was pleased to get information from RCA, but he insisted that all of the equipment used by EMI be designed and built by them in their own factories.[35]

The agreement between RCA and EMI had not prevented Tedham and McGee, who had come to work for EMI in January 1932, from surreptitiously building an experimental "iconoscope" in August 1932. It seems that Tedham, the EMI chemist, had been in contact with Sanford Essig the RCA chemist, and that certain information relating to Zworykin's new single-sided camera tube had been relayed to the EMI laboratories. As a result, Tedham and McGee decided to build such a tube for themselves to see how well it would work. All of the necessary video amplifiers, high-voltage supplies, scanning and pulse circuits, and so on were readily available to them from Browne's EMI cathode-ray tube laboratory.

In the autumn of 1932 an experimental tube was built. Its main feature was that the target was an aluminum-coated disc with a silver and cesium layer. A metal mesh screen was mounted in front of this disc and a pattern of square patches formed through the mosaic. After the mesh screen was removed and the photoelectric layer activated, the tube went through a complicated series of steps of baking and exhausting, similar to the Zworykin process, and was ready for testing.

Tedham and McGee borrowed a signal amplifier from Browne and drove the scanning coils on the neck of the tube in parallel with those of a picture tube from the same scan generator. After a few minor adjustments, to their amazement and satisfaction, it produced a crude picture for a short while. Then the gas from the electron gun proved to be too much for the photo-cathode and the picture disappeared. Somehow, the tube was stored away and survives to this very day.

McGee later claimed many times that this tube had been made independently of RCA and that he and Tedham had followed the ideas of Campbell Swinton (read Zworykin). However, there is abundant proof that EMI had knowledge of the Zworykin iconoscope project and had made a successful copy of it. This in itself was a feat of great significance.[36]

However, it was now time for RCA to officially share the iconoscope with EMI. Zworykin brought with him to England complete specifications, including Essig's method for producing silver mosaics. The agreement between EMI and RCA was modified and EMI was allowed to go into the communications business, including production of Zworykin's camera tube. In September 1933, Shoenberg set up a camera tube laboratory at Hayes with McGee in charge. (Tedham had become quite ill and had to leave the project.) Broadway was put in charge of the picture tube project; Blumlein remained in charge of electrical circuit research and design.[37]

EMI's only competition in England was the Baird Television Company, whose engineering activities were now headed by West. After leaving H.M.V. Gramophone, West had a short period of employment as a sound engineer at Ealing Studios. In May 1933, he became technical director of Baird Television, which had been sold to Isadore Ostrer in January 1932 and was now part of Gaumont-British Pictures. As technical director, West had immediately embarked on a program involving cathode-ray reception. By this time John Logie Baird was no longer playing a major role in the company. In fact, it was alleged that in July 1933 he was paid to stay away from their London research laboratories at the Crystal Palace and to do most of his experimenting at his home in Sydenham.[38]

From London, Zworykin went to Paris, where he presented his paper to the Société des Radioélectriciens on July 26, 1933. There is no record of where or whom he visited while in Paris. However, he did go to the Soviet consulate for a Russian visa, which was stamped on his American passport. Then he left for Berlin, early in August 1933. From there he took a train to the Soviet Union for the first time since he had left in 1919.

This had to be an unusual visit, since the United States and the Soviet Union had no diplomatic relations at the moment. However, RCA and the Soviet Union had an agreement approved in 1929 by the army, navy, and State Department for the exchange of technical information. In 1930, RCA had sent a transmitter engineer, Loren Jones, to Moscow to aid the Russians in their development of broadcast stations. So there was some kind of RCA presence there during these difficult times.[39] Thus Zworykin had no problem with the U.S. State Department for this trip.

It is easy to understand why the State Department was so eager to give allow Zworykin to go at this time. The Soviet Union was a brutal, security-conscious police state. Obviously the chance to introduce a world-renowned scientist into Russia would be of benefit to the United States government. While Zworykin would not be free to visit any place outside of Moscow, he certainly could report back on what he had seen in the various universities and radio manufacturing laboratories while giving his lectures. At any rate, and even though it was an official visit, no news of it has ever appeared in any available Western newspaper or magazine.

According to Soviet sources, he presented his paper on the iconoscope at the Moscow Home of Scientists before a group of Soviet radio experts. Shortly after Zworykin returned home, the United States and the Soviet Union did establish diplomatic relations in November 1933. This was the first time since the 1917 revolution.[40]

From the Soviet Union, Zworykin returned to Berlin, where he presented his paper once again, on September 4, 1933. In Berlin he visited Dr. Fritz Shröter and the laboratories of Telefunken A. G., the German correspondent for RCA. It is certain that Zworykin gave Shröter the same information on the iconoscope that he had given Shoenberg and EMI. However, according to all available evidence, Shröter routinely ignored most of the information passed along by RCA. For example, Telefunken's picture tubes, as displayed in 1933 at the Berlin radio show, were still gas-filled, in spite of the fact that the information on the Zworykin kinescope was available as part of the RCA/Telefunken agreement.

Part of the explanation may also lie in Shröter's opinion, expressed in his 1932 survey of television, *Handbuch der Bildtelegraphie und des Fernsehens,* that did not give Zworykin's methods for producing a camera tube much chance of success. For instance, Shröter stated that Zworykin's and other proposals for a photoelectric screen scanned by an electron beam to be converted into electrical fluctuations "were highly improbable." Shröter was also concerned that the voltage difference would be more a function of the wave-length than the intensity of the light. He went on to state that "perhaps those devices which permit accumulation of charge by photo-effects by the use of individual cells, might be the only solution for the future for transmission of normal brightness with a fine raster." He continued, "None of the above described cathode-ray scanning devices have the chance of being realized in the foreseeable future because the vacuum technology difficulties are very great and the tools for their being overcome have not yet been developed. Neither the potassium or selenium rasters can be provided with the high amount of individual cells with homogeneous inherent sensitivity. Nor can they be produced because of the technical complications or lack of physical research on the basic premise." Yet at this very moment Sanford Essig, in the Zworykin laboratories, had solved that problem with his silver aggregate process, and Zworykin's tubes were actually working quite well.[41]

After his triumphant return from Europe and the Soviet Union, in September 1933, Zworykin had many visitors who came to see what had been accomplished in his laboratories. The most distinguished of them all was Marchese Guglielmo Marconi himself. On October 10, 1933, Sarnoff conducted Marconi through the Zworykin/Engstrom laboratory in Camden. Marconi was given a demonstration of the latest 243-line interlaced RCA television system, and apparently was amazed at the advanced state of television research in the United States.[42] While there is no mention of Zworykin being there, it must have been a rather dramatic moment, this meeting of the "Father of Radio" and the "Father of Television."

Other visitors soon followed. Two radio experts from the Soviet Union, S. A. Veklinsky and A. F. Shorin, came to see RCA's progress. From England, both Wright and Broadway of EMI came in the autumn of 1933, by which time Zworykin's system was using interlaced scanning of 243 lines at thirty frames per second. This was demonstrated to Wright and Broadway, who were amazed to see how effective it was. As proof of its independence from RCA, EMI was still using vertical line scanning rather than horizontal scanning. Their experiments with interlaced scan using a special mirror drum did not work, and all they accomplished was a number of lines moving across the screen at right angles to the scanning lines. When Wright and Broadway told Shoenberg about Zworykin's success, EMI changed over to horizontal scanning early in 1934.[43]

On February 25, 1934, Zworykin reported that he was working on a "super microscope" which featured an iconoscope camera tube with a special quartz window used with an optical microscope. Its main advantage was that the operator was able to observe specimens using either infrared or ultraviolet wavelengths. The storage capability of the iconoscope mosaic allowed the super microscope to retain the image of an object after it had been removed.[44]

The bitter struggle between EMI and Baird Television for control of British television intensified. The BBC had announced that Baird Television's low-definition (thirty-line, twelve-and-one-half frame) service, begun experimentally in September 1929, was to be continued only until March 31, 1934. Since BBC radio did not support itself with paid commercials as did American radio, it was most likely that television service would also be supported by the British government by a similar system, where the operating revenue came from licensing the public.[45] Baird Television was striving hard to persuade the General Post Office, the branch of government which operated the BBC, to start a new television service using Baird equipment. Baird Television had given several demonstrations, but EMI had also been invited to give similar demonstrations and in each case had shown better pictures.[46]

The Baird Television Company under West, which was now allied with Fernseh A. G. in Germany, had made tremendous strides in the reception of cathode-ray television. Without an electric camera tube, they were using the Nipkow disc for both live and film pickups. For studio pickups, Baird was using a flying spot scanner. For outdoor coverage, Fernseh was using an old idea—a new, highly complex "intermediate film" system. This took advantage of the ability of 35mm motion picture film to take pictures with a reasonable amount of light out-of-doors or indoors with artificial lighting. At this time the televising of motion picture film was producing the best results.

The intermediate film system consisted of a 35mm film camera connected to a special high-speed film processor. Large reels of film were fed into the camera, where they were exposed. By forced development, the negative film was processed and ready for use in about sixty seconds. While still wet, this film was then fed into a motion picture projector and scanned by a Nipkow disc. The negative image was converted to a positive image electrically. While this process was both cumbersome and costly, it did work, albeit with rather poor results. However, for the bulk of their programming, Baird used regular 35mm motion picture film. The Baird cathode-ray receivers were all showing excellent 180-line pictures at twenty-four frames per second.[47]

Sydney Moseley of the Baird Television Company was also able to exploit EMI's past ties to RCA, even though EMI was by this time independent of the American company, by claiming that if the BBC chose EMI's system to run a London television station, the control of English television would be in American hands. "It would be a 'public scandal' if the Radio Corporation of America, through one of its 'controlled' companies in London, were to 'march in' through the backdoor of the BBC." Finally John Baird even wrote the king of England, complaining that "the BBC was crushing a pioneer British industry and was giving secret encouragement to alien interests." This claim became Baird Television's battle cry.[48]

In fact, Sarnoff was only too pleased to have EMI start a commercial television service in England. He encouraged Sir Louis Sterling, EMI's managing director, to speed up their laboratory efforts and, although EMI operated independently, RCA continued to give it technical assistance. According to Briggs, Sarnoff told Sir John Reith, director-general of the BBC, that "he was not going to start a regular television service [in the United States] until both competitive pressure and technical experience increased."[49]

At this point, the British General Post Office and the BBC decided to intervene between EMI and Baird Television. On May 16, 1934, a television committee headed by Sir William Mitchell-Thomson, Lord Selsdon, was set up to investigate the various television systems in England, Europe, and the United States. In October 1934, Lord Selsdon sent a subcommittee to visit the United States and examine the state of the art there. They visited the Zworykin/Engstrom laboratories in Camden in December 1934. There they were given demonstrations of 343-line, thirty-frame-per-second interlaced pictures both indoors and outdoors and reported that "very good reproduction was obtained." The subcommittee also visited the Philco laboratories in Philadelphia and reported that the Philco standards were comparable to RCA's and were sufficient to give clear, defined images on the screen. (Evidently Albert Murray had fulfilled his promise to Philco.) For some unknown reason, the subcommittee did not visit the Farnsworth laboratories while they were in Philadelphia, although they had planned to do so.[50]

Meanwhile, at EMI, McGee was able to report that his research group was producing operative camera tubes. Tube number six gave a very presentable picture, and was shown to EMI chairman Alfred Clark and director of research Isaac Shoenberg on January 29, 1934. The early target mosaics were made by a ruling process, but McGee soon switched over to Essig's silver aggregate process. As a result, tube number twelve had greater sensitivity and better picture quality.

McGee's group made such rapid progress that by April 5, 1934, they were able to point a camera out of a window and demonstrate a usable picture. Later in the year, using Ballard's patent, EMI went from 180 lines sequential to a 243-line interlaced picture at twenty-five frames per second. By then, RCA had raised their standard from 243 to 343 lines at thirty frames per second.[51]

On May 22, 1934, the Marconi Wireless Telegraph Company joined forces with EMI to become Marconi-EMI Ltd. This merger combined Marconi's expertise in short-wave radio transmission with EMI's proficiency in cameras and receiving systems. Isaac Shoenberg now sat on the Board of Directors of Marconi-EMI. This powerful alliance left Baird Television with the General Electric Company (not affiliated with General Electric in the United States) and Fernseh A. G. as its only allies.

EMI at this time was spending some £100,000 a year on research, and it appears that they were researching more than television. There have been allegations that the British government was encouraging companies such as EMI and Metropolitan Vickers Ltd., A. C. Cossor Ltd., British Thomson Houston Ltd., the British General Electric Company, and the other large electronic laboratories in England to expand their basic research into other electronic media.

The Radio Research Station at Slough, under Robert A. Watson Watt, had been working with cathode-ray oscillograph tubes since 1922. This was mainly concerned with radio direction finding and observation of atmospherics. While the early tubes were of the Western Electric variety, in the past few years they were mainly procured from the company of E. Leybold-von Ardenne Lichterfelde, Germany, and such reliance on foreign sources made the British government unhappy.

Secret work on radio-location of objects (radar) began in England in February 1935. It became obvious that a broad base of English manufacturing plants and trained personnel would be welcome in case the political situation in Europe, especially with the rise of Hitler, turned into full-scale war.[52]

Progress in television itself continued as well. In May 1934 Hans G. Lubszynski and Sydney Rodda of EMI applied for a patent for an important improvement on the iconoscope, which combined the image section of the Farnsworth dissector tube with the storage plate of the Zworykin iconoscope. This resulted in a camera tube with more sensitivity and fewer shading problems. The new tube was called the "Super Emitron" in England and was designed to replace the old iconoscope.

A similar tube had also been conceived in the Zworykin/Engstrom laboratories by Alda Bedford, so a patent interference was declared between them. Even though the EMI patent had been applied for first, the suit was resolved in RCA's favor due to an earlier conception date. Clearly, when it came to patents and patent interferences, RCA treated EMI just like any other competitor.[53]

Sometime early in 1934, Zworykin was transferred to and made director of the electronic research laboratory of RCA Victor. On May 28, 1934, at the ninth annual meeting of the Institute of Radio Engineers, he was awarded the Morris Liebman Memorial Prize for his work on new television devices using a rapid beam of cathode rays, the first of a series of honors.[54] While Zworykin was being honored for his research, it was Elmer Engstrom who introduced the RCA television system at this same meeting. This was the first time that

RCA publicly acknowledged that Engstrom rather than Zworykin was actually managing the RCA Victor television project. Three major advances claimed for RCA's system were: (1) the ability to pick up and transmit outdoor scenes through the medium of the cathode-ray camera, (2) greater detail, and (3) near-perfect synchronization between transmitter and home receivers. Dr. W. R. G. Baker, now vice president and general manager of RCA Victor, claimed that the system produced detail that was as good as "8mm home movies."

At the same time, Baker acknowledged that great progress had been made but also brought up the question of who was going to pay for starting up a television system in the United States. He suggested a government subsidy like that given to the BBC, but this idea wasn't taken seriously.[55]

Meanwhile a new development had come to Sarnoff's attention. His good friend and RCA stockholder Major Edwin Howard Armstrong had invented a superior method of radio transmission, a wide-band Frequency Modulation (FM) system. It was noiseless and free from both man-made and natural static, offered higher selectivity, and was capable of high-fidelity (full range sound) transmission. However, while experiments showed that the system had merit, Major Armstrong needed a special transmitter to continue his research program.

The two-kilowatt RCA television transmitter at the Empire State Building was just perfect for such tests. It operated in the 44mH region and needed only minor adjustments to transmit a wide-band FM signal. After several meetings and a private demonstration, Sarnoff gave Armstrong permission to use this transmitter for actual broadcast tests. Sarnoff ordered the television transmitter, which was now idle except for an occasional test, shut down. Major Armstrong moved in with his new radio equipment. The necessary changes were made and by June 16, 1934, Armstrong made his first test. A trial receiver was located at Westhampton Beach, seventy miles away. Testing continued throughout 1934 and into 1935.[56]

In the summer of 1934, Zworykin was visited at Camden by the eminent Japanese television pioneer Kenjiro Takayanagi. Takayanagi had been working on cathode-ray television since 1926 and had made great progress. He was shown the complete RCA system in operation and was quite impressed with it.

He then visited Iams's tube laboratory in Harrison, New Jersey, and was given full details of how iconoscopes were built. He was determined to build such a tube when he returned to Japan. Takayanagi also visited Farnsworth's laboratories in Philadelphia, and was quite impressed with their work as well. He reportedly told chief engineer

Arch Brolly that "We already make your receiver set. When I go home we make your transmitter!"[57]

Takayanagi then went on to England, where he visited the laboratories of Baird Television in London, and to Germany to visit the 1934 Berlin Radio Exhibition. In Berlin, even though Germany and Japan had similar goals for world domination, the German authorities denied him access to either the Telefunken or Fernseh television research laboratories. This was in sharp contrast to the friendly reception that Takayanagi had received from the RCA Victor Company, the Philco Corporation, and the Farnsworth laboratories. The Nazi military regime was in control of all television research there, and they, like the British and the Americans, were engaged in highly secret research on radio-location of objects.[58]

At this time, Philo Farnsworth had received new financing from a stockholder, Seymour Turner, and had put together a demonstration unit to prove that his system was still very much in contention. Farnsworth was now using improved photosensitive materials (cesium on silver oxide, the Koller process) in his dissector camera, which gave it more sensitivity. He had added a built-in secondary emission electron multiplier that made it possible to pick up live pictures in ordinary sunlight. He was also using a short coil for his magnetically focused picture tube and was able to present pictures with almost the same amount of brightness as the Zworykin kinescope.

Beginning on August 25, 1934, Farnsworth gave ten days of live television demonstrations at the Franklin Institute in Philadelphia. Since both his camera and receiver were transportable, he was able to present pickups from the institute's roof featuring vaudeville acts, tennis games, and appearances of various politicians. This was the first public demonstration of electronic television ever given. Even though Zworykin had a more sensitive camera, more lines, odd-line interlaced scan, the kinescope, and so on, in short a superior system, in his laboratory, Sarnoff had forbidden any public demonstration of it.[59]

In June of 1934, Baird Television asked Farnsworth to bring one of his image dissector units to London for evaluation, as they desperately needed an electronic camera tube to compete with Marconi-EMI. It is ironic that Baird Television, which had so bitterly complained about the connection between EMI and RCA, should now have to go to an American television company to provide them with an electronic camera so that they could have a competitive cathode-ray system.

Farnsworth's system was installed at the Baird laboratory in the Crystal Palace in London. A satisfactory demonstration of motion picture film only (no live pickup) was given. George Everson claimed that this demonstration was an important factor in the Selsdon committee's findings.[60]

Not only was the latest image dissector tube still a potent competitor, RCA and EMI no longer had exclusive use of the iconoscope. Successful iconoscope tubes had been built by the Philco Corporation in Philadelphia, Telefunken A. G. in Berlin, by a scientist in the Soviet Union named Krusser, and by John R. Hofele in the Bell Telephone labs in New York City.[61]

In the Zworykin/Engstrom laboratories, Alda Bedford applied for a patent on a "shading generator" on October 26, 1934. This complicated electric device made it possible to neutralize the moving charge pattern on the face of the iconoscope and get excellent pictures from it. Finally, and of the most importance, an all-electric interlaced synchronizing and blanking generator had been developed and was in use by the end of 1934. This was mainly the effort of Alda Bedford, with the assistance of Raymond Kell and Merrill Trainer.[62]

At this time the Zworykin/Engstrom television system at Camden was operating on 343 lines at thirty frames per second, interlaced sixty fields per second. Marconi-EMI's system was operating at 243 lines at twenty-five frames per second, interlaced fifty fields per second. This was the beginning of the division of standards between the American and European systems that has lasted to this day.[63]

(Bottom) Zworykin "on camera" at the RCA laboratory in Camden, 1933; (top) actual picture off of the kinescope tube. Courtesy of Les Flory.

Banquet given by Zworykin at the Hotel Walt Whitman in Camden, June 1, 1934. According to the original caption, those in attendance were (left to right) Braden, Luck, Malter, Young, Wolff, Rhodes, Campbell, Beers, Smith, Murray, Iams, Horn, Goldsmith, Zworykin, Engstrom, Vance, Flory, Goldsborough, Painter, Ogloblinsky, Briggs, Tolson, Turner, Roberts, Kurz, and Essig. Courtesy of Harley Iams.

1934 RCA iconoscope camera. Courtesy of the David Sarnoff Research Center.

Zworykin with electron multiplier tube, 1936. Courtesy of the David Sarnoff Research Center.

Les Flory (left) and Zworykin with experimental tube, RCA Victor laboratory, Camden, 1935. Courtesy of the David Sarnoff Research Center.

Zworykin at a meeting of the Congrès de l'U.I.R., Paris, Mar. 1936. (Left to right) Barthélemy, Harbicht, Fortin, Poincelot, Schrötter, Zworykin, Nestel, Le Duc, Karolus, Chevallier, Knoll. Courtesy of Jacques Poinsignon.

David Sarnoff addressing the opening of the RCA exhibit at the New York World's Fair, Apr. 20, 1939. (Top) Sarnoff in front of the camera; (bottom left) rear view showing camera; (bottom right) actual picture taken off the air. Courtesy of the David Sarnoff Research Center.

Zworykin (top) and James Hillier with RCA electron microscope, 1942. Courtesy of the David Sarnoff Research Center.

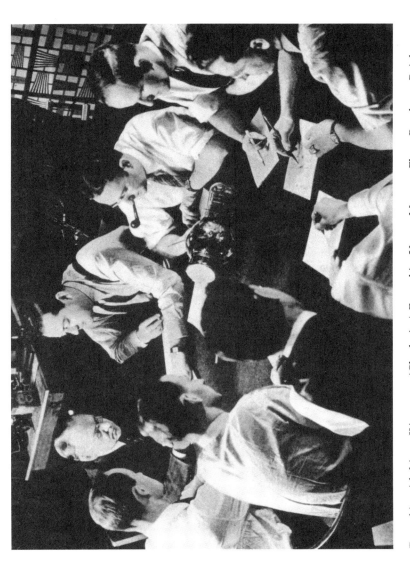

Zworykin with his staff in 1940. (Clockwise) Zworykin, Vance, Morton, Flory, Ramberg, Rajchman, Banca, and Massa. Courtesy of Les Flory.

Sir Isaac Shoenberg. Courtesy of the Royal Television Society.

Allan Dower Blumlein. Courtesy of the Royal Television Society.

J. D. McGee with a display of his tubes, Imperial College, London, 1972. Author's collection.

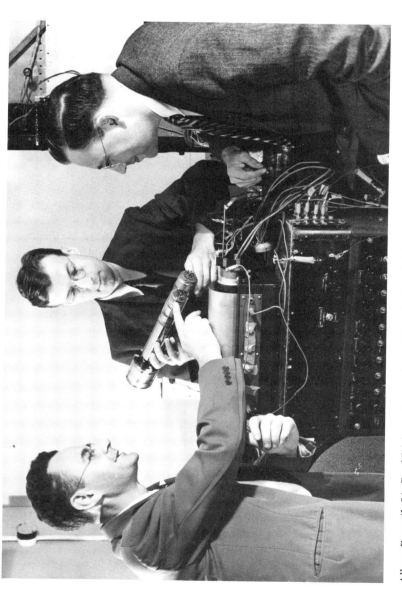

Albert Rose (left), Paul Weimer (center), and Harold Law (right) with early image orthicon camera tube, 1945. Courtesy of the David Sarnoff Research Center.

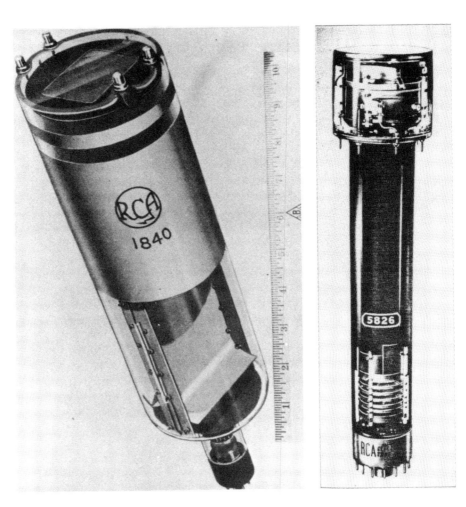

1840 orthicon tube and 5826 image orthicon camera tube. Courtesy of the David Sarnoff Research Center.

Display of color tubes, 1951. (Left to right) E. W. Engstrom, E. W. Herald, H. B. Law, Zworykin. Courtesy of the David Sarnoff Research Center.

Zworykin with display of camera tubes, Princeton, 1954. Courtesy of the David Sarnoff Research Center.

Manfred von Ardenne (left) and Zworykin, 1973. Courtesy of Manfred von Ardenne.

Zworykin and the author, RCA laboratories, Princeton, 1976. Author's collection.

David Sarnoff. Courtesy of Karsh/RCA, David Sarnoff Research Center.

Zworykin, Miami, Fla., 1980. Courtesy of *Television International*.

A Return to Russia

In September of 1934, Vladimir Kosma Zworykin was again traveling back to the Soviet Union. Sometime in 1933, perhaps while he was on his first visit, he had received an invitation from Moscow to give a series of lectures on television. With the resumption of diplomatic relations between the United States and the Soviet Union in November 1933 an extended visit was possible.[1]

In spite of his defection, Zworykin claimed that he was quite highly regarded by the Soviet regime. Earlier, he had received a very flattering invitation to return to the USSR, with the assurance that his past was known and there would be no recriminations. He rejected this offer, stating that he had settled in the United States and had become an American citizen, and had no intention of renouncing his citizenship.

Returning to Russia to present a series of lectures on television was different. Zworykin knew from the literature that there was a great deal of interest in television in the USSR and he was curious to see how much progress they had made. There was always the chance of detention, as his family and friends warned him, and they tried to persuade him not to accept the invitation, but he had had no problems in August 1933 and anticipated none in 1934.

Before making this decision Zworykin decided to consult with Sarnoff, who said that he had no objection to an official visit, as there was always the possibility of doing more business with the Soviet Union in the future. He saw little danger of Zworykin's being detained as long as he had an exit visa, but he left the decision to Zworykin himself.

The State Department also had no objection. In fact, they welcomed the idea of Zworykin, as a trained scientist, touring the closed society of the Soviet Union and visiting laboratories. This visit could provide valuable information, which was hard to get in those days. However, they warned him that they could not protect United States citizens returning to their country of origin. Zworykin weighed all of this and decided, in spite of the objections of his family and friends, to accept the proposal.

So Zworykin returned again, first by ship to Europe and then by train from Berlin, to the country of his birth. At the Soviet border, he was met by an engineer from the Communications Trust who remained with him for the rest of his visit. He was furnished with food coupons and rubles and had to register his money and camera along with his passport. Also, he was warned that taking pictures of certain installations was forbidden. Zworykin got along very well with his guide, even though there were occasions when Zworykin claimed he bent the rules a little.

Zworykin first visited Leningrad, where his two sisters were still living. Here he was met by his brother-in-law, a professor at the Institute of Mines. His first impression of Leningrad was that it hadn't changed at all in the past seventeen years, but that it had been taken over by "country-folks." There were few carriages and almost no automobiles, and the streets were filled with shabbily dressed people.

He was taken to the Astoria Hotel, which once was one of the finest hotels in the city and was where his father had always stayed when visiting Leningrad. It was unchanged outside and shabby inside. He was pleasantly surprised to be given a sumptuous suite of rooms in fairly good condition, which he later found was usually reserved for foreign dignitaries.

On the first day he ordered breakfast from a rather elaborate menu. Since he was paying for it with food coupons, he ordered one of his old favorite dishes, "riabchick," a sort of a game bird. However, he had to settle for something more simple like boiled eggs. He was joined during the meal by his guide and two engineers, who presented him with a rigorous lecture schedule. Lectures were to be given almost every day; and there were to be many visits to noted laboratories and several formal dinners. He then found out that his trip had been sponsored by the Soviet government communications industry, not, as he had assumed, by the universities.

By bargaining a bit, he was able to rearrange his schedule to include some sight-seeing and a visit to his sisters. He also agreed to lecture in Russian rather than English, even though much of the technology had to be described in English terms. He found that the question and answer periods were often longer than the lectures themselves, and he was a quite impressed with the knowledge and discipline of the students.

He also visited several of the electrical laboratories, which were mostly in old buildings and poorly equipped. He was not at all impressed with their facilities. Obviously the Soviet authorities showed him only what they wanted him to see. However, he did see many original experiments and some unfamiliar results.

One of his lectures was at the St. Petersburg Polytechnic Institute, where he met an old friend, Petr L. Kapitza, who had been given his own laboratory at Cambridge University. Kapitza was accompanied by Professor Abram F. Ioffe, another Soviet physicist whose lectures Zworykin had attended as a student in Leningrad. Zworykin asked Kapitza when he expected to return to England, as he planned to visit Cambridge University on his way home. Kapitza was quite noncommittal at the time.

Zworykin also visited his old school, the St. Petersburg Institute of Technology. He was rather disappointed not to find a single person he had known before the war. He noted that the institute had changed considerably.

He also attended the theater and saw several ballets and operas. These he found to be as well performed as when he had left. Zworykin claimed that all inquiries about Professor Boris Rozing indicated that he had been arrested at the beginning of the 1917 Revolution, exiled to Archangel, and died soon after. (Actually Rozing had died in 1933.)

Zworykin visited his sisters, who were well but looked older than he had expected. The fact that his brother-in-law was a full professor at the Institute of Mines and a member of the Russian academy gave the family certain privileges. His older sister Maria had abandoned her medical career and was now working as an illustrator at the institute. She had married and had a daughter, but had lost her husband during the revolution. From her he heard that their mother had died in Murom during the civil war and that the rest of the family had been scattered throughout the region.

About a week later, Zworykin left for Moscow. The train was the Red Arrow, which he found to be in excellent condition. It left Leningrad at midnight and arrived in Moscow at seven o'clock in the morning. The cars were clean and comfortable and hot tea and biscuits were served in the compartments.

Upon arriving in Moscow, he was taken to a comfortable hotel. Moscow had changed more than Leningrad. He found that they were putting up many new buildings and a subway was under construction. Again he had a schedule of almost-daily lectures, attended the theater in the evenings, and visited as many of his relatives as he could locate.

While in Moscow he was asked by the head of the communications industry if RCA Victor would be interested in selling and installing a television transmitter and a number of receivers in Moscow. Zworykin had expected this question to come up and explained that he was only an engineer and would have to take this inquiry back to the head of RCA, David Sarnoff.

From Moscow, Zworykin was flown in a military plane to Kiev, Kharkov, and Piatigorsk. He noted that civil aviation in the Soviet Union was just being organized and planes had to land on a grassy runway. From Piatigorsk he was driven over a military road to the city of Tbilisi, the capital of Georgia, which had been chosen for his visit because his older brother Nicolai lived there. Zworykin's brother had married a second time and was working as a construction engineer. Like many other engineers in the USSR, he had been arrested on various vague charges of sabotage and spent several months in jail. Only when the project he was working on had been finished was he given his freedom and never arrested again.

In Tbilisi Zworykin was installed in the best hotel and wined and dined by a local group of engineers and officials. At one of the dinners he was introduced to Lavrenti P. Beria, who at the time was chairman of the Georgian Communist party and a close friend of Joseph Stalin. Later, in 1938, Beria became head of the NKVD, the dreaded Soviet secret police. Zworykin obviously impressed Beria, who asked him if he would like to visit the Caucasus. Zworykin said he would like to see the Black Sea shore, so Beria arranged to have him flown there by a military plane. In two hours he arrived in Souhumi, on the eastern shore of the Black Sea. He was lavishly entertained for several days before returning to Moscow by way of Sochi. He was driven by automobile to Sochi and flew in a regular plane back to Moscow.

Upon his return, he attended several meetings in which the possibility of purchasing a complete RCA television system was again discussed. After a few days, it was time for Zworykin to leave Moscow. His sister, several engineers he had met on the visit, and his guide all came to see him off at the railway station. Everybody expressed hopes of seeing him again. After a tearful farewell, Zworykin entrained for Berlin.

Zworykin claimed that he never felt any hostility while in the Soviet Union. He was never questioned about the circumstances of his leaving the country in 1918. Yet he confessed a certain sense of relief when his papers were cleared and he was installed in his train compartment.

At the Soviet border, Zworykin felt this sense of security disappear. The camera he had brought with him to the Soviet Union was a new German Leica. At his farewell in Moscow, he had given it to a nephew who had admired it. The border policeman looking over Zworykin's papers and passport noticed that the camera was not there. Zworykin had a box of 35mm projection slides with him and just indicated that the camera was there somewhere. After a short pause, and without a word, the guard stamped Zworykin's papers and left the train.

Zworykin's trip to the Soviet Union had lasted about six weeks. He returned to the United States around November 4, 1934. His return from the USSR was again met with mixed feelings by many of his acquaintances. It was a great relief to his family and friends, as quite a few predicted that he would never be permitted to return. The majority of the White Russian émigré colony hadn't approved of his trip in the first place, on ideological grounds, and looked on his safe return with much suspicion.

A few days after his return, Zworykin reported to Sarnoff about his conversation with the Soviet communications officials and their interest in purchasing RCA television equipment for Moscow. As a result, further contact was made with the Soviet Embassy in Washington and the Soviets placed an order for a complete television system.[2]

Late in 1934, Gregory N. Ogloblinsky was killed in a high-speed automobile crash while on vacation in France. Zworykin had lost one of his best friends and fellow workers, one who was loved and admired by all who knew him. Though "Oglo" was quite shy and demanded no publicity, his accomplishments, though unsung, were to live on long after he had passed away. His role in the Chevallier affair, for example, was never made public. There were many who credited him with most of the success of Zworykin's early television systems. Certainly his tragic death created a void in the Zworykin/Engstrom research division that had to be filled.

While RCA had been carefully hiring such outstanding scientists as Dr. Otto Schade, Dr. R. R. Law, Dr. George Morton, Dr. I. G. Maloff, and Dr. D. W. Epstein, there still remained a gap. This was filled in 1935, when RCA hired Dr. Albert Rose, a brilliant graduate of Cornell University. He was assigned to the Harrison, New Jersey, tube laboratory to assist Harley Iams in his research in camera tubes, including improving or replacing the iconoscope.[3]

Zworykin made another of his several unpublicized visits to the Soviet Union in February 1935, when he supposedly visited a television institute, the All-Union Scientific Research Institute for Television in Leningrad. According to one Soviet source, Zworykin had nothing but praise for their recent accomplishments. Ever the consummate diplomat, he was quoted as saying, "On my first visit here I wanted to acquaint you with some of my new ideas and accomplishments. Now after my second visit [this was actually his third] I leave as a colleague of yours, I fear that when I come the next time I will have a lot to learn from you." Zworykin returned to the United States in March 1935.[4]

With Elmer Engstrom brilliantly directing the RCA television project, Zworykin had long ago expanded his field of research and was quite content to let his colleagues at the Camden and Harrison laboratories continue to improve the RCA television system. Instead, Zworykin was continuing his research into the various electric phenomena that applied to light and electricity, such as electron optics and secondary emission. Specifically, he was now engaged in research on an electron multiplier tube, a tube with a series of electrodes that were bombarded in such a way that each one gave off more electrons than the one before, resulting in a sizable gain of current. On April 28, 1935, he gave a lecture before the Institute of Radio Engineers in Washington, D.C., entitled "The Application of Secondary Emission."[5]

Meanwhile, in England, the report of Lord Selsdon's television committee, given on January 14, 1935, recommended that both Baird Television and Marconi-EMI be given the opportunity to supply television equipment for a series of demonstrations to be given in London at a later date. While several other British companies had been surveyed, including A. C. Cossor, Scophony, Ferranti, and the English General Electric Company, none of them was prepared with either finances or equipment to participate in the upcoming competition.[6]

The promised setting up of a high-definition television station in London did not go unheeded in the United States. On February 26, 1935, David Sarnoff told the American public that "Technical and geographical difficulties make television impossible in the United States at present." However, he stated that "The United States is keeping pace with developments in Europe." He also said that Zworykin's laboratory was working on larger and better picture tubes.[7]

On April 3, 1935, the BBC announced that the first television station in England would be built at the Alexandra Palace on the north side of London. This was the site of a large old exhibition building that offered space for such a development. It was also located on an elevation, which was necessary for maximum television coverage. The General Post Office confirmed that Baird Television and Marconi-EMI would be the competing companies.[8]

Then on May 7, 1935, Sarnoff announced plans for an extensive field test program by RCA. He wanted to assure the world that RCA had not lost its lead to Great Britain. Some of the goals announced were to establish the first modern television transmitting station in the United States, to manufacture and distribute a limited number of receiving sets, and to build studio facilities to develop the necessary studio technique. However, Sarnoff also noted that this program would take some twelve to fifteen months to complete. He was still in no hurry to launch a television service in the United States, especially since RCA could benefit from the lessons learned in setting up a television station in London.[9]

At this time, Armstrong was still conducting a series of field tests of his new FM system using the RCA television transmitter on top of the Empire State Building. The tests were so successful, in fact, that Armstrong demanded of Sarnoff that RCA commit itself to "adopting FM—not as an auxiliary service, but as a substitute for the existing AM system!" This would mean a complete reordering of radio power, a probable alignment of new networks, and the eventual overthrow of the carefully restricted AM radio system on which RCA had grown to power. While quite impressed with the Armstrong system, Sarnoff was not about to tamper with the successful AM radio system, both local and national, that had weathered the Depression so well.

In addition, Sarnoff now needed the Empire State Building transmitter for Engstrom's television field test program. In April 1935, he politely requested Armstrong to remove his equipment from the Empire State Building and terminate his experiments there. Armstrong was deeply hurt by this decision. He was technically correct but economically wrong, in that FM service eventually became the dominant radio medium. This was also the end of the friendship and the start of a deadly feud between the two men.

Dr. W. G. R. Baker, RCA's vice president of engineering, was quite pleased with the results of Armstrong's tests and supported him. This infuriated Sarnoff, and within a few short months, Baker was "promoted" out of his job and went back to work for GE. He was replaced by L. M. Clement, who had been chief engineer of the International Standard Electric Company in London and had previously been employed by both Westinghouse Electric and the Federal Telegraph Company.[10]

On June 7, 1935, the new technical standards for the London television service were revealed. Baird Television was to continue with 240 lines sequential scan at twenty-five frames per second, the minimum requirement of the Selsdon committee. However, Marconi-EMI was going to use an advanced 405-line interlaced system at twenty-five frames (fifty fields) per second. This came as a complete surprise to everyone else experimenting in television, including RCA.

Isaac Shoenberg had decided to stretch his system to its limits. Since RCA was operating on 343 lines at this time, he decided to surpass them and went on to the next highest number using odd-line interlaced scanning. The consensus was that such a high standard could not be achieved at this time, and most experts thought it unnecessarily high for home use. But Shoenberg was confident that his superb team of research engineers, led by Allen Blumlein, could have such a system operating by the time the trials started.[11]

Philo Farnsworth also wanted to let the radio industry know that he was very much in contention. He had continued to perfect his system, and while Sarnoff procrastinated, Farnsworth saw the opportunity to demonstrate his new, improved television system. On July 30, 1935, he gave a demonstration from his Philadelphia laboratories at Wyndmoor, Pennsylvania. The pictures were 240 lines sequential at twenty-four frames per second, and the press reported that the film pickup was presented with "a clarity and sharpness of definition that surprised many observers."

Farnsworth then went on record that now was the proper time to launch a television service in the United States. He proposed building television transmitters in San Francisco, Philadelphia, and New York City. He confirmed his ties with Baird Television in Great Britain and Fernseh A. G. in Germany. He believed that television had advanced to the stage where it had real entertainment value, and that promoting television would not necessarily upset the radio industry but make contributions to it.[12]

By now, not only had the Bell Telephone labs built a successful iconoscope of their own design but John R. Hofele was working on the first practical photoconductive camera tube. This was demonstrated to Bell Telephone executives Buckley, Streiby, and Axel G. Jensen on July 26, 1935. It used a film of red mercuric acid (HG I_2) cooled by carbon dioxide (CO_2) and was viewed on a special cathode-ray tube built by C. J. Davisson. While it used the seventy-two-line standard, Bell's disc system, Hofele planned to go to 240 lines in September 1935.[13]

In July 1935, Sarnoff went to Europe, ostensibly to study the state of European television. Actually he went to negotiate the financial separation of RCA from EMI, which for the past two years had neither needed nor desired any assistance from RCA to provide a television service. EMI's, and especially Shoenberg's, desire to be free from RCA's domination was now finalized. According to the 1935 RCA stockholders' report, RCA sold its EMI stock for some $10,225,917 in cash to investors in England.[14]

No longer could Baird Television or anyone else in England claim that RCA had any financial or technical interest in English television. One of the questions still remaining is where the financing came from. The large sum involved strengthens the case for some kind of subtle monetary intervention by the British government, which certainly went to great extremes to help EMI rid itself of American domination and at the same time to build up a British industrial base of laboratories, manpower, and materials for the future.

The remarks of Alfred Clark, chairman of EMI, at an EMI stockholders' meeting on November 15, 1935, are of extreme interest. He stated, "The position which your company [EMI] has achieved in television is due to the work of its research engineers. In the last decade there has been a notable awakening of the importance of research if carried out under the best of conditions and the large concerns in this country now regard it essential to successful business." He continued, "The development of the various components of television, cathode-ray tubes and electric eye scanners will soon be available in quantity. The cost of this important department has of course, been treated in our books as a 'trading charge' because we are dealing with an essential part of the business upon which our success depends."[15]

What Clark had left unsaid was that EMI's basic system was built around Zworykin's iconoscope (now named the Emitron), Ballard's patent for interlacing, and Zworykin's kinescope (now called the Emiscope). Zworykin could bask in the glory that even though it was totally English built, owned, and operated, EMI's system was mostly based on his inventions.

John Baird bitterly called attention to these facts. His contract with the General Post Office for a television system had expired on March 31, 1934. He complained, "And so, after all these years, we were out of the BBC. The fact that it was the RCA system, imported from America, the scanning used being covered by the RCA-Ballard patent and the transmitter being the iconoscope of Zworykin, did not hinder the Marconi Company proclaiming the system all British. The iconoscope was now called the emitron. Ballard was ignored and in an amazingly short time the Marconi publicity department had established it in the public mind that Marconi had invented television."[16]

Meanwhile, Zworykin continued work on a secondary emission multiplier with the assistance of two RCA engineers, Dr. George

Morton and Dr. Louis Malter. Their secondary emission multiplier, which was similar to one presented by Farnsworth in 1934, had a double row of electrodes extending the tube's length. One row was coated with cesium and acted as targets; the other row supplied electrostatic fields that guided the electrons in the desired path. From a modulated light source, electrons were driven from target to target, building up the current flow. This resulted in a significant amount of electrical gain without increasing the "noise" in the signal. On October 23, 1935, Zworykin presented a paper, co-authored with Morton and Malter, entitled "The Secondary Emission Multiplier— A New Electronic Device" at a meeting of the Institute of Radio Engineers in New York City. This same paper was read again before the Institute of Radio Engineers at Rochester, New York, on November 18, 1935.[17]

In December 1935, the *New York Times* reported that Amtorg, the Soviet trading corporation, had ordered $2,000,000 worth of radio equipment and machinery from RCA for shipment to the Soviet Union. The purchase was approved by the U.S. army and navy and the State Department, and the equipment was to be made in the RCA plants in Camden and Harrison, New Jersey. In 1936, a special commission arrived from the USSR to inspect and accept the television equipment. They remained in the United States until 1937.[18]

On January 4, 1936, Zworykin and Morton demonstrated a new electron-optical device to a meeting of the American Association for the Advancement of Science at St. Louis. In this demonstration, they revealed for the first time details of a new electron image tube that they called a "third eye." This was a tube that received light from the entire optical spectrum. Light striking a photocathode would be accelerated and focused by a system of ring-shaped electrodes where the electron image was thrown onto a fluorescent screen. In other words, there was no scanning involved.

Articles in both *The Times* and *Radio-Craft* claimed that its color sensitivity was such that it could reach up to 13,000 angstrom units in the infrared and down to 1,800 angstrom units in the ultraviolet section of the spectrum. It was asserted that this tube could see through fog, smoke, and darkness and that it made possible the examination of living specimens under infrared light. Zworykin gave credit for this device to Morton and Malter, who worked with him on this project.[19]

Soon after giving this demonstration, Zworykin left for Europe again, on a lecture tour that took him through England, France, Germany, and Hungary. In London on February 5, 1936, Zworykin delivered a paper, "Electron Optical Systems and Their Applications," before the wireless section of the Institution of Electrical Engineers.[20]

He also visited the Alexandra Palace, which was being readied to be the site of the future London television station, and the laboratories of EMI at Hayes, Middlesex. There he found that EMI had built a television system very similar to RCA's. However, he was quite surprised to discover the 405-line standard that EMI had proposed. With this giant step forward, EMI had finally surpassed RCA in the great television race.

He also commented that the Baird television pictures were quite good, considering that they were 240-line pictures. In fact, when it came to motion picture film transmission, the Baird 240-line system was considered better (except for the flicker) than the EMI 405-line interlaced pictures from the emitron film projection system. EMI's iconoscope tube did not react well to sudden changes in light values from scene to scene in motion picture film.[21]

In March, Zworykin traveled to France, where the Postes Télégraphes et Téléphones Service was operating a television station from the Eiffel Tower. It was transmitting on 180 lines at twenty-five frames per second and used a Nipkow lens disc for film pickup. All reception was on cathode-ray tube receivers. A system of 240 lines was being built by the Compagnie pour la Fabrication des Compteurs et Matériel d'Usines à Gaz, which was directed by René Barthélemy, with Chevallier as one of his chief engineers. At the Sorbonne, Zworykin repeated his paper on electron optical systems to the Société des Radioélectriciens. On March 10, Chevallier hosted a reception for Zworykin in Paris. Apparently there was no animosity between them, and Chevallier was quite friendly.[22]

Later in March Zworykin went on to Berlin. On July 12, 1935, the German television service had been put under the control of Hermann Göring, the Reich's minister for aviation. As a result, all television research in Germany was carried out under strict security measures and electrical laboratories such as Telefunken (which was allied with both RCA and EMI), Fernseh (allied with Baird Television and Farnsworth), Loewe Radio, and Lorenz were all under rigorous military control. This bureaucratic secrecy hampered German television research very much. Even Zworykin's own iconoscope and Farnsworth's image dissector were considered top military secrets.

At this time, German research in television consisted mainly of refining and perfecting techniques from both the United States and England, which they did very well within certain constraints. They were still operating on 180 lines sequential scan at twenty-five frames per second.

Rather than television, the large German electric companies were more concerned with radio-locating devices (radar) and communications, all instruments that had direct application to the coming monstrous German war effort. The German propaganda machine did not concentrate on television either, although according to an article in *The Times* the director-general of German broadcasting, Eugene Hadamovsky, stated that it was his aim "to imprint the image of the Führer in the hearts of the German people by television." This was an idle boast. While there was much talk of the impending sale of receivers, none ever appeared on the market. However, the Germans were preparing to televise the 1936 Olympic Games from Berlin.

Zworykin was allowed to visit the laboratories of Telefunken A. G. and gave his lecture at the Technische Hochschule in Charlottenburg where he had enrolled some twenty-two years earlier.[23]

During this visit to Berlin, Zworykin also met Baron Manfred von Ardenne, the noted German television scientist. While the two men had corresponded and knew of each other's work, this was the first time they had met personally.[24]

Zworykin then continued on to Budapest and neighboring Ujpest, where he visited the Tungsram Forschungslaboratorium. He was greeted by the director, Dr. Eugene Aisberg, who showed him through their laboratories. Zworykin was quite impressed with the advanced quality of the work being done by the Hungarians.

Dr. Paul Selenyi showed him a method of storing images by electrostatic charging. A rotating metal plate or drum was covered on one side with an insulating material and was connected to a source of DC potential. This formed the anode of a three-electrode system. By modulating the DC potential according to the picture to be transmitted, a pattern is formed on the insulating plate that can be made visible by dusting with lycopodium powder. This certainly was an early predecessor of the modern copying machine.[25]

While in Budapest, Zworykin was visited by one of the directors of a big television laboratory in Germany. It seems that a very important government official had missed Zworykin's lecture and wanted him to come back to Berlin and repeat it. Zworykin politely refused, as he was not very pleased with the political conditions he had found on this trip to Germany. This upset the director, who apparently was under orders to bring Zworykin back with him, but Aisberg congratulated Zworykin on this decision and confirmed his previous impression of the terrifying situation in Nazi Germany.[26]

Zworykin then returned to London. He called Kapitza at Cambridge University and found that he had never returned from the Soviet Union and that his friends were afraid that he had been detained for an indefinite period. In fact Kapitza had been given the post of director of the Institute of Physical Problems in Moscow, but he was not allowed to leave the Soviet Union for some thirty years. It was 1966 when he again visited England and 1969 when he traveled to the United States. Zworykin also found that his old Russian friend Dr. Serge M. Aisenstein, was living in England and was the director of the English Electric Valve Company research laboratory in Chelmsford.[27]

Early in April 1936, Zworykin sailed home to the United States. On April 30, he gave a final lecture in a series on applied electron optics to a meeting of the American Physical Society held in conjunction with the Institute of Radio Engineers in Washington, D.C.[28]

"The World of Tomorrow"

   Well aware of the progress being made in England in preparing the new London television station, David Sarnoff decided to give a closed circuit demonstration of the RCA system to the press on April 24, 1936. This was presented from the television studio at the RCA Victor plant in Camden, New Jersey, and was RCA's first public presentation of television since May 1932. It used the latest 343-line interlaced thirty frame (sixty field) system. A single iconoscope camera that was inside the studio was first pointed outside at the set, where a staged fire was put out by the Camden fire department. The camera was then trained on the studio for some light entertainment. Finally, a reel of 35mm film was projected.

  A standard television receiver, featuring a cabinet in which the picture tube was upright and the picture reflected through a hinged mirror, was used. The set had thirty-three tubes and fourteen control knobs. The picture measured five by seven inches and had a greenish cast. According to Zworykin, who was present, "Green was chosen because the human eye is more sensitive to that color." After the demonstration, Ralph R. Beal, research supervisor of RCA, commented, "It is safe to say that home television is at least eighteen months away."[1]

  The Empire State Building television transmitter went back on the air on June 29, 1936. This was the occasion of a demonstration of the RCA television system at Radio City for an assembly of 225 RCA licensees. The program consisted of speeches by Major General James G. Harbord, chairman of the board of RCA, and David Sarnoff, president of RCA; and a report by Otto S. Schairer, vice president in charge of patents.

  There was also a short live presentation, consisting of a dance by twenty girls, and a short film of some army maneuvers. The program finished with a short "nod" on camera from the RCA engineering team, who were present in the studio. RCA revealed that three receivers were in operation in the New York City area and that within a short time one hundred more sets would be scattered throughout the city.

Other television demonstrations were given in the United States in June 1936. One was by Don Lee Television under Harry Lubcke in Los Angeles, and another by the Philco Corporation in Philadelphia. Philco's engineering staff at the time included A. F. Murray, H. Branson, P. J. Konkle, F. J. Bingley, W. N. Parker, N. S. Bean, and S. F. Essig.[2]

Even though RCA and Sarnoff indicated that a television service in New York City would not be started in the near future, this seemed like an opportune time for Westinghouse Electric to reclaim control of Zworykin's television patents. On May 22, 1936, the Westinghouse patent department revoked all of the RCA attorneys' (Thomas Goldsborough and H. D. Gregory) powers of attorney, and on June 2, 1936, returned them to Westinghouse Electric. A lawsuit followed.

In July 1936, Westinghouse Electric filed an equity action against RCA asking that the U.S. patent commissioner issue various patents of inventions to Vladimir K. Zworykin. This was in accordance with agreements signed on October 8, 1923, and March 27, 1925, in which Zworykin gave full control of his patent applications to Westinghouse Electric. RCA had acquired these patent applications from Westinghouse Electric on October 8, 1932, as part of the antitrust settlement, a month before a consent decree had been signed between the radio group and the Department of Justice. This transaction was one of the few favorable items that Westinghouse Electric received from this agreement. Both General Electric and Westinghouse felt that somehow they had not received their fair share in their agreement with RCA. This smoldering resentment was to last for many years and in General Electric's case later had dire consequences for RCA.

In the lawsuit now underway, Westinghouse Electric was appealing the decision of February 26, 1936, which gave priority to Harry J. Round of the Marconi Wireless Telegraph Company. Since RCA was the American agent for Marconi, it was now a dispute between Westinghouse and RCA. Therefore, Zworykin found himself in the middle of this altercation. Even though he was now an employee of RCA, he had to testify for Westinghouse Electric, which wanted full control over the successful and valuable iconoscope camera tube. The court battle lasted almost two years.[3]

In August of 1936, the eleventh Olympic Games from Berlin were televised by Deutsche Reichspost, the German post office. However, this was far from successful. The television pictures came from three sources. There were three Telefunken iconoscope cameras, a single Fernseh image dissector, and two intermediate film trucks, but the overall technical quality was quite poor.

Another problem was the lack of opportunity to view the results. The often-promised manufacture and sales of home receivers had never occurred. There were just a few receivers scattered around Berlin, and the few experimental sets that existed were mainly in the Olympic Village itself. Most public viewing was done in some twenty-five specially built "television parlors."

The German television standard of 180 lines at twenty-four frames per second sequential scan had low definition and very pronounced flicker. According to one report, "on an ordinary set the pictures were partly unrecognizable. They resembled a very faint, highly under-exposed photographic film. Many turned away in disappointment."[4]

However, the major event of the year was the debut of the London television station on November 2, 1936. On opening day, both Baird Television and Marconi-EMI demonstrated their systems. Each half-hour program featured speeches by R. C. Norman, chairman of the BBC; C. G. Tryon, postmaster general; Lord Selsdon, chairman of the television advisory committee; Sir Harry Greer, representing Baird Television; and Alfred Clark, of Marconi-EMI. Also included in this first program were a song by Adele Dixon, a comedy skit by "Buck and Bubbles," and a musical number by the BBC television orchestra led by Boris Pecker, as well as a British Movietone newsreel.

According to *The Times,* "the Baird transmission suffered from a tendency to flicker and a curious and uneasy shifting of parts within the image." It concluded that "at present [for Baird Television] entertainment prepared before hand [motion picture film] was more suitable than 'direct vision.'" On the other hand, EMI's transmission was judged to come through with "less flickering and with more sharpness of detail than Baird's." This was an augury of things to come.

The Marconi-EMI 405-line, interlaced twenty-five frame picture was so superior to the Baird system that it was only a matter of time before Baird Television dropped out of the race. This defeat was inevitable, but because Baird Television had a powerful press behind it and Marconi-EMI, with its ties to RCA, had been so secretive about its work, an immediate government decision on EMI's behalf would have provoked a row of immense proportions. The trials continued until the end of 1936, but on February 4, 1937, the BBC announced that they had chosen the Marconi-EMI system over that of Baird Television. Service began on February 8. Baird Television then turned its efforts toward large-screen theater television.

Not only was Marconi-EMI technically superior but the BBC's programming under Cecil Madden was superb. It included game shows, musical numbers, drama, and a variety of outside broadcasts that covered everything from coronations to cricket matches, boxing, and exhibitions. A steady stream of visitors from the United States and elsewhere were amazed at the uniformly high quality of the pictures, the regularly scheduled programs, and the coverage of "outside broadcasts" (remotes). A fitting tribute to Alfred Clark, Louis Sterling, and Isaac Shoenberg and his superb team of engineers, Marconi-EMI's success was the envy of the television world. The only real problem was the high cost of home receivers, around 95 to 150 guineas ($450 to $700). Apparently fewer than three hundred receivers found their way into homes in London at the time.[5]

In January 1937, the Radio Manufacturers Association (RMA) television committee agreed to a new set of television standards for the United States. It was planned to raise the number of lines from 343 to 441, with a channel width of 6mh, in order to prove that the United States was still ahead in the television race.

The Philco Corporation also indicated that it was using a new radio transmission system using single side-band modulation. This made it possible to get more picture detail within the prescribed six megacycle bandwidth. This was a tremendous advance. Under Albert Murray, Philco was making great progress in the field of television and had become RCA's major competitor in the United States, far surpassing Farnsworth Television.[6]

On May 12, 1937, at the twelfth annual convention of the Institute of Radio Engineers in New York City, Zworykin presented a paper on a new RCA projection kinescope, which used a tube developed by Dr. R. R. Law and his associates at the RCA Harrison laboratories. The tube was about eighteen inches long and produced an image one and one-half by two and one-quarter inches on its fluorescent screen. At a demonstration given by RCA, it was able to project pictures eighteen by twenty-four inches with brightness equal to home movies. It was also shown on a three-by-four-foot screen.[7]

Also in May 1937, all of the television equipment, a complete television station including the radio transmitter, ordered from and built by RCA reached Moscow. It was to be installed in their television center there. In June 1937, twelve RCA engineers from Camden and Harrison left to supervise the installation of the television equipment and to instruct the Soviets in its operation. The project manager was Loren Jones, who was accompanied by Thomas Eaton, Henry Koza-nowski, John Read, Waldemar Poch, and several others.[8] Meanwhile a report from Moscow indicated that the Soviets were already oper-ating television equipment made in the USSR patterned after Zwor-ykin's iconoscope. All other transmitters and equipment to be used in the USSR were to be built in Leningrad and Kiev. However, the RCA installation was still not completed in August 1938.[9]

On July 22, 1937, Farnsworth Television and the American Tele-phone & Telegraph Company signed an agreement whereby each had the privilege of using the other's patents. This was quite a feather in Farnsworth's cap. The agreement was nonexclusive; both were free to license other parties.[10]

On October 14, 1937, after his return from another unpublicized trip to Europe, Zworykin hosted a meeting of a select group from the Society of Motion Picture Engineers (SMPE) on the sixty-second floor of the RCA Building in New York City. At this meeting, the audience was treated to a demonstration of television pictures from the Em-pire State Building transmitter, which were claimed to be 30 percent clearer than a year ago and to have almost the same clarity as home movies. This demonstration also featured the new large-screen pro-jector developed by Law, with a three-by-four-foot screen.

This meeting was also the first to officially recognize the special interest that the SMPE (and later the Motion Picture Academy) was taking in the budding television industry. Previously, in August 1937, an SMPE committee had submitted a very pessimistic report about television; they indicated that the pictures were small, the receivers were costly, and program content was mediocre. Zworykin's lecture and demonstration were designed to show how much television had progressed. The picture quality was now good enough that the SMPE could no longer ignore the presence of television, and the commit-tee continued to keep track of its progress.

Newsweek, after relating his early adventures at Westinghouse, stat-ed that, "At present his work with television is finished. Basic diffi-culties have been ironed out. Henceforth, work is in the hands of the engineers." This statement wasn't quite true. While Dr. Zworykin had long ago delegated basic research on television to others, as direc-tor of the electronic research laboratory he was still a very potent force at RCA. This was mainly due to his strong relationship with Sarnoff.

Under Zworykin's guidance, many new television projects were undertaken and successfully completed. His laboratory had a vitality and spirit that contributed to the many advances that came out of it, in spite of his petty insistence that his name be put on all projects and papers and his inclination to interfere with everyone else's projects.

In a sense, Zworykin's demeanor there was a plus factor, since Sarnoff knew the star value of a personality such as his, especially for RCA's public relations. As a result, Zworykin's laboratory never had to worry about funding for its many projects. While not as large, as well financed, or as highly organized as the Bell Telephone labs, it was attracting highly qualified scientists and did not lose the lead in television research until many years later, when RCA was a different company.[11]

In 1937, Harley Iams and Albert Rose had begun several research projects as a means of improving the performance of the iconoscope. Realizing the deficiencies of existing targets, Iams and Rose began to experiment with photoemissive, photoconductive, and photovoltaic targets. They also concentrated on improving Alda Bedford's new tube, the image iconoscope, which combined the image section of Farnsworth's with the storage plate of Zworykin's. This gave them the best results to date.[12]

At the thirteenth annual convention of the Institute of Radio Engineers on June 17, 1938, Zworykin reported on the gains made in getting greater contrast from the kinescope. This was the work of his colleagues Law, Iams, and W. H. Hickok. He also described the progress on his "electrostatic multiplier" tube, a great improvement over the "magnetic multiplier" tube of a year ago.[13]

At this same meeting Iams displayed another television advance with the public introduction of the RCA (Bedford) image iconoscope. This tube, called the super emitron by Marconi-EMI, had been in use in England since 1937, yet in patent interference (no. 74,655), the RCA patent department was able to prove an earlier conception date than that of Marconi-EMI's researchers, Lubszynski and Rodda. So the image iconoscope belonged to RCA. In this regard, even though RCA and EMI were working together, RCA treated EMI as it had both Westinghouse and GE.

However, RCA was not really interested in this tube. While many image iconoscopes had been built for experimental purposes, they had been installed in only one RCA camera, according to Robert Morris, the radio engineer in charge of the Empire State Building transmitter during the 1930s and 1940s.

There were two reasons for this lack of interest. While the image iconoscope was a major advance over the regular iconoscope, it still had most of the problems arising from the use of the high-velocity electron scanning beam. Furthermore, with the lawsuit filed by Westinghouse Electric, RCA was in danger of losing the basic rights to Zworykin's iconoscope. A replacement camera tube that belonged solely to RCA was a necessity.[14]

However, the iconoscope still remained the principal camera tube for RCA/NBC, Marconi-EMI, and Telefunken A. G. The London television service was using the super emitron for outside broadcasts with excellent results. The finest example was the telecasting of British prime minister Neville Chamberlain's return from Munich on October 1, 1938, which included his "Peace in Our Time" speech.

EMI's outside broadcast vehicles were stationed at Heston Aerodrome, and with three cameras the event was televised live as it happened. It was relayed by microwave to the Alexandra Palace, where it was rebroadcast. This was the first time in history that a live news event was telecast as it occurred, which meant another first for the London television service.[15]

On October 20, 1938, Sarnoff finally announced the start of a television service in the United States. The occasion was the annual meeting of the Radio Manufacturers Association (RMA) in New York City. Sarnoff stated that RCA would inaugurate a public television service with the opening of the New York World's Fair in April 1939. He said also that RCA's field tests and manufacturing experience had "convinced us that television in the home is now technically feasible." He continued, "The problems confronting this difficult and complicated art can be solved only from operating experience, actually serving the public in their homes" and noted that there was "no fear that the advent of home television service would paralyze broadcast receiver sales or upset the broadcast industry." He thought of television as a supplementary service, and assumed that homes would have two receiving sets in order to get both services.[16]

There was only minor mention of the ongoing business depression, whose effects the United States and the rest of the world were still feeling. The American economy was recovering very slowly, chiefly because of the rise in defense spending. Only the ominous events in Europe, as it prepared to go to war, were giving the United States any financial stimulation at all.

There is little question why Sarnoff chose this particular time to start a television service in the United States. Technically, there was no problem. The RCA system was certainly equal to the Marconi-EMI system, now operating so successfully as the London television station. Sarnoff also knew that television had not hurt the sale of radio receivers in London, since television receivers were just too expensive to go into the average home. So he did not have to worry about the sales of radios in the United States. Besides, he knew that it would take several years before television could compete financially with radio because of the enormous start-up costs.

Finally, the political situation in Europe was getting precarious. If there was to be a war, it would certainly involve the United States, and no one could predict how many more years would be lost before it was over. In a *New York Times* article published on January 3, 1939, Sarnoff was quoted as saying that "If action were deferred to await near perfection, the medium would be postponed almost indefinitely." A decision had been made; RCA would begin its television service with the opening of the New York World's Fair, appropriately called "The World of Tomorrow."[17]

On December 20, 1938, after the application had spent fifteen years in the Patent Office, Zworykin was finally granted U.S. Pat. no. 2,141,059 for his December 29, 1923, television application. This decision had had to wait for settlement of the Westinghouse suit in the District of Columbia court of appeals. As indicated earlier, the Patent Office tribunals had decided against Zworykin several times. This patent had been involved in no fewer than eleven interferences, had been twice before the United States court of appeals, and had been in litigation for some twelve years.

Since the District of Columbia court of appeals was less technically minded than the Patent Office, its decision was to be made on legal rather than technical grounds. The case was finally resolved on one issue: the construction of the photocell. If in the filing the 1923 application Zworykin disclosed a photosensitive plate of a specific construction, then he was entitled to be awarded priority. The only questions in controversy were (1) whether the potassium of the photoelectric element was in the form of elemental areas and (2) whether it was new matter in the original 1923 application. If these were so, Zworykin was the first inventor.

Westinghouse Electric's lawyers attempted to prove that there was only one way to deposit the potassium on the plate and that when this was done properly it would form elemental areas. Westinghouse brought in several highly regarded scientists, including Prof. Jacob Kunz, Dr. Irving Langmuir, Dr. Saul Dushman, and Prof. Joseph T. Tykociner, as witnesses who claimed that this was what Zworykin had in mind. They all stated that there was only one way to deposit the potassium on the plate, which was to halt the process while the potassium was still in drops, before it coalesced to form a layer. Once this fact was agreed upon, then of course it was not new matter and properly belonged in the patent. This argument convinced the court of appeals and on November 2, 1938, it ruled that the Patent Office should grant the Zworykin patent. It was issued to the Westinghouse Electric and Manufacturing Company.

This was a major victory for Westinghouse's lawyers, who had worked hard to protect the company's interests. It was also a major victory for Zworykin, as he could finally claim that the iconoscope had indeed been invented in 1923.[18] However, it was a setback for RCA, which had actually done all of the basic research and development on the tube. In reality, though, because of the licensing agreements between RCA and Westinghouse, this suit made very little difference. Harley Iams continued to build iconoscopes at the Harrison laboratory for both General Electric and Westinghouse.

On December 29, 1938, Zworykin gave a lecture on his work at RCA on an "ultra" (dark-field transmission) electron microscope to the American Association for the Advancement of Science in Richmond, Virginia. He described two models, one thermionic (with a hot cathode), originally attributed to R. P. Johnson and W. Shockley, and the other the H. O. Müller cold-cathode type. Zworykin demonstrated the RCA version, based on the Müller design. The enlarged images of gas molecules, filterable viruses, and bacteria could be seen on a fluorescent screen. At the same time, Dr. Ladislaus Marton from the University of Brussels was also working in Zworykin's laboratory on a related project, an advanced magnetic electron microscope.[19]

Late in 1938, Dr. V. K. Zworykin was awarded an honorary degree of Doctor of Science from the Brooklyn Polytechnic Institute.[20]

On January 19, 1939, the television committee of the Radio Manufacturers Association announced a new set of technical standards for American television, RMA standard T115. It raised the number of lines to 441 and proscribed the use of single side-band transmission and reception. This was agreed upon by the members of the radio industry. The RMA also decided that a television service should begin with the opening of the World's Fair in New York City, largely at the insistence of David Sarnoff.

Six American manufacturers, American Television Corporation, Andrea Radio Corporation, Allan B. Du Mont Laboratories, the General Electric Company, Philco Radio and Television Corporation, and the RCA Manufacturing Company, announced plans to have television receiving sets on sale by May 1, 1939. Other manufacturers who promised to have television receivers ready were Stewart-Warner, Stromberg-Carlson, Pilot, Garod, Meissner (kits), and Westinghouse, but not Zenith. According to them, "Zenith was ready but television was not." Also conspicuously missing was Farnsworth Television, which was not prepared to manufacture television receivers as it had neither the necessary resources nor financing.[21]

RCA made plans to have an exhibit building at the New York World's Fair that was to feature a "Hall of Television." It was to be equipped with thirteen TRK-12 home receivers and a flask receiver with a special 18-by-24-inch front viewing screen. There was also a special combination receiver for the "living room of tomorrow" that included a radio receiver and a phonograph. In addition, there was a special laboratory model projection type receiver for images six by ten feet in size. At the entrance of the building was a complete television receiver enclosed in a glass cabinet to show the inner workings of the set.

The exhibit would also feature a regular RCA iconoscope television camera, a complete 16mm film projection chain, and the new NBC "Tele-Mobile" remote trucks ready to pick up live pictures from any part of the fairgrounds.

Meanwhile, RCA was overhauling its radio transmitter at the Empire State Building to conform to the new standards by April 30, 1939. CBS promised to have its new television transmitter (built by RCA) operating from the Chrysler Building at the same time. However, this did not happen.[22]

On April 19, the day before RCA dedicated its exhibit, Zworykin told a *New York Times* reporter that "he never dreamed five years ago that he would see television start as an industry almost in his own back yard, Forest Hills, New York."

On April 20, ten days prior to the opening of the New York World's Fair, RCA dedicated its television exhibit building. This ceremony was viewed by several hundred people watching it on receivers inside the building at the fairgrounds and on receivers in the RCA building in Manhattan by means of the NBC tele-mobile unit, which relayed its signal to the Empire State Building.

The program opened from Radio City with an introduction by announcer Graham McNamee. Then the scene switched to Flushing Meadows, where it showed the Trylon and Perisphere (trademarks of the fair). The camera was then moved to the garden behind the RCA Building. The star of the telecast was Sarnoff, who made the dedication speech. He was introduced by Lenox R. Lohr, president of NBC. David Sarnoff stated:

> Today we are on the eve of launching a new industry, based on imagination, on scientific research and accomplishment. . . . Ten days from now this will be an accomplished fact. The long years of patient exper-

imenting and ingenious invention which the scientists of the RCA Research Laboratories have put into television development have been crowned with success. I salute their accomplishments and those of other scientists, both here and abroad, whose efforts have contributed to the progress of this new art. On April 30th, the National Broadcasting Company will begin the first regular public television-program service in the history of our country; and television receiving sets will be in the hands of merchants in the New York area for public purchase. And now we add radio sight to sound. It is with a feeling of humbleness that I come to this moment of announcing the birth in this country of a new art so important in its implications that it is bound to affect all society.

Others who appeared before the single camera that afternoon were Major General James G. Harbord, chairman of the board of RCA; F. J. Nally, first vice president of RCA, and Dr. Vladimir K. Zworykin, director of electronic research and inventor of the iconoscope or radio "eye" around which the television system was built. In Zworykin's few words before the camera, he hinted that "New wonders are still developing in the laboratory."

Following the ceremonies, the audience witnessed a boxing match staged at the studio. This has often been called the first sports event ever televised in America. For Sarnoff and Zworykin it was the climax of eleven years of unparalleled cooperation. This was the moment they had waited for so long.

RCA had installed four types of television receivers at Radio City, where a group of more than one hundred newspaper and magazine writers were gathered to witness this momentous event. The *New York Times* claimed that the quality was excellent with every detail "distinct."[23]

Ten days later, on Sunday, April 30, television made its quasi-formal debut in the United States. At 12:30 P.M. the camera focused on the Trylon and Perisphere in the distance and the public was treated to pictures of a panoramic view of the fair buildings. There was an opening parade and a shot of Mayor Fiorella La Guardia as he entered the president's box. Images were then broadcast of President Franklin Delano Roosevelt, speaking at the opening ceremonies. RCA had scattered dozens of developmental receivers in various department stores all over New York City. Witnesses claimed that the pictures were clear and steady.

However, observers from the BBC were amazed at the nerve of the Americans in having only one camera on the scene, according to the article published in the *New York Times* on May 7. They said that they would have used at least three or four cameras for fading to scenes from different angles to gain variety. Then, too, they wondered, "If the single camera burned out at a crucial moment, what would they do?" By this time the British had almost two and a half years of experience running a daily television service, so they were highly amused at the efforts of RCA/NBC to present such an event in such a crude fashion. At the moment, of course, this was the best that RCA/NBC could offer. Multicamera remotes in the United States would come later in the year.[24]

RCA/NBC wasn't the only television manufacturer at the World's Fair with an exhibit. General Electric, Westinghouse Electric, and the Crosley Radio Corporation (in conjunction with the Du Mont laboratories) also had exhibits dedicated to television. Both General Electric and Westinghouse had studios with live cameras for interviews and a half-dozen receivers. Crosley had several sets on display that mainly picked up the NBC telecasts. So did the Ford Motor Car Company exhibit, which had several receivers located in an executive lounge. While there was enthusiasm for the new medium, the various department stores involved—Macy's, Bloomingdale's, and Wanamaker's—all reported that while there were many inquiries about the receiving sets, they had made few or no sales.[25]

On June 7, 1939, RCA's Iams and Rose revealed the details of a new television camera tube called the "Orthicon" (*orth* for its linear electrical output and *icon* from iconoscope) to the New York section of the Institute of Radio Engineers. It was the first in a long line of camera tubes using low-velocity electron beam scanning and the forerunner of a new breed of television camera tubes that would revolutionize and vitalize the industry.

This radical new tube was described as being twenty inches long and four inches in diameter. The image plate (target) was two by two and one-half inches wide. Its sensitivity was claimed to be some ten to twenty times greater than the iconoscope. It had no shading problems (flatness of field) and picture resolution was 400 to 700 lines. Iams and Rose claimed that it was in the final stage of development and would be ready for use by the first of the next year.[26]

The orthicon tube was the result of a developmental program by Iams and Rose started in 1937 using low-velocity scanning beams. This proved to be a long and arduous task. Their main objective was to operate the tube with a low-velocity beam at (zero) cathode potential in order to eliminate secondary emission. The electron beam struck the target at low velocity, thus eliminating the shading problems of the iconoscope and improving its efficiency. Their experiments showed that for improved operations, these novel tubes needed a uniform axial magnetic field that would keep the beam well focused and at a consistent angle of incidence as it approached the target.

In addition, a two-sided target was needed in order to make the shape of the tube practical. This was made of a thin mica sheet. The side facing the beam was covered with a mosaic of photosensitive elements, while the side away from the beam was coated with a translucent conducting film of metal; this was known as the signal plate. The optical picture was focused on the photosensitive side of the tube through the translucent signal plate. This created a severe loss of light and electrical signal as it went through the mica sheet.

When Iams and Rose finally produced an operable tube and applied for patents, the RCA patent department found that Philo Farnsworth had many of these features covered in his patents on the image dissector and that he had applied for the first patent on a low-velocity beam scanning tube in April 1933. While he had never built or operated such a device, he had the priority and RCA had to come to terms with him. Iams and Rose continued to improve the tube's performance and prepare it for studio use.[27]

Zworykin himself went on a lecture tour in 1939, beginning with a speech on electron optics in Columbus, Ohio, on February 6. He gave a talk entitled "Electron Optics as Applied to Television" in Connecticut on March 15 and in Pittsburgh on March 21. He concluded his lectures in the United States in the summer of 1939, when he gave an address on electron optics before the Pacific Coast section of the Institute of Radio Engineers in San Francisco on June 28.[28]

In July 1939, Zworykin celebrated his fiftieth birthday at his new summer home in Taunton Lakes, New Jersey. He had more than one hundred friends at his housewarming. There he became good friends with Boris Polevitsky, who had been the mayor of Murmansk, and his wife Katherine. Zworykin's friends, who included David Sarnoff, stated that they had many marvelous parties and good times at his Taunton Lakes home. He spent his weekends there even after the opening of the RCA laboratories in Princeton in 1942.[29]

He then continued his lecture tour in Europe, his last trip abroad before the war. He planned to visit various laboratories and universities in Rome, Zurich, Paris, London, and Dundee, Scotland. He left New York City on July 28, aboard the SS *Saturnia* bound for Naples, with the expectation of spending several days in Egypt.

During the crossing he received a radiogram from the RCA representative in Palestine requesting him to come to Tel Aviv and give a talk on television there. As an inducement, he was offered a tour around the country and a chance to see some of the archeological finds. He was met in Alexandria by this gentleman and flew with him to Tel Aviv. Here he gave his lecture before an enthusiastic group of engineers.

At that time, Palestine was under British rule and it was necessary to get a special permit to travel through the country, so Zworykin went to Jerusalem to get the papers. He desperately wanted to see the Dead Sea as he was very interested in King Solomon's copper mines. On his way back to Damascus, he could not resist the temptation to have a swim, or rather a float, in the Dead Sea.

When Zworykin arrived in Damascus, he was handed a telegram from the American consul telling him to clear out of the country as soon as possible. It was obvious that war was imminent. He packed his bags and headed for the airport at Beirut. Here he found that all planes were full and only by bribing certain officials did he finally get a seat on a plane for Rome.

In Rome, the news was more alarming. Italy had declared a full mobilization, and all transportation out of the country was canceled. From past experience, Zworykin knew that the concierges in the big hotels were most likely to be helpful to travelers so he approached the concierge at his hotel. He was told that he had a difficult problem, but the concierge promised to help if he could. He was told to go to his room and have a good night's sleep.

At midnight he received a call stating that for a large monetary premium he could have a seat on the morning plane to Paris. Zworykin agreed and went to the airport the next morning. Here he was met by a certain individual who bypassed all of the main passageways and took him behind one of the hangars directly to a heavily loaded plane, which he boarded without difficulty. In a few hours he landed at Le Bourget Airport in Paris. Here, although the excitement was not as high as in Rome, normal passage was not available. However, by again bribing the right people, Zworykin managed to get a plane to London that same evening.

Zworykin found that London was much calmer than Europe, that the Italian mobilization had been canceled, and that there would be no armed conflict for the time being. He stayed in London for a couple of days before departing for Dundee. Here he found that the conference was going ahead as planned, and on August 31, 1939, he spoke to the British Association for the Advancement of Science on a new scanning microscope, a device capable of 100,000 times magnification. It appeared to be an advanced model of the device he had introduced in 1934.[30]

The next day, September 1, 1939, Nazi Germany invaded Poland,

and the rest of the conference was called off. The organizing committee announced that they had secured accommodations on the SS *Athenia* for all participants who wished to go home. It was to sail for the United States from Liverpool the next day.

This created a problem for Zworykin, who had arranged to have his luggage shipped by way of American Express to London. From past experience, he knew that to travel on a British ship without proper dinner clothes was "impossible." So he decided to return to London and try his luck there.

The next day, to his horror, he read that the *Athenia* had been torpedoed and sunk with considerable injury and loss of life. Clearly Zworykin's traditional good luck had stood by him, though he was marooned in London for several days. It had now become a quite different city from the one he had just arrived some four days earlier.

While trying to arrange passage home, Zworykin wired Sarnoff in New York and sought permission to speak with certain military officials in Great Britain about secret work being done in the RCA laboratories for military purposes. This included work on flying bombs with television sights and measuring of distances by means of ultrashort radio pulses (radar).

Permission was granted and Zworykin met with Sir Charles Galton Darwin, director of the National Physical Laboratories at Teddington, and two military aides from the Royal Navy and the British war office. Zworykin was surprised to discover that they seemed to be quite disinterested in these proposals. He was told that Great Britain was already at war and therefore did not have time to embark on such a long-term research project, and that the best thing Zworykin could do was to go back to his laboratory. But they promised to help him get home. As a result, the next day he acquired a berth on the SS *Aquitania* to New York. It is interesting to note that later in the war, when Darwin visited the RCA laboratories, Zworykin was told that both projects they had discussed earlier were top secret and could not be discussed with him.

When Zworykin arrived back in New York he found that the war in Europe had barely affected life in the United States. His laboratory work continued as before except that certain military projects were getting more and more attention. Sarnoff was quite concerned about the war in Europe and immediately made plans to divert RCA's great technical resources for the coming war effort. A part of this plan called for centralizing all the RCA research laboratories.[31]

The vicious attack on Poland by Nazi Germany had many tragic consequences. While there were no actual hostilities between Germany and Great Britain at the onset (the so-called phony war), one of the first casualties was the BBC's London television station at the Alexandra Palace. Without advance warning, it had been ordered to shut down on at noon on September 1, 1939, while showing a Mickey Mouse cartoon. Mickey said, "Ah tank ah go home now" (à la Greta Garbo) and the station went off the air. There were approximately 19,000 to 23,000 television sets in the London area, all of which were now rendered useless.

Three reasons were given for closing down the Alexandra Palace on this day: (1) it was a luxury item, enjoyed only by a minority; (2) its highly trained staff was needed for work on radar; and (3) its radiation could be detected by the enemy. Later, however, there were rumors that the high cost of operating the London Television station was the primary reason for this abrupt closing.[32]

Whatever the reason, its closing created a gap that took many years to be filled again. The BBC had shown how a modern television service was to function and had blazed a path both artistically and technically. All the cameras and technical gear were carefully packed away for the day when the war would be over.

This left the United States alone in the field of broadcast television. It is true that Nazi Germany continued to televise certain programs for a short while, but they were of minor consequence. There were only a few experimental television receivers scattered around Berlin and broadcasting was an exercise in futility. The German propaganda machine had utterly failed to realize the potential of the new medium and did nothing to exploit it.[33]

While RCA/NBC tried to sustain interest in television with a series of special events, they were not done in the style and spirit of the BBC. The remotes (outside broadcasts) were sporadic, unimaginative, mostly covered by a single camera, and did not arouse much interest. Only the carefully staged multicamera studio programs were of better quality. Sales of television sets were very disappointing. Three months after the World's Fair inaugural, only eight hundred sets had been sold and five thousand were stacked up in dealer and distributor warehouses. CBS was stalling for time; it had no desire to start a television service and, for reasons that were disclosed later on, made every excuse not to go on the air.[34]

With the start of commercial television in the United States, RCA finally had to come to terms with Philo Farnsworth. In anticipation of this, Farnsworth Television and Radio Corporation had hired E. A. Nichols, former head of RCA's licensing division to negotiate with Otto Schairer, head of RCA's patent department. After months of hard bargaining, RCA signed an agreement in September 1939 for continued use of Farnsworth's patents. For the first time in RCA's history, they were paying royalties.

By the terms of the agreement, RCA acquired a nonexclusive license under the patents of the Farnsworth Corporation for television transmitters and receivers and other radio and sound recording and reproducing apparatus. Farnsworth acquired a standard nonexclusive license for broadcast and television receivers and electrical phonographs under RCA patents. The agreement was announced on October 2, 1939. Later, in April 1940, at an FCC hearing, Sarnoff finally publicly acknowledged Farnsworth's contributions to television by describing him as "an American inventor who I think has contributed, outside of RCA itself, more to television than anyone else in the United States . . . , he has made significant inventions."[35]

By this time, Sarnoff's efforts to get a television service started in the United States were faltering. Not only was there a lack of interest from the public but the other radio/television manufacturers, especially Philco, Du Mont, and Zenith, were voicing their objections to the use of the technical standards agreed on by the RMA television committee, mainly because they resented the dominance of RCA in this area. They were determined not to let RCA gain the control in television that it had in radio.

In addition, Major Edwin Howard Armstrong was claiming that RCA was trying to block the adoption of his new FM radio system. This offended the FCC chairman, James L. Fly, who ordered an inquiry into the matter. As a result, the FCC ordered forty new FM channels allocated, taking away Television Channel One and awarding it to Major Armstrong for FM use. This was a bitter defeat for Sarnoff and RCA.[36]

In addition, the introduction of Rose and Iams's new orthicon camera tube, in September 1939, proved less satisfying than promised. Although it was three to ten times more sensitive than Zworykin's iconoscope and free of spurious signals (shading problems), the orthicon introduced a few problems of its own. The main one was a tendency to "charge up" when exposed to sudden bright lights such as photographers' flashbulbs. This upset the tube, cutting off its picture, and it took a while until the tube stabilized itself. This also occurred with certain television scenes in which parts of the scene were in bright sunlight and the rest in deep shadows. Due to its linear output, pictures failed to have a full range of light values. It also presented problems with its inefficient two-sided target. Not only did it absorb a great deal of light from the scene but, as Zworykin and Ogloblinsky had found in 1930–31, construction of a practical two-sided target presented an enormous challenge.[37]

However this last problem had finally been solved. In the Harrison laboratory, Rose had arrived at the solution, a near-perfect two-sided target consisting of an extremely thin piece of semiconducting glass. The light from the scene would strike the front of the glass, which was coated with a suitable photoelectric layer. With the proper parameters, it allowed the electrons to go through the glass, where they would be discharged by the electron beam from the rear. This created the picture signal. While still a laboratory experiment, it paved the way for future improvements in television camera tubes.

On September 20, 1940, Albert Rose filed for a patent disclosing the new two-sided glass target. However, the War Department decided that Rose's patent was a military secret and withheld it until the end of the war.[38]

In February 1940, Zworykin published his now-classic book *Television: The Electronics of Image Transmission,* written in collaboration with George Morton, a research engineer at RCA. It was a monumental book that described every technical aspect of television to date.[39]

In May 1940, the RCA radio research laboratories at Camden disclosed an electron microscope for general use. It had been designed and constructed under Zworykin's supervision by Ladislaus Marton with a group of co-workers including Arthur W. Vance, M. Charles Banca, and J. F. Bender. At this time, Dr. James Hillier joined Zworykin's electronic research laboratory in Camden and was soon working with them to develop the first commercial electron microscope for industry.

The commercial electron microscope was first described in November 1940, at a lecture given at the American Institute of Electrical Engineers winter convention in Philadelphia, and again on January 27, 1941. Zworykin's lecture was entitled "An Electron Microscope for Practical Laboratory Service." Zworykin was determined to produce an electron microscope that could be used by the average technician with a minimum of training or technical knowledge.[40]

In June 1940, NBC/RCA televised the Republican national convention from Philadelphia, using both iconoscope and orthicon cameras. The pictures were transmitted by coaxial cable to New York City. The signals from the Empire State Building on station W2XB were also relayed to a special receiver in the Helderberg Mountains 129 miles away. Here they were sent a mile and a half away to the General Electric transmitter W2XBS and rebroadcast to Schenectady, Troy, and Albany, New York. The Philco Corporation also televised parts of the convention but transmitted it only in the Philadelphia area.[41]

The FCC announced that a sponsored (paid) television service was to be permitted after September 1, 1940. This prompted RCA to announce a great sale of receivers at reduced prices in March 1940. The FCC reacted by asking for new hearings to find out if RCA was unduly retarding higher technical standards. These hearings resulted in an FCC ruling that it would allow full commercialization of television only when there was a consensus within the industry about standards. The National Television Systems Committee (NTSC) was set up to solve the problem. The committee, chaired by Dr. W. R. G. Baker of General Electric, consisted of representatives of all the major television manufacturers and associated companies. It held its first meeting on July 31, 1940.[42]

In September 1940, CBS revealed that it had a color television system operating in its laboratories. Under the direction of Dr. Peter Goldmark, CBS's chief television engineer, they had produced a remarkable television system based on a rotating color wheel at both the transmitter and receiver. At first, it could only show color motion picture film and slides. The development of this system was one of the reasons why CBS had shown such reluctance to start a television service earlier. Introducing such a system at this time had two significant effects. First, it insured that the adoption of any new technical standards would have to take color television into consideration, and second, it protected CBS Radio's interest in AM radio broadcasting, since it meant a possible delay in the adoption of television standards.

RCA now had CBS as well as Philco to contend with in the development of television. CBS had long been NBC's chief competitor in network radio broadcasting and was now to become a formidable factor in the laboratories. The only other rivals were the much smaller Allan B. Du Mont laboratories and the television laboratory of Philo Farnsworth.[43]

On January 27, 1941, the NTSC submitted its report on proposed new television standards for the United States and announced on May 2 that its new standards had been adopted officially by the FCC. Commercial television could begin on or after July 1. According to the new standards, the number of lines was raised from 441 to 525. (This 525-line, thirty-frame per second interlaced standard is still in use in the United States today.) They also specified the use of FM for the sound channel but kept open the question of color television, since it required further testing and might be able to co-exist with monochrome television.[44]

On March 5, 1941, David Sarnoff announced that RCA intended to build the world's largest radio laboratory at Princeton, New Jersey. Otto S. Schairer was named vice president of the RCA laboratories. Other officers were to be Ralph R. Beal, research director, who would have general direction of all research and development; Dr. C. B. Jolliffe, in charge of the RCA frequency bureau, as chief engineer; Elmer W. Engstrom, director; and both Zworykin and Browder J. Thompson as associate directors.[45]

Interestingly enough, as the July starting date for a television service approached, Sarnoff showed very little enthusiasm for the coming commercial service, telling the *New York Times* that "We have lost none of our enthusiasm for television, nor our faith in its eventual success, but no one company can do it all." He also expressed doubt that advertisers would be interested because of the low number of sets, optimistically stated at two to three thousand, all of which had to be converted to the new FCC standard, operating in the New York area. Finally, he added that national defense was causing a strain on both his facilities and technical experts. Due to government orders, neither raw materials nor production capacity could be diverted to television receivers.

Because of a presidential order on May 27, 1941, RCA closed down its television receiver assembly line and had to be content with the ample supply in their warehouses. According to Robert Sobel, "only 10,000 [this figure is in dispute] of the promised 25,000 receivers were manufactured and sold, and these stood idle most of the time." However, these receivers were being modified to conform to the new technical standards so as to be available to potential customers. Bilby later claimed that only a few thousand sets were in the public's hands.

David Sarnoff knew this was not the time to introduce such a radical service to the American public. At the moment they could have cared less. However, there was nothing to do but continue in the hope that somehow the United States would not be involved in the war, either in Europe or in the Far East.[46]

The official opening of the American television service on July 1, 1941, was a rather interesting affair. As the only station with paid programming, NBC/RCA fulfilled its promise on opening day. This included a test pattern clock with the Bulova Watch logo, a news broadcast sponsored by Sun Oil, a television version of a radio show for Lever Bros., and a quiz show for Proctor and Gamble. NBC also televised a Dodgers-Phillies baseball game and a USO program in the evening—all in all, an impressive start. Sarnoff had learned well from the London television service.

CBS, on the other hand, which began its television programming reluctantly, was beset by technical troubles. It televised a dance lesson, a newscast, and an exhibit of art in the evening. Du Mont, also troubled technically, televised some live and film programs in the evening. For the United States, it was a rather inauspicious beginning for the most powerful communications medium in history.[47]

On October 8, 1941, Zworykin was awarded the Rumford medals of the American Academy of Arts and Sciences, for "outstanding contributions to the subject of light." He was honored specifically for his work on the development of the electron microscope. In the *Proceedings of the Institute of Radio Engineers,* he was quoted as saying that "Electrons play the role of light, their much shorter wavelengths making it possible to see details and objects twenty to fifty times smaller than those which can be resolved with a light microscope."[48]

There was only minor television programming in the United States for the rest of 1941. Out of twenty-two licensed stations across the country, only seven were actually on the air. They were NBC/RCA (W2XBS, New York City), CBS (W2XAX, New York City), Du Mont (W2XWV, New York City), Don Lee (W6XAO, Los Angeles), Zenith (W9XZV, Chicago), Philco (W3XE, Philadelphia), and General Electric (W2XB, Schenectady).[49]

The bombing of Pearl Harbor by the Japanese navy on December 7, 1941, quickly ended most television activities in the United States. A limited amount of television, connected with civilian defense, continued briefly in New York and Chicago, but this was soon stopped. The United States went to war with both Japan and Nazi Germany and for the next four years almost all television activity was confined to the laboratory, where it was put to military use.

CHAPTER 11

World War II

With the treacherous attack on Pearl Harbor, American industry quickly converted over to total war production. Sarnoff telegraphed President Franklin Roosevelt that "All our facilities and personnel are ready and at your instant service. We await your commands." In fact, RCA's scientists and engineers would play a major role in the development of radar and sonar and of the electronic navigation systems known as loran and shoran. They also worked on walkie-talkies and other forms of communications. RCA poured out over two thousand types of radio tubes, including 20,000,000 miniature tubes, and did the basic work on the highly secret proximity fuse.

When it came to the use of television for war purposes, Sarnoff stated that "The potentialities of television-directed weapons seem to be of the greatest importance." He pointed out that RCA was the most qualified producer and "the only presently qualified supplier and the only one able to solve the remaining problems." Most of the wartime advances in television weaponry were in fact spearheaded by RCA in collaboration with the Office of Scientific Research and Development of the armed forces.[1]

The consolidation of RCA's research laboratories into one facility (now called the David Sarnoff Research Center) in Princeton had started in August 1941 and was finished in September 1942. Zworykin moved his staff and research facilities there. He had rented a home at 91 Battle Road in Princeton in anticipation of such a move, in addition to the summer home he maintained in Taunton Lakes.[2]

During the war, Zworykin served as a member of the Ordnance Advisory Committee on Guided Missiles, three important subcommittees of the National Defense Research Committee, and later on the scientific advisory board to the commanding general of the United States Army Air Forces. His main research involved aircraft fire control and infrared image tubes for the sniperscope (a carbine-mounted aiming device) and the snooperscope (a portable telescope using infrared light for illumination). His work also included effective improvements in radar systems and storage tubes and several projects involving the use of television for reconnaissance and in guided missiles.[3]

On April 25, 1934, Zworykin had submitted a secret proposal to Sarnoff for a "flying torpedo with an electric eye." Basically, it outlined a program for the control of guided missiles by means of television. Sarnoff was so impressed with the scheme that he immediately set up a meeting with the navy and war departments in Washington at which the inventor explained his radical concept to skeptical members of the armed forces. Zworykin met with admirals Ernest J. King and Harold R. Stark and generals Westover and Tschappat. Out of this meeting came a pledge of technical and financial support for Zworykin's ideas.[4]

In 1935, under the able direction of Ray Kell, who was later joined by Waldemar Poch and Henry Kozanowski, RCA began work on a lightweight airborne reconnaissance television system that was successfully demonstrated in 1937. Equipment was built using a model 1850 iconoscope camera that was installed in a Ford trimotor airplane. Two years later, work was commenced on still smaller and lighter equipment. By 1940, Zworykin's flying bomb design, incorporating radio control, won Washington's authorization to proceed with advanced aeronautical design. The new equipment was flight-tested in 1940–41.[5]

On March 6, 1940, some of this experimental reconnaissance equipment was demonstrated to the public by RCA and NBC. A twin-engine Boeing Mainliner owned by American Airlines took off from La Guardia Field and circled the World's Fair Grounds, the Empire State Building, and the Statue of Liberty at an altitude of two thousand feet for about forty-five minutes. The pictures from the plane were so clear that experimenters felt there "were possibilities of adapting such a tele-airplane in wartime, especially for reconnaissance flights, bombing operations and map making." The March 17 report in the *New York Times* mentioned that army and navy officials at Radio City and other outposts watched the demonstration with keen interest. This was the only indication that this exercise was part of the guided missile program begun by Zworykin in 1935.[6]

The first model of what was known as Block I television was flight-tested in April 1941 from an army B18 bomber at Wright Field, Ohio. In November 1941, with a grant from the National Defense Research Council, three RCA-built television guided missiles were tested successfully at Muroc Dry Lake in California. With the start of the war, work was intensified in the design of television equipment for use in both aircraft and guided missles.[7]

The RCA laboratories developed three basic television missile systems during the war. The first was the "Block" system, a lightweight (under one hundred pounds) missile system used by the army in the GB-4 radio-controlled glide bomb, and later in war-weary old B-17s, and by the navy in the TDR-1 drones and the GLOMB radio-controlled glide bomb. These missiles were designed to operate unattended and were considered expendable.

The second was the "Ring" system, started in November 1942, which was designed to provide high-resolution pictures for reconnaissance and other long-range viewing. It was devised for attended operation with two or more cameras and was not considered to be expendable. The first tests were made using two standard 1840 orthicon cameras and their associated equipment in a PBY-4 Catalina Flying Boat at a naval air station in Florida.

The third system, "MIMO," named for a new miniature camera tube developed in the RCA laboratories, was similar to the "Block" system except for the smaller camera tube. It was to be mounted in an army missile called the "ROC," a high-angle radio-controlled bomb made by the Douglas Aircraft Company.

By 1944, RCA had delivered more than 4400 cameras and associated transmitting equipment, including 500 "Block" television cameras, to the armed forces for reconnaissance purposes and to guide pilotless bombs.[8]

Zworykin's work on the infrared image tubes that led to the sniperscope and snooperscope had its amusing side as well. More than once, he and his associates were apprehended by the local Princeton police while testing models of his infrared imaging system by driving automobiles at night without headlights. Apparently the police thought they were spies riding around the RCA laboratory trying to get secret information.[9]

While acting as technical advisor to the various camera tube and guided missile projects, Zworykin was also constantly striving to improve the electron microscope. By 1942, his associate, James Hillier, had succeeded in building a scanning electron microscope that could be used by unskilled workers in the average research laboratory.

Early in 1943, Zworykin went on a lecture tour aimed at describing the various uses of the electron microscope in chemistry and metallurgy.[10] Also in 1943, he was elected to the National Academy of Science.[11]

In June 1943 Zworykin was asked by a group of members of the

Russian War Relief, Inc., to accept the chairmanship of their New York chapter. His first reaction was negative, since from his first days in the United States he had stayed out of the political arguments of the Russian émigré colony, and, furthermore, he was too busy with his experimental laboratory to do any outside work.

However, it was pointed out to him that his very neutrality made him the ideal candidate for the position. In addition, he was told that Eleanor Roosevelt, wife of the president of the United States, and Henry Wallace, the vice president, were members of this same group. At the time it seemed a worthy cause as the Soviets were valiantly defending their country against the Nazi onslaught. In spite of their political differences, there was a common bond between the Soviet Union and the other Allies. Zworykin finally agreed on the condition that the chairmanship would be in name only and he would not be asked to give any of his valuable time.

What Zworykin should have realized was that the FBI under J. Edgar Hoover were suspicious of any organization with ties to the Soviet Union, however innocent. As a result, the FBI automatically opened a file on each and every member of such organizations, including Mrs. Roosevelt and Vice President Wallace. For that matter, the FBI had already opened a file on Zworykin for a variety of reasons. For example, the Russian War Relief, Inc., ranked quite high on the FBI's list of subversive organizations and was the object of several clandestine raids on its headquarters. Zworykin claimed that he had only attended a few of this organization's luncheons. He also became a member of the science committee of the National Council for American-Soviet Friendship in December 1943. This organization had long supported the political aims and goals of the USSR.[12]

Directing one of the many research groups at the new RCA laboratories in Princeton was Zworykin's associate and codirector, Browder J. Thompson. They had become very good friends; in fact, they even shared Zworykin's house. Thompson was engaged in research in the use of radar and went overseas (although not required to do so) to evaluate its performance. When he arrived in Italy, Thompson insisted on going on actual bombing missions to evaluate the system's performance. On one of these flights, on the night of July 4–5, 1944, his plane was shot down by an enemy aircraft and both he and his pilot were killed. This was quite a blow to Zworykin and the entire staff of the RCA laboratory.[13]

Other similar tragedies had occurred earlier in both England and France. On June 7, 1942, Alan Blumlein, C. O. Browne, and A. Blythen, all from EMI, were killed in the crash of a Halifax bomber.

They had been checking out the highly secret H2S airborne radar system, which they had helped develop. Their plane began to have engine trouble and overshot its landing; it crashed and burned, killing all on board. This was such a tragic loss that it was not reported to the English public until the war was over.[14]

Another calamity had occurred in Occupied France. Professor Fernand Holweck of the Édouard Belin and Madame Curie Radium laboratories (whom Zworykin had met in Paris in 1928 and again in 1939) had joined the French underground and was imprisoned by the Nazis in December 1941. He was shot and killed while in custody, while they claimed he had committed suicide.[15]

In 1943, after the capture of Paris, the Nazis began a television service from the Eiffel Tower. The Vichy government arranged for the Nazis to undertake television transmissions in Paris on the German standard of 441 lines, twenty-five frames interlaced, under the auspices of the Luftwaffe. The purpose of the television transmissions was to entertain wounded German airmen in the various hospitals in Paris.

A new television studio, "Magic City," was built at 180 Rue de l'Université in Paris. It had seats for 250 people and a large control room at the back with facilities for mixing up to six live cameras. There was also an elaborate telecine projector which was equipped with orthicon tubes. The technical equipment was furnished by both German and French companies. The cameras and receivers were made by Telefunken-Allgemeine Elektricitäts-Ges (A. E. G.), and Fernseh A. G. German television engineers were joined by the Compagnie pour la Fabrication des Compteurs et Matériel d'Usines à Gaz under the direction of René Barthélemy, who was collaborating with the Nazis on this project, under a contract with the German Post Office (DRP).

The Nazis would not permit French companies to pursue work of industrial importance but would allow them to improve the French television system under their guidance. In a September 1945 article in *Electronic Engineering,* Pierre Hémardinquer claimed that "The transmission was poor and the synchronizing varied considerably during transmission. The French public, both on moral and material grounds, did not take much interest in Programmes under enemy control, but nevertheless this did allow certain manufacturers to develop receivers and undertake useful research."

"Magic City" remained on the air until August 16, 1944, at which time the Nazis abandoned Paris. They took with them all of the studio equipment but left the telecine (motion picture) transmitters.

There was general destruction of the Eiffel Tower transmitter itself, with fragile apparatus smashed and the oil-filled transformer hit by bullets. The aerial and feeder apparatus were supposed to have been blown up but were saved due to the haste with which the Nazis left Paris.[16]

On March 10, 1944, Colonel David Sarnoff was called to active duty, to participate in the coming Allied invasion of France. He was scheduled for overseas service with the Signal Corps in London. He took a leave from the presidency of RCA and was made a special assistant on communications to General Dwight D. Eisenhower.

Colonel Sarnoff did a superb job of organizing the radio coverage of the landings for the Allies. He then covered France, North Africa, Italy, saw the liberation of Paris, and finally ended his Signal Corps duty with an assignment in Germany. About this time, General Harbord, who had taken over Sarnoff's duties at RCA while Sarnoff was in Europe, had become ill. It was time for Sarnoff to return home.

Sarnoff sailed for the United States on October 21, 1944, and arrived back in New York City on October 28, still on active duty. On December 7, he received his long-awaited promotion to brigadier general.[17]

Five days later, on December 12, 1944, Sarnoff and Zworykin were honored at a meeting of the Television Broadcasters Association (TBA) at the Hotel Commodore in New York City. Sarnoff was the guest of honor that evening. As a special award he received the title "The Father of American Television" and a medal, "For his vision of television as a social force, and the steadfastness of his leadership in the face of natural and human obstacles in bringing television to its present state of perfection." The award was given by Paul Raiburn of Paramount Pictures.

On December 13, Sarnoff was quoted by the *New York Times* as saying that "The nation which established television first will undoubtedly have the first great advantage in establishing its designs, its patterns, and its standards in the rest of the world and will thus gain a great advantage in the export markets."

Zworykin was given the first award in engineering in recognition of his development of the iconoscope and the storage principle of picture pick-up, resulting in the first practical television pick-up equipment. Several other significant figures in the development of television were honored that evening, among them Philo T. Farnsworth, Dr. Peter C. Goldmark, Frank J. Bingley, Dr. Allan B. Du Mont, Dr. W. R. G. Baker, and Dr. Alfred N. Goldsmith.[18]

Early in September 1944, General Henry H. "Hap" Arnold, commanding general of the U.S. Army Air Forces, met with the renowned aeronautical scientist Dr. Theodore von Kármán of the California Institute of Technology of Pasadena, who was also a member of the Aerojet Engineering Corporation and later of the Jet Propulsion Laboratory (JPL), to set up a committee to study the future of military aviation. In October 1944, at the Eglin Field proving ground in Florida, von Kármán consulted with three trusted colleagues, Hugh Dryden of the National Bureau of Standards, Frank Wattendorf, director of Wright Field, and Zworykin, who had long been consulting with von Kármán on RCA's guided missile project, about this committee. Collectively they laid down a structure and drafted names for General Arnold's board of aeronautical scientists.[19]

On October 26, 1944, the headquarters of the Army Air Forces recommended that Zworykin be appointed as an expert consultant to the commanding general (Arnold) of the Army Air Forces on scientific matters. Zworykin was to be assigned to von Kármán's office. On November 9, Brigadier General Donald Wilson, U.S. Army assistant chief of Air Staff Operations, Commitments and Requirements, stated that Zworykin had been issued an investigation certificate from RCA laboratories permitting him to work on classified material. It was valid until May 1, 1945.[20]

This was noted by the FBI, which had actively been looking into Zworykin's background since 1943. Military intelligence reported that Zworykin had been a member of the executive committee of the science committee formed by the National Council for American-Soviet Friendship since December 1943. It was alleged that he "has been reported a Communist member and is active in front organizations of the C.P." This information supposedly came from an informant, whose identity was known to the bureau, on April 7, 1943.[21]

The FBI files also indicated that on June 9, 1944, Zworykin, as head of the Russian Technical Book Committee (part of the science committee of the National Council for American-Soviet Friendship), had written at the request of Soviet authorities to Dr. Ernest Orlando Lawrence at the University of California, Berkeley, requesting a list of books in major scientific fields that would be most helpful to the Russians. He also requested authority to use Professor Lawrence's name as a sponsor for his committee. This immediately caused alarm, as it was noted that Lawrence was engaged on the highly secret DSM (atomic bomb) project, something that Zworykin may have known.[22]

This immediately triggered a massive investigation into Zworykin's activities. On August 19, 1944, the FBI first reported that Zworykin was head of the research laboratory of the Radio Corporation of

America in Princeton, and that he maintained a summer home at Taunton Lakes, New Jersey, and a year-round home in Princeton. He was separated from his wife Tatiana, who was believed to live on Long Island. The report noted that no RCA officials had been contacted, because it was known that "he is an extremely close friend of David Sarnoff, President of the Radio Corporation of America and a member of the Board of Directors. Sarnoff, it is known, has on numerous occasions visited at the subject's home at Taunton Lakes."[23]

On October 27, 1944, it was alleged that Zworykin "on his own initiative in May 1944 undertook at the laboratory conducting experiments in nuclear physics (Akin DSM Project)—a fact which is unknown to RCA Board of Directors and a closely guarded laboratory secret." This report also mentioned that Professor Lawrence was presently employed on work of a highly confidential military development about which the Soviet government was exerting efforts to obtain information.[24] Information from Zworykin's draft board came up with the amazing fact that Zworykin was Jewish.[25]

There began an immediate discreet investigation to determine Zworykin's identity, background, and activities, as well as discreet physical surveillance and immediate consideration of the feasibility of technical and microphone surveillance. This report, whose date is blanked out, was sent by J. Edgar Hoover himself to the Communications Section.[26]

A later report, dated November 15, 1944, related that "the Army's secret weapon presently being developed, known as the DSM Project, is based primarily on nuclear physics experiments."[27] On November 18, 1944, another report related that Zworykin, "unknown to the officials of the RCA where he is employed is independently engaged in research in nuclear physics related to the DSM project. . . and has discussed the 'breaking of the atom' with someone in the Office of Scientific Research and Development in Washington, D.C."[28]

On November 21, 1944, Carl A. Holloman, a CID agent in army intelligence, reported that one of Zworykin's co-workers at the RCA laboratory, Lloyd P. Smith, had earlier worked on the DSM project and then returned to RCA. The report noted that Smith "is presently employed there." Holloman believed that Smith was a good friend of Zworykin and was working in the same laboratory with him. He advised that "Smith does know the entire process of the project and could, for that reason, furnish any information." Another informant was quoted as saying that he believed that Smith was a loyal American and had not furnished any information about the subject to any unauthorized personnel.[29] This report was brought to the FBI's attention on December 20, 1944.[30]

On November 27, 1944, J. Edgar Hoover informed the U.S. attorney general that through "the Bureau's investigation of the Comintern Apparatus, which includes known espionage agents, it has been ascertained that Vladimir Kosma Zworykin, 91 Battle Road, Princeton, New Jersey . . . [the rest blanked out]. Respectfully, John Edgar Hoover." This apparently was the excuse for the actual physical surveillance of Zworykin which started on this date.[31]

On November 30, 1944, S. K. McKee requested that Zworykin's name be placed on the National Special Censorship Watch List, so that copies of all letters addressed to and from him would be sent to the bureau.[32] This request was confirmed on January 9, 1945, and he was put on Special List No. 46.[33]

On December 8, 1944, Zworykin's phone at his Princeton residence was tapped.[34] Since it was noted that Zworykin spent his weekends at Taunton Lakes, a phone tap there was also authorized on December 9, 1944.[35] Zworykin was now classified as Internal Security-R(ussia), so technical surveillance was authorized provided full secrecy was assured.[36]

On December 19, 1944, the FBI requested that the State Department notify them in case Zworykin applied for a passport or indicated that he planned to travel abroad. On December 21, 1944, Zworykin swore an oath of allegiance to the United States as part of his appointment to the von Kármán/General Arnold U.S. Army Air Forces project.[37]

On December 27, 1944, the FBI was told that RCA expected to send "a five-man commission to the Soviet Union in the near future. Information relative to this matter has been previously furnished to Military authorities who have now advised that the final approval for this mission to go to the Soviet Union rests with the Department of State."[38]

Another obvious attempt to connect Zworykin with the DSM project was disclosed on January 3, 1945. This FBI memo noted that "you will recall that Zworykin is the chief of the research laboratory for RCA in the field of electronics. [Text blanked out.] It is to be noted that the DSM project involves this type of experimentation."[39]

Even Zworykin's visits to von Kármán's office were logged. He visited von Kármán on January 9, 1945, and spent the day at the Pentagon. Another report revealed that on January 20, 1945, Zworykin "met in Philadelphia with an 'unidentified blonde woman' probably Mrs. Boris Polevitsky of Yeadon, Pennsylvania, with whom he apparently had frequent contact." Zworykin knew nothing of this surveillance.[40]

With the war ending in Europe, von Kármán decided to send a scientific delegation abroad to collect aeronautical data, interview foreign scientists, and integrate the findings with American experience. He immediately organized a group of specialists—including George S. Schairer, chief aerodynamicist of Boeing; H. S. Tsien, California Institute of Technology; Lee A. DuBridge, University of Rochester; Frank Wattendorf, director of the Wright Field wind tunnel; Hugh Dryden, National Bureau of Standards; and Zworykin—to go to Europe. Among the German scientists to be interrogated by von Kármán's group were Dr. Fritz Shröter and Professor Wilhelm Runge of Telefunken, Dr. Max Knoll, Ernst Ruska, and Prof. August Karolus, all of whom Zworykin had known for many years.[41] Von Kármán's mission was code-named "Operation Lusty," and was in addition to an army secret mission, code-named "Alsos," headed by Dr. Samuel A. Goudsmit and led by Colonel Boris T. Pash. The Goudsmit-Pash mission was already in Europe looking for signs of a German atomic bomb project, specifically scrutinizing facilities and scientists.[42]

On January 12, 1945, an FBI informant reported that Zworykin was told by an unknown person that the NDRC (National Defense Research Committee) was quite concerned, "as there is a meeting at Douglas [Aircraft] next week and they want 150 *memo* [MIMO] units right away." (These were television cameras made by RCA for the ROC guided missile project manufactured by Douglas Aircraft.)

The FBI also learned that Zworykin had accepted a second year as chairman of the Russian relief drive connected with the Red Cross, although he would not be able to devote any time to it, mentioning that he was going to Europe in the latter part of March. The FBI assumed he was referring to his proposed trip to Russia in connection with his work on what they considered to be an unidentified committee to study postwar aviation.[43]

The FBI report also noted that on January 22, 1945, Zworykin had visited the radiation laboratory at Massachusetts Institute of Technology, where he appeared to be well known. A report of the investigation of the Manhattan engineering district was included, in spite of the fact that Zworykin was conducting experiments in high-velocity electronics intended to reduce the length of electronic impulses. This of course would be part of a research program in radar development, which Zworykin was probably involved with at RCA. However, the report stated that Zworykin was "reportedly not concerned with disintegration of atoms as such."[44]

Another part of the investigation into Zworykin's activities was to identify and check the background of Mrs. Boris Polevitsky of 835 Connell Street, Yeadon, Pennsylvania, with whom he had frequent telephone communication between January and July 1944.[45]

On February 19, 1945, J. Edgar Hoover wrote a letter to Lt. Col. John Lansdale, Jr., of the U.S. Engineers Office of the Manhattan District to the effect that Zworykin was not engaged in the fields of atomic and nuclear physics. "It has now been determined that Zworykin's activities in this report should not be technically classified under either of these terms as he is not primarily concerned with the disintegration of the atom or in isolating the nuclei of atoms."[46] This should have ended the FBI's investigation of Zworykin, but didn't.

Zworykin's ease of entry into the Pentagon and other military installations irritated the FBI. It was reported that Zworykin "had received an Army Temporary Duty Pass valid for thirty days on November 9, 1944, but returned this pass on November 22, 1944, at which time he received the Army's Gold Border Pass which entitles the bearer to enter all war localities without question." He turned in this pass on December 20, but was issued a new pass on January 9, 1945, which was still outstanding at the time of the FBI report.[47]

On March 7, 1945, Zworykin visited von Kármán at his office in the Pentagon and spent the whole day there.[48] On March 8, the Newark Field division indicated to J. Edgar Hoover that Zworykin intended to make a trip to Europe, possibly to Russia, sometime in March. This was followed by a March 24 letter to the Washington field division, indicating that they should be particularly alert for any information obtained through confidential informants relative to proposed travel by Zworykin or members of the aviation committee to the Soviet Union.[49]

The FBI agent in charge revealed that on March 14, 1945, a "highly confidential and reliable source having access to the premises of the subject at Taunton Lakes was able to procure photographs of the contents of subject's residence."[50]

It was also noted on March 26 that "there are no secrets in the radar field or scientific air force developments not available to the subject, since he [Zworykin] has the confidence of Army and Navy officials as much any one individual in the United States."[51]

On March 27, an FBI report disclosed that much of the information about Zworykin and his activities came from an individual of Russian descent who had been acquainted with Zworykin and his associates.[52] As we saw earlier, many of Zworykin's Russian "friends" did not like or trust him, and perhaps envied his fame and fortune. So it must be assumed that this testimony was tainted. Still, the FBI put great stock in informants of this kind who did not mind ruining someone else's reputation for a variety of personal reasons.

The FBI had bugged Zworykin's Princeton home and were making recordings on "sound scriber" disks. Typical of the recordings were intimate conversations between Zworykin and Katherine Polevitsky. Unfortunately for the FBI, all of their conversations were in Russian and presented problems in translation.[53]

On April 3, Zworykin spent the day with von Kármán at the Pentagon. He stayed overnight at the Carlton Hotel in Washington, where RCA had a suite of rooms, and spent the next day with von Kármán as well.[54] During the period from March 23 to April 8, the FBI noted that "several contacts were made with Mrs. Boris Polevitsky, his paramour."[55]

Alleged information about Mrs. Polevitsky indicated that "the Spring Bulletin of the Philadelphia School of Social Science and Arts which is sponsored by the Communist Political Association carried the name of Mrs. Katherine Polevitsky as an instructor in this school." She was employed by the bacteriology department of Evans Dental Institute, at the University of Pennsylvania. The report also indicated that Zworykin visited this Medical Arts Building on April 11, 1945.[56]

Theodore von Kármán of Cal Tech and George Morton and Vladimir Zworykin of the RCA laboratories all received top secret clearances from the AC/AS, Intelligence, Counter Intelligence Division, Station Intelligence Branch on April 11. Zworykin and Morton were members of the scientific advisory board for radar, communications, and weather. Zworykin also served on the physics, radar, and television panel.[57]

On April 18, 1945, in a letter from Hoover to Lansdale, it was claimed that, relative to a visit by Zworykin to the radiation laboratory at M.I.T. on March 26, "No information is contained therein indicating an interest on his part in the DSM project."[58]

On April 19, the foreign duty unit of the War Department requested the State Department to issue a special passport to Vladimir K. Zworykin, "who is proceeding to England, France, Belgium, Holland/ Germany, Switzerland and Sweden on official business." (There was no mention of Russia.) This was followed by a letter from the Air Force on May 1, requesting that Zworykin's services be extended for presentation of a special report on "Television" to General Arnold.

Going abroad for Zworykin required considerable preparation. Along with the special passport, it meant ordering an army officer's uniform, as he had been assigned the assimilated rank of full colonel.[59]

On April 22, 1945, an FBI informant advised the bureau that "Zworykin has received final clearance from military for European trip as member of a commission of about 6 persons headed by Doctor Theodore von Kármán of the Army Air Forces." The informant guessed that the entire group would first go to England, using an army plane, and that Zworykin, alone or with others, would then go on to Russia.[60] This was confirmed in a letter from agent Strickland to D. M. Ladd on April 23, 1945.[61]

Von Kármán and his group left secretly for Europe on April 26, 1945, from Gravelly Point, Virginia, in a C54 transport, without Zworykin. *He was the only one of the group who had been denied permission to go.* Von Kármán's mission started in London on April 28. Here they met with such top-notch British personnel as Sir Robert Watson-Watt, Professor Melville Jones, Sir James Chadwick, Lord Cherwell (Prof. Frederick Lindeman), Sir Henry Tizard, Wing Commander Frank Whittle, and Air Vice Marshal John W. Jones.[62]

Von Kármán's group went on to Paris on May 1. They then proceeded to Germany, returned to London, then went on to Switzerland and back to Paris in June 1945. Von Kármán then went by himself to Moscow and to Hungary, and returned to Paris in July 1945 before finally returning home.[63]

On April 27, the bureau was advised that "Subject [Zworykin] informed unidentified woman that he was leaving quote Town unquote Monday next and that he expected to be gone for two months. Subject has been observed to be presently dressed in uniform of U. S. Army Officer."[64] On April 28, J. Edgar Hoover noted that "subject had applied for passport to visit Soviet Union. No record of subject being commissioned as US Army Officer. Bureau desires to continue surveillance to point of embarkation."[65]

Also on April 28, Strickland of the FBI reported to Ladd that "Mr. Frederick Lyon of the State Department has requested a summary of the information appearing in Bureau files relative to Vladimir Kosma Zworykin who has applied for a passport to visit the Soviet Union." Strickland noted that Zworykin "was to accompany a mission headed by Theodore Von Karman to study scientific developments relative to lenses and electronics by German scientists. His passport application was actively sponsored by a Colonel Glantzburg [*sic,* Col. Frederick E. Glantzberg was deputy director, military, of the von Kármán scientific group], a Colonel McHugh [Lt. Col. Godfrey Temple McHugh, executive officer of the same group], and a Major Hammond of the Army Air Forces." Strickland enclosed a blind memorandum stating that Zworykin "had been a sponsor of Russian War Relief, Inc., . . . and had been a patron of the Congress of American Soviet Friendship Council and is a member of the National Council for Soviet American Friendship, Inc." [66]

At this time Colonel Glantzberg, who was endorsing the issuance of Zworykin's passport, ascertained that it was being held up at the insistence of a "civilian intelligence agency." The FBI had given an adverse report about Zworykin to both the State Department and military intelligence. The State Department had then held up his passport pending "clear up of the security feature."[67]

Another FBI report noted that their informant at Newark had "overheard a telephone conversation between Colonel Lansburg of Washington [*sic,* Glantzberg] and Zworykin's secretary, in which the Colonel advised that Zworykin's visa or passport for his proposed trip to Europe had been cancelled. The Colonel was obviously quite disturbed because those credentials had been cancelled. He advised the secretary they had been cancelled at the insistence of another government bureau and not the Army." According to the report, "the other members of the proposed commission of which Zworykin was traveling as a member are leaving on Monday as planned. . . . [Actually, the rest of the group had already left on April 26.] At 3:00 p.m. today Zworykin was still not aware that his credentials had been cancelled, as he could not be reached by his secretary." The report explained that Zworykin "had left his home this morning, wearing his new Army uniform and went to Philadelphia, where he intended to meet Mrs. Polevitsky, his paramour." The informant said that Zworykin would probably suspect that his activities were being checked and he was told that the bureau would protect him carefully.[68]

It appears that Zworykin came back to the RCA laboratories on Monday, April 30, where he first went into James Hillier's laboratory in Princeton in a resplendent full colonel's uniform and informed Hillier that he was being sent to Europe. Zworykin stated that he was leaving that evening. He gave no reasons for his trip. Everyone in the laboratory said their farewells and wished him well.

His secretary then informed him of the fact that his trip had been canceled over the weekend and that she had not been able to contact him. He then went to Washington that same day to find out what had happened.[69]

On April 30, a teletype from agent S. McKee was sent to the Washington SAC (special agent in charge) indicating that "Zworykin Internal Security-R registered at the Carlton Hotel, DC where he plans to remain possibly to Wednesday, attempting to have his passport approved. Surveillance to resume when he arrives back in Princeton."[70]

Agent McKee later wrote to Hoover that this same informant "has advised that the RCA and David Sarnoff, its president, are starting an investigation to find out why Zworykin was not allowed to go to Russia." Another informant reported that Zworykin "was of the opinion that something 'very strange' was going on." Obviously the second informant had very close connections to both the headquarters of RCA in New York and the laboratory at Princeton.[71]

A letter from S. S. Alden to Ladd on May 1 quoted a Mr. Reynolds as saying that the only derogatory information available was that Zworykin had sponsored a meeting of a "front" organization.[72]

In spite of this report, military intelligence (G2) was determined not to have the State Department issue a passport to Zworykin. On this very same day, Zworykin had been cleared by the Air Force. In a classification review of Zworykin it was agreed "(1) to an extension of appointment, (2) that he continue with the work that he originally appointed, and (3) the work to be performed is not of a classified nature." Such a review indicated that it there would be no problem for Zworykin to finally go abroad. However, this did not happen.[73]

Zworykin went to the passport division of the State Department with Maj. Godfrey Temple McHugh of the Army Air Forces (von Kármán's executive officer) on this same day. Here they were notified that his passport was still being held up. They were told that it was necessary for Zworykin to clear himself completely with the military authorities and that there was nothing that the State Department could do. On May 2, he again went to von Kármán's office for several hours. At noon on May 3, with this pressure from G2, the Army Air Forces finally withdrew their request for Zworykin to go abroad.[74]

He left Washington on May 3, and the next day appeared back in his office in civilian clothes at his regular time. Without a word about the previous events, Zworykin continued to follow his regular office routine.[75]

But now even his position with RCA was in jeopardy. Through the services of John Cahill, an attorney for RCA, the FBI had succeeded in planting a "spy" in the RCA laboratories to watch what was going on there. A Mr. McGrady, special assistant to the undersecretary of war and labor advisor to RCA, was aghast that Zworykin had access to the most secret army and navy material, material "so secret that other people in RCA not actually in the laboratory know nothing about it." McGrady wanted Zworykin removed. He informed Cahill of the situation. Cahill "was disturbed at the legal responsibility of RCA if they continued to employ Zworykin knowing that he could not get a security clearance from the Government to leave the United States." After hearing the details from McGrady, Cahill agreed that Zworykin was to be told that "it would be necessary for him to clear himself completely with the military authorities and that there was no matter pending before the Department of State concerning him."[76]

In a letter from Hoover to Lansdale on May 2, Hoover advised that "Zworykin [still] expects to depart from the United States in the near future to visit the European Theatre of Operations with a military mission." He added that "copies of all pertinent reports have been furnished to the Office of Military Intelligence."[77]

On May 12, a request was made to "determine the identity of a Colonel Glantzburg [*sic*], who has been pushing the passport application of Zworykin." It turned out that Glantzberg had been a veteran of the murderous Ploesti oil fields bombing raids.[78]

On May 19, an FBI report on Zworykin said that on May 14, the RCA Corporation and David Sarnoff, its president, were starting an investigation to find out why Zworykin was not permitted to go. "In this respect it should be noted that David Sarnoff presently holds a commission as a colonel [actually he was now a brigadier general] in the U.S. Army Signal Corps."

The same report indicated that information on Zworykin appearing in bureau files was furnished to the State Department and the Office of Military Intelligence. The FBI reported that Zworykin requested not only permission to visit occupied countries but a visa for the Soviet Union and Turkey as well. It was noted that Zworykin expected to visit the Soviet Union to receive a medal for his scientific accomplishments some time in the summer of 1945. The State Department had refused to issue a visa to Zworykin to visit the Soviet Union although considerable pressure on Zworykin's behalf was exerted by representatives of the Army Air Forces.[79]

On May 29, 1945, S. S. Alden wrote to Ladd that a special agent "had discussed this entire matter with General C. W. Clarke, Military Intelligence Service . . . to ascertain what action would be taken by the War Department concerning Zworykin's trip. General Clarke has advised [name blanked out] that as a result of his recommendation Zworykin's trip has definitely been canceled. General Clarke pointed out to [name blanked out] that he was well aware that Zworykin exercised all sorts of pressure to be permitted to make this trip, but he felt nevertheless, that due to the information in the Bureau's files, that Zworykin should not be permitted to accompany this mission."[80]

On June 1, 1945, the investigation into Zworykin's activities continued. It is known that the FBI had gained access to Zworykin's summer home at Taunton Lakes. However, access had not been gained to his home in Princeton or his office at the RCA laboratories, where they presumed that he had concealed documents that would directly involve him with the Russian espionage system. It was obvious that they had not actually found any adverse information on Zworykin to this date. Later that year (October 1945), while Zworykin was away on vacation in Florida, the FBI searched his Princeton home. The results "of this thorough search were completely negative in value so far as this investigation is concerned."[81]

His phones were still being tapped, his Taunton Lakes home had been broken into, listening devices had been installed at his home, his mail was still being intercepted, and a certain amount of physical surveillance was still taking place. In spite of all of this effort, no such incriminating evidence was ever found. On June 4, 1945, J. Edgar Hoover confirmed that "Zworykin's trip to Europe in behalf the War Department has now been cancelled."[82]

All of Zworykin's immediate efforts to find out when he could get his passport or why it was not ready were to no avail. Later he was told that his passport had been denied by the State Department because of his being a member of the Russian War Relief, Inc. Believing that the organization was legal and included such prominent personalities as Eleanor Roosevelt and Vice President Henry Wallace, Zworykin came to the conclusion that the real reason for his troubles must be his Russian origins. In this he was mistaken, since in fact the Russian War Relief, Inc., and the National Council of American-Soviet Friendship were high on the FBI's list of subversive organizations.

While Zworykin, who was sometimes quite naive, may have believed that lending his name to organizations like these at this time carried no real significance, there were many government officials and agencies (the FBI, for instance) who took such affiliations quite seriously. It hardly mattered that during the war the Soviet Union was fighting for its very existence, against a common enemy.

And of course the FBI was correct, the Soviet Union was doing its best to infiltrate certain projects (especially the DSM). But as the Klaus Fuchs case proves, the FBI's harassment of people such as J. Robert Oppenheimer often blinded them to the real culprits. Being regarded with such suspicion was a bitter pill for Zworykin to swallow after so many years and so many contributions to the United States. He felt that he had been confined in some sort of political cage.

Zworykin claimed in his manuscript that he had immediately resigned from von Kármán's committee and had planned to resign from RCA, as he would no longer be able to work on secret projects. However, when Sarnoff heard of the problem, he had the RCA legal department investigate the matter and persuaded Zworkyin to stay on.[83] This is actually another piece of information in Zworykin's favor, since Sarnoff was a dedicated anticommunist cold warrior and would have dismissed Zworykin out of hand if there had been the slightest doubt as to his loyalty. It is also known that Sarnoff enjoyed a good relationship with J. Edgar Hoover and certainly intervened on Zworykin's behalf. At any rate Zworykin remained at RCA. The records also indicate that Zworykin was still on von Kármán's committee in June 1946.[84]

On October 16, 1945, Zworykin again applied for a passport to visit England, Holland, France, Switzerland, and Sweden. This application too was rejected.[85] While he was on a two-week vacation in Miami Beach, Florida, Zworykin's leased home in Princeton "was thoroughly searched with completely negative results."[86] Finally, the FBI noted that Zworykin purchased Dr. E. A. Lowe's home at 103 Battle Circle around November 1, 1945.[87]

In 1946, the Russian War Relief, Inc., was dissolved and Zworykin's membership was terminated. On April 11, 1946, the FBI requested that "all Offices sever the outstanding leads in this case and unless information is developed which would indicate that subject Zworykin is acting as an agent of the Russian Government or is engaged in espionage activities, the Newark Office should submit a closing report."[88]

On September 17, 1946, the FBI admitted that "The file on subject was reviewed and it was found that all leads have been covered. *THIS CASE IS BEING CLOSED UPON THE INSTRUCTIONS OF THE SPECIAL AGENT IN CHARGE.*" In a memorandum of November 1, the FBI advised the Passport Division of the State Department that "this office has no further interest in this matter." Zworykin later claimed that in 1947, "on the basis of public information that the National Council of American Soviet Friendship was suspected of being subversive, I resigned from that organization."[89]

So it appeared that Zworykin's ordeal was finally over. But this was not really true because, while actual FBI surveillance abated, the information in their files was always there to be used against him when the occasion arose, which would happen several times. However, Zworykin's uncanny ability to get out of sticky situations again served him well. No charges were ever filed against him or brought out in public; he retained his top secret clearance; and his status with RCA and Sarnoff remained untainted.

World War II had actually benefited the American electrical industry because of tremendous production contracts for research and development from the Office of Scientific Research and Development. The largest portion went to Western Electric, General Electric, RCA, and (in order) Du Pont, Westinghouse, Remington Rand, Eastman Kodak, Monsanto Chemical, and Zenith Radio. They had all come out of the war with advanced technologies that the government had paid for, and RCA was no exception.[90]

Work done on the image orthicon camera tube, designed and built in 1944 by Albert Rose, Paul K. Weimer, and Harold B. Law of RCA for guided missiles using television, gave RCA one of the most essential pieces of equipment needed for postwar television. Although Zworykin had not actually participated in its development, acting only as the director and advisor, the camera tube had come out of his research laboratory and he took at least partial credit for a tube that was so superior to any other television camera tube that it assured the supremacy of RCA when television resumed after the war.

The project began with a two-sided glass target, which Rose had sought a patent for in September 1940. The target was considered so valuable to the war effort that the Department of Defense held up the patent application until the war was over. Instead, under an Office of Scientific Research and Development contract, Rose, working with Weimer and Law, started a top secret project to incorporate this target into a new orthicon camera tube for military purposes.

By 1944, the basic orthicon tube had been tremendously improved by the addition of an image section (similar to that of the image iconoscope) and an electron multiplier based on earlier work by Zworykin, Farnsworth, and others. Its sensitivity was about one hundred times that of the regular orthicon. In addition, it was completely electrically stable at medium and high light levels. The earliest model was the LM15, built in Lancaster, Pennsylvania. This was a large tube, almost seventeen inches long and three inches around, and had been quite complicated to build. However, it clearly showed great promise.[91]

A second project was started to make a smaller version of this tube to save both weight and space. The resulting tube, developed by Weimer, Law, and Stanley V. Forgue, was only 9 inches long and 1½ inches in diameter. It was designed for use in the ROC missile and was called the MIMO (*m*iniature *im*age *o*rthicon).[92]

While the tube may have been top secret in the laboratory, RCA could not suppress its elation. The first indirect hint of the image orthicon camera tube appeared in an ad in *Electronics* in May 1944, which bragged, "Due to research, conducted at the RCA Laboratories, a new super-sensitive television camera tube, rivaling the human eye in its ability to see under conditions of poor light is in prospect for the post-war world."[93]

In January 1945, on a transatlantic visit, Professor J. D. Cockcroft of British Air Defense Research met with Zworykin. He was given a demonstration of "a camera tube of such sensitivity that studio technique would be completely changed." Zworykin considered that it would take four to five years to make real progress on commercial use of the image orthicon, and this would include progress in color.[94]

Although the image orthicon camera tube was the most complicated and expensive electron tube to date, it was as important to the development of television as Zworykin's iconoscope some fourteen years earlier because it filled the need for a tube that could be built into a lightweight camera for use in the studio and out of doors. RCA added two essentials to the camera, an electronic viewfinder that had first been developed by the Du Mont laboratories and a rotatable four-lens turret.

The tube was first revealed publicly in New York City on the night of October 25, 1945, at the Waldorf-Astoria Hotel's Herald-Tribune Forum. There was a demonstration of its ability to pick up scenes in dimly lit interiors and close-ups lighted only by a single candle. This demonstration was followed by one at the Army-Navy football game in Philadelphia on December 1.[95]

The new tube did have problems at first; its pictures were "noisy" (grainy), it had pronounced "halos" (or black edges) around bright objects, and its gray scale (tonal rendition) made faces look a bit unreal. It was also almost impossible to match the gray scale of any two tubes and get them to look alike, a problem typical of the earliest 2P23 camera tubes. However, these faults were slowly overcome, and for the next nineteen years, the image orthicon camera tube became the standard pickup tube in television cameras all over the world.[96]

The end of the war in Europe found most of the Continent in ruins. There were efforts made to return to peacetime activities. At Champ-de-Mars in France, the Compagnie pour la Fabrication des Compteurs et Matériel à Gaz under René Barthélemy demonstrated some of the equipment they had designed and built under the Nazi occupation, including a modern iconoscope tube, called the Isoscope, and an orthicon tube that was not working. They also showed a quite modern looking camera with an electronic viewfinder. There was even a demonstration of a 1050-line system on a fifteen-inch tube with a 15mH bandwidth that gave pictures of extremely good definition and contrast. According to Delaby, the quality of the images was "comparable with that of an ordinary cinema." There were plans to go back on the air on a new French standard of 450 lines interlaced at fifty frames per second, and a test transmission from the Eiffel Tower was sent out on October 1, 1945.[97]

Late in 1945, Zworykin published his third book, *Electron Optics and the Electron Microscope,* co-authored by George A. Morton, Edward G. Ramberg, James Hillier, and Arthur Vance.[98]

CHAPTER 12

"Ad Astra per Video"

On January 10, 1946, the *New York Times* announced that Dr. John von Neumann, a world-famous scientist at Princeton University's Institute of Advanced Studies, and Dr. V. K. Zworykin of the RCA laboratories had advised the U.S. Weather Bureau and the army and navy weather services that they were developing a supercalculator that might solve the mysteries of long-range weather forecasting. There were many obstacles to be overcome and there were no indications that such a project was under immediate consideration. Between May and July of 1946, von Neumann negotiated a contract with the navy to construct such a device. Despite the publicity, Zworykin had no role in the design and construction of such a computing machine.

It is known that earlier von Neumann realized that a computer would need some type of device, such as delay lines and cathode-ray tubes, to store information. Even Zworykin's iconoscope, which was a storage device, was considered. Zworykin probably first met von Neumann when he approached RCA in search of such a device.[1]

Another of von Neumann's ventures at the time involved work on an advanced electron microscope to study bacteriophage until the computers then under development became available. He proposed to used scaled-up electromagnetic centimeter waves to simulate X rays. Because this plan called for an engineering group with radar experience, he discussed it with Zworykin and RCA. Zworykin agreed to participate. A proposal for funding was sent to Mina Rees of the Office of Naval Research, but the money did not materialize and the project was canceled.[2] But Zworykin was engaged in other projects at this time, such as work on a device for guiding blind persons, weather research including a device for charting hurricanes at sea, and a guidance control system for automobiles.[3]

Meanwhile, on June 7, 1946, the BBC London television station reopened. Maurice Gorham was director of television, and his staff included Cecil Madden, program organizer; Jasmine Bligh, announcer; Douglas C. Birkenshaw, engineer in charge; and Tony Bridgewater, engineer in charge of outside broadcasts. It resumed with the same technical standards and equipment that had been so carefully stored during the war. After much debate, it had been decided that it was prudent to go ahead with old standards rather than cause delay while seeking "dubious" improvements.

One of the programs that opening day was the same Mickey Mouse cartoon that had been the last televised item in 1939, but the first major event telecast was the victory parade the next day. This meant reviving a television cable linking the West End with the Alexandra Palace that had not been used since 1939. This event came off quite well. Sadly, six days later, on June 14, 1946, John Logie Baird died at the age of fifty-eight. Unfit for military service and denied a chance to serve among the "Boffins" in secret research work, he died tired but undefeated.[4]

With the return of peace, Sarnoff reaffirmed his faith in Zworykin. In March 1947, he made him a vice president and technical consultant at the RCA laboratories. At the same time, Zworykin received the Franklin Medal, the War Department Certificate of Appreciation, and the Navy Certificate of Commendation in recognition of his distinguished achievements during the war.[5]

One of the first things Zworykin did after his promotion was to preside over the first public demonstration of a simultaneous (that is, not sequential) large screen color television. RCA hosted the demonstration on April 30, at the Franklin Institute in Philadelphia, using a seven-and-a-half- by ten-foot screen. For the occasion, Zworykin gave an address entitled "All-Electronic Color Television."

The system used a flying spot method for film and slides (no live camera existed) to create three registered images that were displayed simultaneously. This was done by means of three cathode-ray tubes with colored phosphors that were projected onto the large screen. RCA had first given a private demonstration of this system using three picture tubes in a cluster called the trinoscope on October 30, 1946, at the Princeton laboratories.

Until this time, RCA had been experimenting with a color television system using a rotating (sequential) color wheel similar to that developed by Peter Goldmark at CBS, using the new image orthicon tube. The Franklin Institute demonstration, however, was the beginning of RCA's efforts to provide a color system that would later fit into the 6MHz television channels and thus be compatible with the existing monochrome standards. All color television research was under the direction of Raymond D. Kell.[6]

RCA planned the first demonstration of American television in Europe in June of 1947. They also planned for Zworykin to deliver a paper on television before the Academy of Science in Rome in celebration of the fiftieth anniversary of Marconi's invention of radio, during the weeks between August 30 and September 15. He was then to give a lecture, "Applications of Electron Microscopy and Electron Diffraction to Metallurgy," during the week of September 28 to October 5 in Liège. On this trip he was to represent the National Academy of Sciences, the Institute of Physics, and RCA.[7]

On March 20, 1947, RCA wrote the State Department requesting a passport so that Zworykin could go to Europe in August. Zworykin applied for a passport on that same day. On April 22, he received notice that his passport application was pending. On June 19, Zworykin received a special passport no. 2438 that permitted him to leave on August 15, on the SS *Queen Elizabeth*. With his passport renewed, Zworykin felt like a free man again.[8]

In 1948, Zworykin was awarded the presidential Certificate of Merit and was made a chevallier of the French Legion of Honor. On October 28, he won the Poor Richard Club's gold medal of achievement in recognition of his invention of the electric scanner in 1934. According to the *New York Times,* "Without this invention, television would still be a laboratory toy." The presentation was to take place on the anniversary of Benjamin Franklin's birthday, January 7, 1949, at the Franklin Institute in Philadelphia.[9]

On March 26, 1949, the American Institute of Electrical Engineers announced that its 1948 Lamme Medal had been awarded to Zworykin for his "outstanding contribution to the concept and design of electronic apparatus basic to modern television." The medal had been established by Benjamin G. Lamme, one-time chief engineer of the Westinghouse Electric and Manufacturing Company. According to the *New York Times*'s account of the award ceremony on June 22, in Swampsscott, Massachusetts, Sarnoff called Zworykin "the scientist extraordinary of this age."[10]

In 1949, Dr. Zworykin published his fourth major book, *Photoelectricity and Its Applications,* an updated version of his 1932 book with E. D. Wilson, *Photocells and Their Application.* This time Zworykin's co-author was Edward G. Ramberg.[11]

In July 1949, the FBI noted that Zworykin was being considered for a position in connection with classified navy contracts. They noted that in February 1948 he had been doing research for the navy, working on a guided missile project for the army, and had been head of television experimentation for RCA. The FBI report contained much positive information about Zworykin's background. For instance, an informant alleged that Zworykin as a young man had been a revolutionist and had been jailed for the "Cause." However, since the revolutionary leaders did nothing for him while he was in jail, he decided that he would "never carry the torch" for anyone again.

The report continued by noting that Zworykin had been associated with the Siberian provisional government of Russia as an engineer and technician in 1918, and that the Zworykins, as White Russians, had to flee from their homeland to escape the wholesale liquidation of prominent persons who had served the overthrown government. The same source indicated that "Zworykin has never spoken about Communism and is in fact an out and out Capitalist." He also stated that Zworykin had expressed reluctance to participate in relief affairs and even annoyance at these requests. Finally the report stated that a source close to the army claimed that around 1939 or 1940 Russian military intelligence considered Zworykin to be hostile to the Soviet regime.[12]

On February 28, 1949, Paul K. Weimer of the RCA research laboratories applied for his first patent on a photoconductive camera tube, the first practical alternative to photoemissive targets such as those found in the iconoscope or orthicon. RCA had undertaken the active development of such a tube in 1947. Early research focused on using CdS (cadmium sulfate) and other high-gain photoconductors such as ZnSe (zinc selenide). Researchers began by evaporating crystalline selenium. They found that red "amorphous" (noncrystalline) forms of selenium were very photoconductive and had many advantageous characteristics. One form was glassy red in color with a high dark resistivity, giving a clean black level. In addition, its sensitivity was exceptional (in excess of 1000 microamperes per lumen) and its resolution was over 600 lines. Finally, its speed of response (lag characteristics) was excellent. All this in spite of the standard textbook theory that red amorphous selenium was supposed to be a nonphotoconducting insulator.

RCA's first tubes were six inches long and one inch in diameter and had a target area of three-eighths by one-half inches. The use of a small target combined with low-velocity beam scanning was almost as important as the photoconductor in achieving satisfactory response times for moving scenes. The sensitivity was also great enough to do away with the need for signal multipliers, and image sections, thus simplifying the construction of the tube and an associated camera. The inventors thanked both Zworykin and Rose for their continued interest and advice during the course of this work.[13]

On March 7, 1950, Paul Weimer, Stanley V. Forgue, and Robert Goodrich of RCA introduced this revolutionary new camera tube at the national conference of the Institute of Radio Engineers in New York City. Zworykin and his staff at the RCA laboratories were credited for developing the tube, called the Vidicon, for closed circuit, industrial, and educational use. This was the beginning of a new generation of camera tubes.[14]

On April 15, 1950, at a meeting of the North Atlantic region of the Institute of Radio Engineers, Zworykin disclosed a new RCA industrial television system capable of presenting three-dimensional images, based on the vidicon camera tube's high resolution at normal light values. He explained that while the images could not be broadcast, they had innumerable uses in science and industry.[15]

The next year, on March 21, 1951, Zworykin received the I.R.E.'s Medal of Honor for "his outstanding contributions to the concept and development of electronic apparatus basic to modern television. His scientific achievements that led to fundamental advances in the applications of electronics to communications, to industry and to National Security." The award was presented at the group's annual banquet in New York City. Zworykin was quoted as saying, "I take this as a challenge to those of us of the older generation to look ahead and join with younger members of our profession in blazing new paths in the effective use of our knowledge and facilities."[16]

Also in 1951, a major change occurred in Zworykin's private life. His marriage to Tatiana Vasilieff Zworykin had never been a success, and their marital relations had deteriorated long ago. In fact, they had been separated for over twenty years. On June 28, 1951, Tatiana filed for a divorce on the grounds of desertion. The divorce was granted on November 13.[17]

The next day (November 14, 1951), Zworykin married Dr. Katherine Polevitsky, whose husband Boris had passed away on December 5, 1950. She and Zworykin had been friends and neighbors at Taunton Lakes for many years. Although she had a medical degree, she had turned to biological research. They were quite compatible and lived a rewarding life together.[18]

For the next three years, Zworykin was kept busy as a consultant at the RCA research laboratory in Princeton. During this period, a heated battle took place between CBS and RCA over whose color television system would be adopted in the United States. The problem was that CBS's mechanical system, which operated at 343 lines at sixty frames/120 fields and was quite capable of producing brilliant color pictures, was incompatible with the television standards set up by the National Television Standards Committee in 1941.

Sarnoff was determined that the newly developed RCA simultaneous "mixed-highs: dot-interlaced" color system, which was compatible with existing monochrome FCC standards (that is, it could be received on existing black-and-white sets), should be the one adopted. A crash color program was started at the Princeton laboratories to improve its performance. No expense for personnel or equipment was spared and bonuses were to be paid for any technological equipment breakthroughs. All noncolor projects were temporarily shelved. An important part of this project was to design and build a practical tricolor (in one envelope) picture tube that could display either monochrome or color pictures. RCA vice president Elmer Engstrom was put in charge of this tremendous effort.[19]

However, impressed by the quality of the CBS color system, the FCC adopted the "incompatible" CBS method on October 10, 1950, over the objections of most of the television industry.[20] As a result, in November 1950, it was decided to set up a second National Television Standards Committee to resolve the color issue, and the committee convened on June 18, 1951. CBS continued to conduct field tests of its system through the summer of 1951 to prove its merits, but the tests were a Pyrrhic victory and CBS soon realized that it had a losing battle on its hands. CBS finally used the Korean war as an excuse to give up its color system.

NTSC testing of the "compatible" system continued through the next two years. In addition to RCA's pioneering efforts, great accomplishments in color circuitry were made by the Hazeltine laboratories, Philco, Zenith, and others. Finally, in December 1953, the FCC chose the new NTSC color system, which combined many of the basic features of the RCA dot-interlaced system but was vastly improved by the entire television industry. This then became the basis of all future color systems. Variations of it were used in countries all over the world. In addition, the tricolor, shadow-mask picture tube developed by the RCA laboratories led the way to brighter, sharper color display devices.[21]

In 1954, toward the end of the red-baiting McCarthy era, doubts about Zworykin's loyalties surfaced again. On February 2, 1954, Zworykin applied for a passport to travel to Spain, Italy, Switzerland, and Sweden. In order to avoid any of the problems of 1945, the RCA legal department wrote a letter attesting to Zworykin's loyalty. In it,

RCA vice president and general attorney Robert L. Werner stated categorically that he had "never come across anything which to my mind raised doubts as to his loyalty to the United States or his adherence to the American form of government." This was accompanied by a sworn statement from Zworykin attesting to his loyalty to the United States. It concluded with his assertion that "On my escape from Communist captivity I came to the United States and I have nothing but hate for the Russian government and for all that it stands." Two days later, on March 3, 1954, his passport was issued.

He had planned a pleasure trip, departing on March 20, 1954, but for some reason, Zworykin left New York City on April 7 for Italy and Sweden. He arrived home from Sweden on May 26. On the day he arrived home, his passport was again taken from him. However, it was later returned, when he visited Japan in May 1955.[22]

What was ironic was that Dr. Hsue-shen Tsien, who had been a member of the von Kármán scientific Air Force advisory board with Zworykin, was being deported in 1955. He had been arrested in September 1950, when he was about to return to China with a trunk full of documents, none of which were actually incriminating. He had been detained by the U.S. Immigration Service for some five years, charged with having been a communist agent since 1936. (None of these charges were ever proven!) Yet Tsien had been allowed to go to Europe with von Kármán in May 1945 while Zworykin had been denied permission to go abroad. Even von Kármán himself had been investigated by the FBI in relation to his earlier service as a government minister during the Communist régime of Béla Kun in Hungary. The FBI's broad brush touched everyone and spared no one.[23]

In July 1954, Zworykin reached sixty-five years of age and due to RCA corporate policy was forced to retire on August 1. Shortly thereafter, the board of directors of RCA elected him an honorary vice president—the first such appointment in the history of the company. This was another way for David Sarnoff to thank him for his efforts on behalf of RCA.

In honor of his retirement, a seminar entitled "Thirty Years of Progress in Science and Technology" was held in McCosh Hall at Princeton University on September 18, 1954. The panelists included Dean Hugh Scott Taylor, who discussed synthetic materials; Dr. J. C. Hunsaker, who spoke on aeronautics; Dr. Isadore I. Rabi, who commented on high energy particles; and Dr. James Hillier, who examined electronics and vision in medicine. Theodore von Kármán had been invited to speak on aviation but declined on the grounds that he would be in Europe on that date. All of this was of course quite a compliment to Zworykin.[24]

After the seminar, Dr. Elmer W. Engstrom, who was now executive vice president of RCA laboratories, introduced the main speaker of the evening, Brigadier General David Sarnoff, chairman of the board of RCA. Sarnoff gave a speech lauding Zworykin's achievements. He received a great laugh when he repeated the story of how Zworykin had convinced him many years ago that he could produce a new, revolutionary cathode-ray tube for some $100,000. Sarnoff stated, "Look at Zworykin, how he deceived me. Before we got a dollar back from television, we spent fifty million dollars!" It was obvious from the speech how much Sarnoff respected his longtime colleague. Certainly his faith in Zworykin had never been betrayed.[25]

On September 27, 1956, A. H. Belmont of the FBI reported that the investigation of Zworykin had "failed to disclose any espionage activities or membership in the Communist Party," but recommended that he be interviewed. "It is believed that sufficient time has elapsed [material blanked out] that an interview of subject, if handled carefully, would not compromise [name blanked out]." Belmont further recommended that "if he is cooperative . . . the interview be expanded to cover his past affiliations and associations."

The recommendation to inteview Zworykin was accepted, but according to a subsequent report the interview was a failure. "The subject appeared ill at ease, breathed somewhat heavily, and followed a policy . . . [of] carefully answering all questions, volunteering little or nothing, and reducing the interview to the shortest time and extent possible."

In response to a further effort, in late November 1956, to make an appointment to show him photographs, Zworykin was increasingly testy. According to the FBI report, "he interjected that he did not want 'to be bothered by the police.' It was explained to him that the FBI were not 'police' and this was just a matter in which his cooperation was being requested as a United States citizen. The subject responded that he 'left Russia to get away from state police' and that he 'didn't want to be bothered.'"

Finally, on January 23, 1959, the FBI conceded that "due to the attitude of the subject it is not possible to make additional contacts with him. In view of the above, no further investigative actions will be taken in this matter and the case is being closed." Apparently both his advanced age and his retirement, along with the absence of evidence against him, had finally convinced the FBI that they had no case. Still, they continued to keep a record of his activities until at least September 19, 1975, the date of the last entry in the FBI file sent to me.[26]

In 1958, Zworykin published his fifth and final book, *Television in Science and Industry* by V. K. Zworykin, E. G. Ramberg, and L. E. Flory.[27]

From June to early August 1959, Zworykin again made a trip to Europe. He visited England first. On July 1, he delivered the James Clerk Maxwell Memorial Lecture, "The Human Aspect of Engineering Progress," at the Cavendish laboratory. He was then made an honorary member of the British Institute of Radio Engineers, which honored him for his efforts "in promoting an international body for the further application of electronic science in the fields of biological research and diagnosis."[28]

From England, Zworykin went on to the Soviet Union on July 6. In Moscow, he was an honored guest at the first post–cold war cultural and scientific exchange between the United States and the Soviet Union, organized by President Dwight D. Eisenhower and Soviet Premier Nikita Khrushchev. This was also the occasion for an American national exhibition at Moscow's Sokolinki Park, part of which was a working demonstration of color television consisting of two RCA color cameras and an Ampex color videotape recorder. With Zworykin in Moscow were Waldemar Poch of RCA, Philip Gundy, vice president of the Ampex Corporation of Redwood City, California, and two of his top engineers, Joseph Roizen and William Barnhart. Roizen, an engineer with Ampex for many years, was chosen for his engineering acumen as well as his ability to speak Russian.

The world-famous "kitchen debate" between Richard Nixon and Khruschev occurred on July 24, 1959. Sixteen and a half minutes of this rough and tumble debate were recorded live by Roizen. He turned the videotape over to Philip Gundy, who smuggled it out of the USSR that same night. It was immediately flown back to the United States, where it was broadcast the next day, first by NBC and then by ABC and CBS. The Soviets were furious at first, claiming that it was to be broadcast simultaneously in both the USSR and the U.S. They threatened an international incident just short of World War III, but, the incident was soon smoothed over with no lasting consequences.[29]

In May 1962, Zworykin attended the Second International Television Symposium at Montreux, Switzerland. He was one of several distinguished pioneers who received awards "in recognition of their outstanding contribution to the advancement of television." The others were Dr. Fritz Shröter of Telefunken; Sir Isaac Shoenberg, director of EMI; Georges Valensi, former director of the International Telephone Consultative Committee of the International Television Union; and Dr. H. Yagi, who had been president of the Musashi College of Technology in Tokyo until 1960.

Zworykin was chairman of the section devoted to color television. He reviewed the present situation in the United States, referring to a new RCA color tube that was six inches shorter than the present 70 degree tube. He remarked that while he questioned the value of color television for broadcasting, it was more important for educational and medical purposes.[30]

In July 1962, Zworykin and his wife Katherine visited Moscow for the International Cancer Congress, which was held from July 22 through July 28.[31]

For a period of years after his formal retirement, Zworykin was the director of the Medical Electronics Center at the Rockefeller Institute in New York. In this capacity and as national chairman of the Professional Group on Medical Electronics of the Institute of Radio Engineers, as founder-president of the International Federation for Medical Electronics and Biological Engineering, and as a member of the board of governors of the International Institute for Medical Electronics and Biological Engineering in Paris, he promoted the use of electronic methods in medicine and the life sciences.

In his work on behalf of interdisciplinary cooperation of physical and life scientists, he built up those organizations dedicated to the advancement of medical electronics and biological engineering. In addition he made significant direct scientific contributions, such as in the development of the radio *endo-sonde,* a radio pill that could be swallowed by the patient and transmit data from inside the stomach and intestinal tract. He also helped developed an ultraviolet color-translating television microscope.[32]

On February 6, 1967, Zworykin, with ten other prominent scientists, was presented the National Medal of Science by President Lyndon B. Johnson. Zworykin was given the award "For major contributions to the instruments of science, engineering and television, and for his stimulation of the application of engineering to medicine." This is the most prestigious, highest scientific honor given in the United States.[33]

Other honors followed. He was given the coveted Founders Medal of the National Academy of Engineering in Washington, D.C., on April 25, 1968, for "inventing the iconoscope, the first practical picture transmission tube, and the kinescope, the receiving tube in TV sets." Zworykin's response, as quoted by the *New York Times,* seems to sum up his attitude toward his life and his work very well. He noted that he considered himself an engineer, "even though I have repeatedly oscillated across the tenuous boundary between pure science and applied science. I have learned to appreciate the satisfactions of both vocations. . . . There is an inescapable nexus between the pure and empirical sciences; neither can effectively survive long without the other. Moreover, each must be motivated by a benevolent interest in humanity, and each must be in league with the future."[34]

On September 10, 1969, Zworykin was given a testimonial dinner by RCA in honor of his eightieth birthday. By then, Robert Sarnoff, David Sarnoff's son, was president and chief executive officer of RCA. The chief speaker was Dr. Erik Walker, president of both the National Academy of Engineering and Pennsylvania State University. There were also statements by Dr. George H. Brown, executive vice president in charge of patents and licensing at RCA, and Zworykin's former research assistant, Dr. James Hillier, now executive vice president in charge of research and engineering for RCA. Hillier's tribute included the remark that "Dr. Zworykin is a warm, friendly, dynamic man who has devoted the totality of his talents to the betterment of his fellow man." This was to be the last time that RCA so honored him.

At this party, RCA announced that it was establishing a Vladimir K. Zworykin Award for $5000 to be administered by the National Academy of Engineering. It would be presented annually for five years, beginning in 1970, for "outstanding achievement in electronic science and technology."[35]

Missing from the proceedings was Zworykin's old colleague and employer David Sarnoff, who was ailing and could not attend. Sarnoff had turned the company over to his son in 1965, in hopes of continuing a dynasty that had begun in 1930. But this was not to be.

David Sarnoff had become ill in August 1968 and on December 11, 1971, he died at the age of eighty. NBC and the other television networks took notice of his passing in news broadcasts. It made front-page headlines in just a few of the leading newspapers and magazines on the following day. An era had come to an end.[36]

Zworykin, however, was blessed with good health and a sharp mind, and his lifestyle changed little after his so-called retirement. As Hillier, who knew him well, commented, "He has the same qualities which contributed to his success as a younger man. He has the same perseverance and crusading spirit in overcoming obstacles, whether man made or technological, the same grandiose planning of technological revolutions, and the same enthusiasm." He was not without his faults, however, characterized by a "certain spontaneity and transparency." Another was his failure to realize that he had a heavy Russian accent, so that many people found themselves "translating" for him while he was speaking perfectly good English.

Zworykin had a wide range of nontechnical interests and exhilarating friendships with writers, artists, musicians, philosophers, and

politicians. Many were leaders of the Russian immigrant community. He excelled in the social graces typical of the well-to-do class of
tsarist Russia. He continued to entertain people at home, loved a
small crowd, and always proved to be a charming and gracious host,
with a mischievous twinkle in his eye and a penchant for telling hu-
morous anecdotes about himself. He loved to swim and play tennis,
and he was a dedicated sportsman, who was always accompanied at
home by a well-trained bird dog. He also loved to travel, and brought
back many amusing anecdotes.[37]

As the years passed, the cold winters in Princeton became too much
for him and he started spending the winters in Miami Beach, where
he kept a modest home. He made friends with many of the teaching
staff of the University of Miami and became a visiting professor at the
Center of Theoretical Studies and the Institute for Molecular and
Cellular Evolution. There he worked with the brilliant Dr. Sidney Fox,
who was the director as well as a research professor. But Zworykin
really missed the "good old days" at the RCA laboratories in Princeton. He often complained that retirement was "like being a gelding;
you can get to run with the horses but you don't get to play their
games."[38]

In spite of his noble purposes, he sometimes became involved in
ventures that did little to heighten his personal prestige. Allegedly
he was sometimes quite naive and was taken in by individuals who did
not mind using his name for their own selfish purposes. For instance,
in 1974 he became involved in an electronic acupuncture scheme
promoted by a nonprofit company called Acupulse. Their system
located the acupuncture points in the body by measuring electric
current with probes and a meter. Then, instead of inserting needles,
the acupuncturist applied electric pulses to the body at varying
speeds. Supposedly the electronic approach made it easier to locate
the acupuncture points, thus paving the way for widespread use of
acupuncture. Unfortunately, Acupulse soon ran into financial problems and ceased operations.[39]

In 1977, Zworykin was inaugurated into the National Inventors
Hall of Fame, an honor which was long overdue. Yet this is still a
confused matter. The award cited U.S. Pat. no. 2,139,296 "Cathode
Ray Tube" and stated that "On November 18, 1929, at a convention
of radio engineers, Zworykin demonstrated a television receiver containing his 'kinescope,' a cathode-ray tube with the principal features
of all modern picture tubes. That same year, Dr. Zworykin joined the
Radio Corporation of America in Camden, New Jersey. As the director of their Electronic Research Laboratory, he was able to concen-

trate in making critical improvements to his system. Zworykin's storage principle is the basis of modern TV." The citation is puzzling. First, U.S. Pat. no. 2,139,296 was neither for the Kinescope (U.S. Pat. no. 2,109,245) nor the Iconoscope (U.S. Pat. no. 2,021,907). It was filed February 23, 1934, and issued December 6, 1938. The main feature seemed to be a concave photoelectric surface. There is no mention of storage. In fact, it was simply an improvement on Zworykin's U.S. Pat. no. 2,021,907, filed November 13, 1931, and issued November 26, 1935. (This was the patent for the first real iconoscope with a single-sided target, and why it was not chosen as the basis for the award is a mystery.) Second, as has been indicated earlier, there was no demonstration of the kinescope on November 18, 1929, though the citation stresses that the kinescope that was Zworykin's prime contribution to the technology of television. In any case, his initiation into the National Inventors Hall of Fame was more than justified for the enormous contributions he made to television.[40]

On October 15, 1980, Zworykin received his last major honor, the first Rhein Ring of the Eduard Rhein Foundation of Germany. Rhein, a German writer, publisher, and physicist, personally presented Zworykin with this award, which he had established as a "prize of honor" for persons "who have made significant contributions to audio-visual techniques."[41]

Also in the fall of 1980, Zworykin honored the 122d technical conference of the Society of Motion Picture and Television Engineers with a brief personal address and recorded a taped interview on early TV developments.[42]

On his ninety-second birthday, July 30, 1981, Zworykin clearly showed his disillusionment with his invention. He claimed that he had never "dreamed that TV would become such a pervasive force worldwide, the technique is wonderful. The color and everything is beyond my expectation." However, he added that "I would never let my children even come close to this thing. It's awful what they are doing."[43]

It has been alleged that the last six months of his life were marred by family squabbling. While the decisions about his estate were well established, there was much bitterness about his competence and ability to take care of himself, including claims of undue influence by a male companion who went with him to Florida each year. Yet he never lost the lust for life. Those who knew him during his last years were surprised at the vivacity of a ninety-year-old man who had the vision and sparkle of a wide-eyed college graduate. Vladimir Kosma Zworykin died in Princeton on July 29, 1982, just one day short of his ninety-third birthday. His passing received very little publicity from the nation's newspapers and magazines, and it is unlikely that many editors remembered who he had been or what he had accomplished.[44]

Some confusion continued even after his death. His granddaughter, Nina Melinda Lee, ordered his body cremated, allegedly without consulting either his wife Katherine or his priest, Father John Turkevich. Father Turkevich had been scheduled to perform the funeral service, but the Russian Orthodox Church forbids a service with cremated remains. Instead a memorial service was held in the Princeton University chapel on August 3, 1982. Zworykin's ashes were scattered over Taunton Lakes near the summer home he had so dearly loved.[45]

Also, a few months after Zworykin died, someone at the RCA laboratories cleaned out his office. All of his papers, books, and artifacts disappeared, and all inquiries as to their whereabouts received negative replies. It is to be hoped that they are being safeguarded somewhere.

Katherine Zworykin died on February 18, 1985, at the age of ninety-seven, and the house on Battle Road Circle was sold. All of Zworykin's precious belongings were sold at random. However, a scholarship was set up in his name at Purdue University, and a fitting monument for him is planned by Katherine Zworykin's granddaughter, Iris Blackmore, and her husband. Meanwhile, a memorial plaque in his honor was officially displayed in the city of Morum, his birthplace, on July 25, 1989, his 100th birthday. Finally, as of July 1994, his first wife, Tatiana, who was born on January 15, 1891, was still alive and well in a rest home in California.[46]

Zworykin and his benefactor and longtime colleague David Sarnoff were more responsible for the growth of television than anyone else in the world. Together, they made RCA and television synonymous. This is not intended to denigrate the efforts of such other pioneers as Farnsworth, Shoenberg, Takayanagi, Belin, von Ardenne, Baird, and countless others who worked so hard to put together the pieces of the puzzle entitled *Television*.

Unfortunately, even before David Sarnoff handed over the reins of RCA to his son Robert in 1965, the company had gone into a steep financial decline. So many unsound business decisions had been made that RCA simply was unable to keep up with either the American competition or the younger Japanese companies that were intent on taking over the entire consumer electronics industry. RCA's rash ventures into data processing, publishing, real estate, and so on were so costly that only its success in color television kept it financially alive.

Also, RCA's lead in television technology was slowly slipping from its hands. While it continued to play a major role in the space and defense programs, it was losing out in broadcast technology. The first major blow came from a small West Coast tape recorder manufacturing company, the Ampex Electric Corporation of Redwood City, California. In April 1956, Ampex introduced the world's first practical videotape recorder. Founded by another Russian émigré, Alexander Poniatoff, Ampex had started a research program into video recording in December 1951. Led by Charles Ginsburg and a brilliant team of engineers including Ray Dolby and Charles Anderson, it produced a machine that completely changed the nature of video recording and television broadcasting itself while other would-be contenders such as Bing Crosby Enterprises, the General Electric Company, the time-honored RCA laboratories under the guidance of Dr. Harry F. Olson and even the prestigious BBC research laboratories were all floundering around with unimaginative, high-speed, prosaic devices such as VERA.

Then RCA lost the television camera market with the introduction of the new N. V. Philips Company's Plumbicon tube in 1962. This new lead oxide vidicon tube was so superior for color rendition that it quickly rendered obsolete both RCA's vidicon and its workhorse, the seventeen-year-old three-inch model 5820 image orthicon. At the same time, RCA also lost its major share of the home color receiver market.

Stung by the success of the new Ampex video tape recorder, in September 1969 RCA announced a new method of video recording called "Selectavision: Holotape," which used holograms. It was a total failure and was soon abandoned. Finally, in desperation, RCA embarked on a costly research program to produce a practical low-cost mechanical (capacitive electronic) video disc player. Selecta-Vision: Video Disc was introduced in March 1975, but, after RCA had spent over 580 million dollars on this project, it too was abandoned.

RCA then had a measure of success by adopting the Matsushita VHS (Video Home System), Japanese Victor Company's video tape recording and playback system, which it marketed under the RCA name. But the competition from home and abroad, especially Japan, was just too much for RCA.[47]

It was a blessing that neither Sarnoff nor Zworykin were around in the latter part of 1987, when RCA was destroyed, not by the Japanese electronic industry but by RCA's founder, the General Electric Company. What General Electric had created in 1919, General Electric destroyed in a "friendly" corporate takeover. It is hard to believe that

GE never forgave RCA and especially David Sarnoff for wanting and getting its independence in 1929–30, yet GE seems to have harbored a smoldering resentment for all those years. I can think of no other reason for the act of filicide that occurred.

Financial and business authorities agreed that RCA could have survived alongside General Electric, but this was not to be. RCA was totally destroyed; only the NBC television network was retained. The radio network was sold; the magnificent RCA research laboratories, where so many of these significant developments had taken place, were turned over to the Stanford Research Institute and made to stand on their own. Everything went, including the very name of RCA. This was sold to Thomson CSF, the powerful French electronics concern along with many of RCA's assets. Thomson CSF continued to turn out home television receivers and portable cam-corders using the familiar RCA name and logo.

At this moment (July 30, 1994), General Electric is quietly destroying what is left of David Sarnoff's NBC television network. According to Grant Tinker, former president of NBC, "GE accelerated the deterioration of NBC into a company that is seen to have no higher purpose than making a buck." Tinker concluded, "GE's philosophy toward NBC seemed to be, 'if it ain't broke, break it!'" The wanton destruction of this entity leaves the United States weaker in its efforts to keep up with competition both foreign and domestic. A rather sad end to what had been an outstanding American success story.[48]

Yet, in July 1989, on the one hundredth anniversary of Vladimir Zworykin's birthday, the deep space craft *Voyager 2* was sending back television pictures of the planet Neptune and its moon Triton with magnificent clarity of detail. These images came from two vidicon cameras spawned in the RCA/Zworykin laboratory, though actually built by General Electro-Dynamics of Texas, which even after twelve years in outer space were daily transmitting pictures of awesome definition. So Zworykin's dream that his "progeny" was born for better uses than mundane, banal television has certainly come true.[49]

As *Voyager 2* continues its amazing journey out of our solar system, every picture returned from it is a monument to this dauntless visionary and everything he stood for.

Notes

Material Obtained under the Freedom of Information Act

As part of my research into Zworykin's life I decided to use the resources of the Freedom of Information Act. I have contacted six different government agencies. The agencies and the dates of my requests and receipt of information are as follows:

1. Department of Justice, Federal Bureau of Investigation (FBI). Applied for Feb. 16, 1990; material received Oct. 27, 1993.

2. Department of State. Applied for Aug. 27, 1990; material received July 14, 1992.

3. Department of the Army, Army Intelligence and Security Command. Applied for Oct. 26, 1990; material received Nov. 16, 1993.

4. United States Air Force. Applied for Sept. 21, 1992; material received Jan. 5, 1993.

5. National Archives and Record Services. Applied for Nov. 9, 1990; material received July 4, 1992.

6. Central Intelligence Agency. Applied for Aug. 21, 1992; no material received to date.

In each case, the request was for information about Vladimir Kosma Zworykin. Since I have listed the dates of application and receipt above, this information will be omitted from each citation. Each agency is abbreviated as follows: FOIA/FBI, FOIA/State Dept., FOIA/Air Force, FOIA/Nat. Archives, FOIA/Army, and FOIA/CIA.

Chapter 1: "Father of Television"

1. Zworykin's relationship to Farnsworth has been discussed by several authors. One of the earliest is Steven F. Hofer in "Philo Farnsworth: The Quiet Contributor to Television" (Ph.D. diss., Bowling Green State University, 1977). Hofer was determined to prove that Farnsworth had "priority for a working system of electronic television" (p. 84). See also S. Hofer, "Philo Farnsworth: Television's Pioneer," *Journal of Broadcasting* 23 (Spring 1979),

153–65. Hofer's whole thesis is based on the fact that Farnsworth won the famous patent interference no. 64,027, Philo T. Farnsworth v. Vladimir K. Zworykin, U.S. Patent Office, Washington, D.C., 1933. At issue was Farnsworth's fifteenth claim from his original 1927 patent application, concerning the forming of an "electrical image." This case only proved that Zworykin's iconoscope did work differently than Farnsworth's dissector tube. Based on this, Hofer claims that "these initial (dissector) devices provided the basis for modern television transmitters and receivers" ("Philo Farnsworth: Quiet Contributor," abstract, p. ii). This is not true, as will be proven in this book. Farnsworth's victory is then further misconstrued by Hofer when he shows in "Appendix E" to his dissertation two tubes titled "Image Dissector Tube No. 1428" which are actually Zworykin iconoscopes. A final point, Farnsworth's original 1927 patent application doesn't even show a cathode-ray tube as a receiver.

The heart of the problem is that while most television historians make much of the conflict between the iconoscope and dissector, Zworykin's principal contribution was his picture tube, the kinescope, the major breakthrough that made modern electronic television possible. While it is theoretically possible to have a television system without a camera tube, using either motion picture film or the flying spot scanner, it is not possible to have television without a picture tube. This is the device which is in millions of homes today. And without a doubt, every modern picture tube in use stems from Zworykin's device.

However, I do agree with Hofer that Farnsworth deserves much more credit than he has received (see esp. "Philo Farnsworth: Quiet Contributor," pp. 87–96). He sums up Farnsworth's many contributions to television technology, especially his many basic patents covering low-velocity beam scanning, synchronizing, generating the high voltage from the horizontal scan frequency, and maintaining a constant black level ("Philo Farnsworth: Television's Pioneer," p. 163). But without Zworykin's picture tube and camera tube to build on, these are meaningless.

There is also a two-part article by P. Schatzkin and B. Kiger entitled "Philo T. Farnsworth: Inventor of Electronic Television" (*Televisions* 5 [1977], 6–8 and 17–20). The authors are reckless with their facts, and the article is actually a misguided effort to prove that RCA had stolen most of Farnsworth's ideas and that Farnsworth had been cheated out of his claim to have been the inventor of electronic television. As I indicate in this biography, Farnsworth was indeed ahead of Zworykin with an all-electronic system and great credit is given him, but not to the extent that they assert.

However, Farnsworth was not "first." Zworykin had already built and operated an electronic camera tube in 1924–25, when Farnsworth had not even begun his experiments into television. Even the image dissector had already been invented in Germany in 1925 by Max Dieckmann and Rudolf Hell. And, sadly, the basic system Farnsworth created was an inferior method that did not survive the test of time. As I have often stated, there are plenty of "firsts" to satisfy everyone. There is no question that our modern system of television owes much more to Zworykin than it does to Farnsworth.

A third effort to make Farnsworth's case is Joseph Udelson's *Great Television Race: A History of the American Television Industry 1925–1941* (University of Alabama Press, 1982). According to Dr. Albert Rose, who wrote me about his correspondence with Udelson, he started out to solve the problem of who was first, Zworykin or Farnsworth, but Udelson could not resolve that problem and went on to write a limited history of American television. Without a technical background he was simply lost. For instance, he had no understanding or appreciation of the work of Charles Francis Jenkins and called him a "tinkerer" (p. 24). Yet Jenkins was America's first television pioneer and made many significant contributions—for instance, Jenkins was ahead of Baird, who is often cited as giving the first demonstrations of television. Jenkins's patents were quite ingenious, and he was among the first with a patent on the storage principle, which was a crucial development in television. (See Chapter 7, pp. 92–93.) Sadly this label has stuck to Jenkins and has been quoted in several subsequent books.

For a more successful consideration of Farnsworth's struggle for his rightful place in the history of television, see the two-part article by Frank Lovece, "Zworykin v. Farnsworth" (*Video* 9, no. 5 [Aug. 1985], 68–72, 117–18; no. 6 [Sept. 1985], 96–98, 135–38). According to Lovece, Farnsworth was the victim of RCA's insistence that Zworykin and RCA had virtually invented television single-handedly. Both pioneers get their due in a well-researched and well-written article.

See also, Albert Abramson, "Pioneers of Television—Philo Taylor Farnsworth," *Journal of the Society of Motion Picture and Television Engineers* 101 (Nov. 1992), 770–84.

2. Each of these pioneers will be treated in turn.

3. "Sarnoff Named 'Father of American Television,'" *New York Times,* Dec. 12, 1944, p. 25:6. (See Chapter 11, p. 176.)

4. Much of this material came from four interviews with Zworykin, two in July 1976 in Naples, California, and two in Princeton in July 1977 and 1978. More details came from extensive interviews and correspondence with Al Pinsky, the public relations manager for the David Sarnoff Research Center, from 1976 through 1982, and others of Zworykin's co-workers. These included Les Flory, Arthur Vance, Harley Iams, George Morton, and Dr. W. D. Wright, among others. I am particularly grateful to the material I received from Donald and Iris Blackmore of Los Alamitos, California (Iris Blackmore is Zworykin's second wife's granddaughter) for more details about his later years and his passing. Unfortunately, I received no cooperation from his surviving daughter, who declined to talk to me.

5. "TV Trends Displease Inventor," *New York Times,* Sept. 19, 1964, p. 30:3; "Notes on People: Vladimir K. Zworykin," *New York Times,* July 31, 1974, p. 39:2; R. Haitch, "TV Pioneer, 85, Finds Retirement Is Not to His Taste," *New York Times,* Nov. 17, 1974, p. 97:1; "Vladimir Zworykin: Father of Electronic

Television Tells What He Likes Most about Television, the Turn-off Switch," *TV International,* 24 (Feb. 1980), 15–17; "TV Turns Off Its Father," *New York Times,* July 31, 1981, p. 16:3.

6. M. Weil, "Vladimir Zworykin, 92, Father of Television, Dies," *Washington Post,* July 31, 1982, p. B4A.

Chapter 2: Growing Up in Tsarist Russia

1. The bulk of this material comes from Zworykin's unpublished manuscript, which he kindly gave to me on July 21, 1976. It is 105 pages long and was obviously dictated by him to his secretary at RCA. It will be known as VKZ MS; see esp. pp. 1–10. Other information comes from the English translation of the third edition of the *Great Soviet Encyclopedia,* 31 vols. (New York: Macmillan, 1973), referred to hereafter as *Sov. Enc.* Vladimir Kosma Zworykin's date of birth is given as July 30, 1889 (vol. 9, p. 702). However, there is some question about the date, as I found out from a Freedom of Information Act request to the U.S. State Department. I had written them about Zworykin's travels. In FOIA/State Dept. PO34, Zworykin gave the date of his birth as June 13, 1889, and in FOIA/State Dept. PO33 he gave it as July 7, 1889. When asked about this by the State Department on Aug. 19, 1937, he claimed that the difference was due to use of the old Russian calendar, which had a difference of thirteen days (FOIA/State Dept. PO31, letter from Zworykin to Shipley, Passport Division, Aug. 20, 1937). However, FOIA/FBI -180, a report from a special agent (whose name has been blanked out) that is dated Oct. 3, 1945, states that Zworykin's certificate of naturalization shows his date of birth as July 17, 1888.

2. Murom is the official Russian name of Zworykin's birthplace (*Sov. Enc.,* vol. 17, p. 702). In French, the town is called Mourum, and that version is the one most widely used, including by Zworykin. I have decided to avoid confusion about places and names by using the transliterations found in the *Soviet Encyclopedia.* An interesting fact is that today Murom makes industrial diesel locomotives and refrigerators, while nearby Aleksandrov makes radio apparatus and television sets (*Sov. Enc.,* vol. 1, p. 225).

3. VKZ MS, pp. 1–10.

4. For further information on the situation in Russia at the time, see Michael T. Florinsky, *The End of the Russian Empire* (New Haven: Yale University Press, 1931) and *Russia: A History and Its Interpretation,* vol. 2 (New York: Macmillan, 1953); *Outline History of the USSR* (Moscow: Foreign Language Publishing House, 1960); N. V. Riasonovsky, *A History of Russia,* 2d ed. (New York: Oxford University Press, 1969); Albert Rhys Williams, *Journey through Revolution* (Chicago: Quadrangle Books, 1969); Victor Serge, *Year One of the Russian Revolution* (Chicago: Holt, Rinehart and Winston, 1972); and Adam B. Ulam, *Russia's Failed Revolutions* (New York: Basic Books, 1981).

5. VKZ MS, pp. 16–23; for information on Abram Fedorovich Ioffe (1880–1960), see the *Soviet Encyclopedia.* Ioffe was quite revered in the Sovi-

et Union, and the article is full of praise for him (*Sov. Enc.,* vol. 10, p. 364). Zworykin later claimed that one of Ioffe's lectures "fascinated me immediately and I thought: Physics—this is what I want, this is my life" (V. K. Zworykin, "Engineering—Service to Humanity," Third Founders Lecture, presented before the Academy of Engineering, April 25, 1969.

  6. VKZ MS, pp. 23–24.

  7. Boris L'Vovich Rozing was born on May 5, 1869. He taught at the St. Petersburg Institute of Technology until 1918, then conducted research at the Leningrad Experimental Electrotechnical Laboratory from 1924 to 1928 and at the Central Laboratory for Wire Communications from 1928 to 1931. In 1892, Rozing expounded a theory on the existence of a molecular field that causes the spontaneous magnetism of ferromagnets. He started work on his television project in 1904 and continued work on it until he died. In 1931, he was exiled to Archangel, where he worked at Archangel's Forestry Technology Institute. He died there on Apr. 20, 1933 (*Sov. Enc.,* vol. 22, p. 313). See also P. K. Gorokhov, *B. L. Rozing: Founder of Electronic Television* (Moscow: Hayka, 1964).

  In 1961 Zworykin told a different story of how he met Rozing. See V. K. Zworykin, "My Most Exciting Moment," *Popular Mechanics* 115 (June 1961), p. 80.

  8. For the early history of television, see Albert Abramson, *The History of Television, 1880 to 1941* (Jefferson, N.C.: McFarland, 1987).

  Rozing's patents include Russ. Pat. no. 18076 (applied for July 25, 1907; issued Oct. 30, 1910), DRP Pat. no. 209,320 (applied for Nov. 26, 1907; issued Apr. 24, 1909), Br. Pat. no. 27570/1907 (applied for Dec. 13, 1907; issued June 25, 1908). It is interesting to note that the British patent, which was applied for last, was the earliest to be issued.

  9. This information comes from a variety of sources. See esp. Robert Grimshaw, "The Telegraphic Eye," *Scientific American* 104 (Apr. 1, 1911), 335–36. See also "Prof. Rozing's 'Electric Eye'—A New Apparatus for Television," *Scientific American Supplement* 71, no. 1850 (June 17, 1911), 384; "An Important Step in the Problem of Television," *Scientific American* 105 (Dec. 23, 1911), 574; E. Ruhmer, "Rosing's System of Telephoty," *Zeitschrift für Schwachstromtechnik* 5 (Apr. 1911), 172–73; and E. Ruhmer, "Der Rosingsche Fernseher," *Die Umschau,* 1911, pp. 508–10.

  There is a picture of Rozing's television apparatus in Gorokhov, *B. L. Rozing,* p. 55. There is also a picture of his laboratory showing the various details described (p. 71). See also "Inventors of Television: Boris Rosing," *Radio-Electronics,* Apr. 1966, 62, and VKZ MS, pp. 24–25.

  10. Max Dieckmann, "Fernübertragungseinrichtunger hoher Mannigfaltigkeit," *Prometheus* 20, no. 1010 (Mar. 1909), 337–41. See also Max Dieckmann, "The Problem of TeleVision—A Partial Solution," *Scientific American Supplement* 68, no. 1751 (July 24, 1909), 61–62; and B. Von Czudrochowski, "Das Problem des Fernsehens," *Zeitschrift für Physik und Chemie* 4, (July 1909) pp. 261–65.

11. E. Ruhmer, "Das elecktrische Fernsehen," *Zeitschrift für Schwach-stromtechnik* 3 (1909), 393–95; see also A. Gradenwitz, "Un appareil de télévision," *La Nature* 38 (Jan. 29, 1910), 142–43.

12. H. D. Varigny, "La vision à distance," *L'Illustration* no. 3485 (Dec. 11, 1909), 451–52.

13. Grimshaw, "Telegraphic Eye," pp. 335–36.

14. A. A. Campbell Swinton, "Presidential Address," *Journal of the Röntgen Society* 8, no. 30 (Jan. 1912), 13; "Distant Electric Vision," *The Times* (London), Nov. 15, 1911, p. 24b; A. A. Campbell Swinton, "Distant Electric Vision," *Nature* 78, no. 2016 (June 18, 1908), 151

Alan Archibald Campbell Swinton (1863–1930) was a well-known British electrical engineer. It was he who introduced young Guglielmo Marconi to William Preece of the General Post Office, who encouraged his efforts in radio-telegraphy in Great Britain. For Campbell Swinton's obituary, see *Nature* 125, no. 3149 (Mar. 8, 1930), 356–57. See also J. D. McGee, "The Contribution of A. A. Campbell Swinton, F.R.S., to Television," *Royal Society, Notes and Records* 32 (July 1977), 91–105; T. H. Bridgewater, *A. A. Campbell Swinton,* Royal Television Society monograph 1 (London: Television Society, 1982); and H. Winfield Secor, "Television or the Projection of Pictures over a Wire," *Electrical Experimenter* 3, no. 4 (Aug. 1915), 131–32, 172–74.

15. VKZ MS, pp. 24–27.

16. On Paul Langevin (1872–1946), see *Grand Dictionnaire Encyclopédique Larousse* (Paris: Librairie Larousse, 1984), vol. 6, p. 6122.

17. VKZ MS, pp. 27–28.

Chapter 3: World War I

1. VKZ MS, p. 29. This is one of the mysteries in Zworykin's life. It is hard to understand why a scientist would be enlisted as a private, in view of the shortage of such trained men. Nor do I know why he was not sent immediately to Officers Radio School for training.

2. VKZ MS, pp. 29–35. Ilia Emmanuel Mouromtseff (1881–?) was born in St. Petersburg and graduated from the Engineering Academy there in 1906 with an M.A. He received a Diploma-Ingenieur degree from the Institute of Technology in Darmstadt in 1910. He joined the radio laboratory of the Russian Signal Corps in 1911 and was assistant chief of the Officers Radio School until 1917. He then came to the United States in 1917 as a member of the supply division attached to the Russian embassy in Washington. He seems to have joined Westinghouse Electric in 1923. In 1947 he began teaching at Upsala College in East Orange, N.J.

3. VKZ MS, p. 35. The April 17, 1916, date is given in "Vladimir K. Zworykin" in *Current Biography: Who's News and Why,* ed. Anne Rothe (New York: H. W. Wilson, 1949), p. 656. But in FOIA/State Dept. PO14, a passport application dated July 12, 1950, he gives the date as April 1917. This is just one example of the many problems I have had with his manuscript. I am

sure that Zworykin was dictating from memory and had very little in the way of notes to help him.

4. VKZ MS, pp. 36–37. See the *Soviet Encyclopedia* for entries on Nikolai Dmitrievich Papaleksi (1880–1947) (vol. 15, pp. 223–24) and Leonid Isaakovich Mandel'shtam (1879–1944) (vol. 15, p. 411). Interestingly enough, the former manager of the Russian wireless telegraph and telephone company, which was founded in 1908 by the Marconi Company, was none other than Isaac Shoenberg. Shoenberg had left Russia just before the war, to seek employment in England. Although there have been several attempts to connect Zworykin to Shoenberg at this time, there is no evidence that they ever met before 1933 (see Chap. 7, n. 63). After Shoenberg left, Serge M. Aisenstein (1884–1962) was director and technical manager of the Russian wireless company until 1917. After the revolution he settled in England. See S. M. Aisenstein, "A History of Some Foundations of Modern Radio-Electronic Technology," in the "Letters" section, *Proceedings of the Institute of Radio Engineers* 46 (Apr. 1958), 780–81.

5. VKZ MS, p. 37. In addition to the written material given me by Zworykin, I have also collected many of his patent applications and interferences and his correspondence with the U.S. Patent Office. Such material is cited as VKZ Pat. File, plus a description of the nature of the material. See Zworykin's testimony in the interference proceedings between him and Farnsworth in VKZ Pat. File, Pat. Int. no. 64,027 (Farnsworth v. Zworykin), Record for Zworykin. On p. 65 of this document, Zworykin claims that this occurred while he was working for the Russian Wireless Telephone and Telegraph Company "in 1917 as the actual beginning of his work on television." In Pat. Int. nos. 64,027 and 73,200, he also claimed that he discussed work "on this problem and proposed to the Director S. M. Eisenstein and Chief Engineer A. M. Cheftel to start this development. This moment I think can be called the actual beginning of this work." Also in Pat. Int. no. 73,203 (Iams v. Ballard v. Farnsworth) he stated that "when the war was over" he would work in their laboratories. This was meant to give him a priority of 1917 for the conception of his idea for an all-electric television system.

6. Jesse D. Clarkson, *A History of Russia* (New York: Random House, 1961), p. 432. This was Dec. 16/17, 1916 (old style), Dec. 29/30 (new style).

7. VKZ MS, pp. 37–43.

Chapter 4: Revolution and Escape

1. VKZ MS, p. 44.

2. Florinsky, *Russia*, vol. 2, p. 1380.

3. VKZ MS, p. 45. Guchkov was a very powerful, ambitious industrialist from Moscow and the leader of the Octobrist party. After the revolution, his position as minister of war and the navy was taken over by none other than Leon Trotsky. See Adam B. Ulam, *A History of Soviet Russia* (New York: Praeger, 1976), pp. 25–40.

4. VKZ MS, pp. 44–47
5. VKZ MS, p. 47.
6. VKZ MS, pp. 47–59; VKZ Pat. File, Pat. Int. no. 64,027 (Farnsworth v. Zworykin), Record for V. K. Zworykin, p. 66.
7. VKZ MS, p. 59.
8. VKZ MS, p. 59–66. Zworykin's escape route is very confusing. I have accepted his version most of the time, but there are times when I have had to deal with facts that just didn't fit. For instance, on page 60 he said that he was going to try his luck by going through the northern mining region and finally reached the "Nadejdinski Mines." According to the map, however, this is not north in the Urals but far to the west near Irkutzk, which is north of the route of the Trans-Siberian Railroad. So it is possible that Zworykin is recalling the route he took from Vladivostok to Omsk after he had returned from the United States. Clearly, his manuscript must be used carefully, though it must also be admitted that in the middle of a civil war it is quite hard to go directly to one's destination and Zworykin never makes it clear whether he was trying to reach Archangel or Vladivostok.

One thing is certain, he was trying to get to Omsk, which was still the headquarters of the provisional government. Here Zworykin expected to get orders permitting him to leave the country. See S. V. Novakosky, "100th Birthday of V. K. Zworykin," *TKT* (Cinema and Television Technique) no. 7 (1989), 64–68. Novakosky notes that Zworykin was sent on an official mission that enabled him to leave the country but does not mention that he was wanted by the police for failing to register as a former officer of the Russian army.

According to the *Soviet Encyclopedia,* Tolmachev explored the "Tunguska Coalfields," also near Irkutzk, between 1915 and 1917 (vol. 26, p. 424). While there is no reference to his being in the Urals, there is no reason why he could not have been on a scientific expedition as Zworykin's manuscript indicates.

9. VKZ MS, p. 66; see also Harrison E. Salisbury, *Russia in Revolution, 1900–1930* (Chicago: Holt, Rinehart and Winston, 1978), p. 465. According to Florinsky, the Japanese landed in Archangel in April 1918 and the British, French, and Americans arrived in August 1918. According to Salisbury, the Japanese landed in Archangel in April 1918 and the British, French, and Americans prepared for armed intervention. According to Williams, Vladivostok was also occupied by French, British, Japanese, and American troops in June 1918 (*Journey through Revolution,* p. 57). The American embassy left Vologda on July 25, 1918, and was set up in Archangel on August 3. It seems that the Soviets had asked Ambassador Francis to move his embassy from Petrograd to Moscow, but he refused. In addition, Francis had given a rousing anticommunist speech on July 4, 1918, which upset the Soviets no end (*New York Times,* Aug. 13, 1918, p. 4:2). The American vice consul was J. K. Caldwell (*New York Times,* May 6, 1918, p. 2:6). Zworykin had certainly arrived at the right time.

10. VKZ MS, pp. 67–69.

11. VKZ MS, p. 70.

12. VKZ MS, pp. 70–72. The date of his return to New York from his second trip is given in VKZ Pat. File, Pat. Int. no. 64,027 (Farnsworth v. Zworykin), Record for Zworykin, p. 73.

Chapter 5: An Invitation from Westinghouse

1. VKZ MS, p. 73; VKZ Pat. File, Pat. Int. no. 64027, Record for Zworykin, p. 74. Kolchack was outlawed by the Bolshevik government on Aug. 16 (*New York Times,* Oct. 30, 1918, p. 2:6) and captured on Nov. 15 (*New York Times,* Nov. 16, 1918, p. 5:3. For details, see Harrison E. Salisbury, *Russia in Revolution, 1900–1930* (New York: Holt, Rinehart and Winston, 1978), p. 226.

2. VKZ MS, p. 73.

3. This comes from a variety of sources. See esp. Gleason L. Archer, *History of Radio to 1926* (New York: American Historical Society, 1938), pp. 128–89, and Gleason L. Archer, *Big Business and Radio* (New York: American Historical Society, 1939), pp. 4–13.

4. David Sarnoff (1891–1971) played the major role in the history of television. As will be seen in this book, his ambitions for RCA coincided with his drive for power. His determination and his support for Zworykin made our modern television system possible. Still, though Zworykin spent over fifty years with RCA, there is no evidence of any great personal relationship between him and Sarnoff. Sarnoff's humble origins were in sharp contrast to Zworykin's birth into the landed, ruling class, so perhaps their backgrounds were too diverse to allow them to be friends. Yet they complemented each other quite well and shared a common purpose: to develop television to its highest degree. As we shall see, they succeeded beyond their wildest dreams. Books about Sarnoff include Carl Dreher, *Sarnoff, an American Success* (New York: Times Books, 1977); Kenneth Bilby, *The General: David Sarnoff and the Rise of the Communications Industry* (New York: Harper and Row, 1986); Robert Sobel, *RCA* (New York: Stein and Day, 1986); and his official biography by Eugene Lyons, *David Sarnoff* (New York: Harper and Row, 1966). In addition to published sources, I have used Elmer Bucher's "official" history, "Television and David Sarnoff," part of a thirty-volume history of David Sarnoff entitled "History of Radio and Television Development in the U.S.A." (unpublished manuscript, 1956, David Sarnoff Library, Princeton). Television is covered in part 12, "Black and White Television," chaps. 1–14. Elmer Bucher was a longtime associate and personal friend of Sarnoff and his account is often a "romantic" (read biased) version of television history. According to Sobel, it is "wholly admiring, even adoring" (*RCA,* p. 41). I was unable to check Bucher's sources as the Sarnoff papers were not available to me. Only his speeches seem to be on file. (I did not go into other aspects of Bucher's history as I was only concerned with RCA's contribution to the history of television). Bucher's account of Sarnoff and RCA will be referred to hereafter as Bucher, "Television and Sarnoff."

5. Archer, *History of Radio,* pp. 110–13.

6. Archer, *Big Business and Radio,* pp. 3–13.

7. VKZ MS, p. 79; Archer, *History of Radio,* pp. 191–204.

8. VKZ MS, pp. 73–74.

9. Archer, *History of Radio,* pp. 210–11.

10. VKZ MS, pp. 74–77. See V. K. Zworykin, U.S. Pat. no. 1,643,734 (filed June 19, 1922; issued Sept. 27, 1927). This was for a "thermocouple" and was assigned to C and C Developing Company. See V. K. Zworykin, U.S. Pat. no. 1,484,049 (filed Aug. 22, 1922; issued Feb. 19, 1924) for "High Frequency Signaling Apparatus."

11. VKZ MS, p. 77. This was an unusual, if not illegal, arrangement, but it was one way for Westinghouse to keep Zworykin's patents out of the patent pool. See VKZ Pat. File, Pat. Int. no. 64,027 (Farnsworth v. Zworykin), Brief for Dr. Zworykin, p. 44.

12. V. K. Zworykin, U.S. Pat. no. 2,141,059 (filed Dec. 29, 1923; issued Dec. 20, 1938). See also VKZ Pat. File, patent application.

13. Edvard-Gustav Schoultz, Fr. Pat. no. 539,613 (filed Aug. 21, 1921; issued June 28, 1922); Boris Rtcheouloff, Russ. Pat. no. 3,803 (filed June 27, 1922; issued Oct. 31, 1927). For Campbell Swinton's discussion of an electric camera tube, see his "Presidential Address," p. 10. In 1928, Campbell Swinton took notice of Zworykin's 1925 patent ("Television by Cathode Rays," *Modern Wireless* [London], 9 [June 1928], 595–98). (See n. 33 below.) In this article he has nothing but praise for Zworykin, though he does state that "where a gauze grid is employed, and a large number of small photoelectric cells, similar to my suggestions of 1911 and 1924, the cells are composed of minute globules of specially prepared potassium hydride in an atmosphere of argon" (p. 596). There seems to be no question in Campbell Swinton's mind that it was indeed akin to his earlier ideas. He was unaware at the time of Zworykin's original Dec. 1923 patent application, which was still in the Patent Office, that specified a layer of photoelectric material, not minute globules.

14. See Marcus J. Martin, *The Electrical Transmission of Photographs* (London: Sir Isaac Pitman & Sons, 1921), diagram on p. 103. Since Campbell Swinton's plan had been revealed first in 1911, then in 1915, and finally in 1921, it is hard to believe that Zworykin, as an ardent television researcher, had not come across it before 1923. This could explain why Zworykin described a target with a continuous film, which was different from Campbell Swinton's target, which was a mosaic of rubidium cubes. Of course, this layer of photoelectric material on the target was to cause Zworykin great problems in getting his 1923 patent application through the U.S. Patent Office. For the charge that Zworykin was influenced by Campbell Swinton's design, see G. R. M. Garret and A. H. Mumford, "The History of Television," *Proceedings of the Institute of Electrical Engineers* (London) 99 (1952), pt. 3a, 25–42 (paper presented Apr. 28, 1952). Zworykin's reply was printed as a corrigendum in the same issue (p. 24). Here Zworykin states that he could not have had any inspiration from Campbell Swinton's "1925" (it was actually 1924, see n. 22 below) paper as he had applied for U.S. Pat. no. 2,022,450

in Dec. 1923. This was not true, for this patent was a division (filed Nov. 21, 1931) of the original 1923 patent application. See V. K. Zworykin, U.S. Pat. no. 2,022,450 (originally filed Dec. 29, 1923; divided on Nov. 21, 1931; issued Nov. 3, 1935). This had been amended to specify "globules," which was not part of the original application. This patent was cited as the 1923 patent application which was not granted until Dec. 20, 1938. Another mystery here is that this important patent apparently was never granted anywhere else in the world, especially England, France, or Germany; at least I have never been able to find a record of such a patent. It is possible that it was filed overseas and was rejected or dropped for a variety of reasons.

15. On the other hand, as we will see, Zworykin was certainly not adverse to using and improving the ideas of other researchers.

16. VKZ Pat. File, Pat. Int. no. 64,026, Record for V. K. Zworykin, p. 4. Here he claims that he did reduce it to practice between Jan. 1 and June 2, 1924. I believe that this indicates the beginning of the camera tube project. It is almost certain that Campbell Swinton's April 1924 articles provided the impetus for this.

17. Édouard Belin (1876–1963), French engineer and physicist, was born in Vesoul, France, and died in Territet, Switzerland. He applied for his first facsimile/television patent in 1903 and sent his first successful facsimile picture between Paris and Lyon on Nov. 9, 1907. See Fr. Pat. no. 339,212 (applied Dec. 8, 1903; issued Jan. 10, 1905) and "The Problem of Television," *Scientific American Supplement* 63, no. 1641 (June 15, 1907), 26,292. For a detailed description of Belin and his life's work, see Bernard Auffray, *Édouard Belin: Le père de la télévision* (Paris: Les clés du monde editeurs, 1981). This book was kindly furnished me by Jacques Poinsignon, the ranking French television historian. See also "Belin Shows Tele-Vision," *New York Times,* Dec. 2, 1922, p. 1; Robert E. Lacault, "The Belin Radio-Television Scheme," *Science and Invention* 10, no. 12 (Apr. 1923), 1166–1217. This article probably did the most to rekindle interest in television during this period after the war.

Belin applied for his first real (as opposed to facsimile) television patent in 1922. Until this time there was very little distinction made between a slow scan picture system (facsimile) and a rapid scan system (television) in the patent offices around the world. See, for example, Édouard Belin, Fr. Pat. no. 571,785 (filed Dec. 27, 1922; issued May 23, 1924); Georges Valensi, Fr. Pat. no. 577,762 (filed Dec. 29, 1922; issued Sept. 10, 1924); Alexandre Dauvillier, Fr. Pat. no. 592,162 (filed Nov. 29, 1923; issued July 28, 1925); Denes von Mihaly, Fr. Pat. no. 546,714 (filed Jan. 22, 1922; issued Nov. 22, 1922); Dionys von Mihaly, DRP Pat. no. 422,995 (filed Jan. 25, 1923; issued Dec. 25, 1925). See also Nicholas Langer, "A Development in the Problem of Television," *Wireless World and Radio Review* 11 (Nov. 11, 1922), pp. 197–201.

18. C. F. Jenkins and T. Armat, U.S. Pat. no. 586,953 (issued July 20, 1897); Charles Francis Jenkins, U.S. Pat. no. 1,544,156 (filed Mar 13, 1922; issued June 30, 1925). See also C. Francis Jenkins, *The Boyhood of an Inven-*

tor (Washington, D.C.: privately published, 1931), and Albert Abramson, "Pioneers of Television—Charles Francis Jenkins," *Journal of the Society of Motion Picture and Television Engineers* 95 (Feb. 1986), 224–38. This article covers Jenkins's work on the early motion picture projector as well as his early work on television in the United States.

Charles Francis Jenkins (1867–1934) was the first American pioneer of television and had a working system before John Logie Baird did, yet for some reason his exploits have gone unsung. My 1987 attempt to have Jenkins installed in the National Inventors Hall of Fame was rejected because I couldn't even get enough votes to have him considered. On Jenkins's 1923 demonstration of his system, see Hugo Gernsback, "Radio Vision," *Radio News* 5 (Dec. 1923), 681–824; and Watson Davis, "The New Radio Movies," *Popular Radio* 4 (Dec. 1923), 436–43.

19. Charles Francis Jenkins, *Animated Pictures* (Washington, D.C.: privately printed, 1898); Charles Francis Jenkins, *Vision by Radio: Radio Photographs: Radio Photograms* (Washington, D.C.: Jenkins Laboratories, Inc., 1925); Charles Francis Jenkins, *Radiomovies: Radiovision Television* (Washington, D.C.: Jenkins Laboratories, Inc., 1929); and Charles Francis Jenkins, "Radio Movies," *Transactions of the Society of Motion Picture Engineers* no. 21 (Aug. 1925), 7–11. In this paper, which was presented on May 18, 1925, Jenkins thanked both General Electric and Westinghouse Electric officials.

Jenkins tried to sell several of his television patents to General Electric but Dr. Ernst Alexanderson, head of GE's television research program, thought he could build a better system without purchasing any of them. However, at one time Alexanderson was interested in Jenkins's U.S. Pat. no. 1,521,189 (applied for Aug. 30, 1922; issued Dec. 30, 1924) for the transmission of motion picture film, though GE never purchased it. I was allowed to inspect the Alexanderson file at the Collections of the Schaffer Library, Union College, Schenectady, N.Y., in July 1980. This will be known as the Alexanderson File.

See Alexanderson file, Alexanderson to E. W. Allen, Dec. 10, 1924; Alexanderson to Huff (patent dept.), Apr. 8, 1929; and Alexanderson to Lunt, Oct. 28, 1927, about the McFarlan Moore neon lamp. McFarlan Moore was at the incandescent lamp department of General Electric at Harrison, N.J.; see Alexanderson file, McFarlan Moore to Roberts (general manager, GE lamp dept.), Oct 18, 1927.

Dr. Ernst F. W. Alexanderson (1878–1975) was born in Sweden and educated there and in Germany. He started work with General Electric in 1904, and became their chief engineer. His greatest accomplishment was the high-frequency radio alternator that bears his name.

20. The quote is from David Sarnoff, *Looking Ahead: The Papers of David Sarnoff* (New York: McGraw-Hill Book Company, 1968), p. 88. According to Lawrence Lessing, "As early as 1925 David Sarnoff had set his company's sights on being first in the commercial development of television" (*Man of High Fidelity: Edwin Howard Armstrong* [Philadelphia: J. B. Lippincott, 1956], p. 231). As we now know, it was as early as April 1923.

21. See J. L. Baird and Wilfred E. L. Day, Br. Pat. no. 222,604 (applied for July 26, 1923; issued Oct. 9, 1924), and J. L. Baird, "An Account of Some Experiments in Television," *Wireless World and Radio Review* 14 (May 9, 1924), 51–65. This article was originally a paper that Baird presented on Mar. 26, 1924.)

For information about John Logie Baird (1888–1946), see any of the following biographies and his autobiography. His legend in Great Britain is such that even though his biographers all tell the same basic story, their emphases differ depending on the author's viewpoint. Baird has become a folk hero in the history of English television, and there is little distinction made between his real and imagined contributions to the development of television as a viable instrument of communication. Baird's contribution to television was summarized best by British television historian Tony Bridgewater, who worked for Baird from 1928 to 1932. According to Bridgewater, "His contribution was to show the possibility of television. To get it going and to get it known, to interest the BBC. He was a stimulus a catalyst and, as a personal achievement of course, it was pretty stupendous to get it going at all" (quoted from Bruce Norman, *Here's Looking at You: The Story of British Television, 1908–1939* [London: BBC, 1984], p. 141). In my opinion, Baird was the first to popularize the concept of television in England, though Campbell Swinton was twenty-five years ahead of the times and never lost the opportunity to promote electric television, though without ever producing anything to further his ideas. Baird certainly tried hard to build working equipment for demonstration purposes, and, to keep financial backers happy, his partners made sure that the public was always informed of his various schemes. This also stimulated other inventors to make progress along different lines. Baird's allegiance to mechanical television was his downfall, but his sincerity and integrity have never been questioned.

The standard biographies of Baird are Ronald F. Tiltman, *Baird of Television: The Life Story of John Logie Baird* (London: Seeley Service, 1933); Sydney Moseley, *John Baird: The Romance and Tragedy of the Pioneer of Television* (Long Acre: Odhams Press, 1952); Margaret Baird, *Television Baird* (Pretoria: Haum, 1973); and Geoff Hutchinson, *Baird: The Story of John Logie Baird, 1888–1946* (privately printed, 1985). See also John Logie Baird, *Sermons, Soap and Television: Autobiographical Notes* (London: Royal Television Society, 1988).

22. A. A. Campbell Swinton, "The Possibilities of Television with Wire and Wireless," *Wireless World and Radio Review* 14 (Apr. 9, 1924), 51–56; (Apr. 16, 1924), 82–84; and (Apr. 23, 1924), 114–18. (The paper was presented on Mar. 26, 1924.)

23. The earliest memo pertaining to television that I can find in the GE files is that of Alexanderson to Dunham (of the patent dept.), entitled "Television by Radio," July 24, 1924, Alexanderson File. See also Alexanderson File, Alexanderson to Lunt, Jan. 17, 1925. The first television patent that I can find from General Electric is C. H. Hoxie, Br. Pat. no. 240,463 (filed Sept. 23, 1924; issued Nov. 25, 1926). There were three important television patents filed for GE in 1925, all coming from Camille Sabbah, a research

engineer. They covered the three important variations of electric camera tubes. See C. A. Sabbah, U.S. Pat. no. 1,706,185 (filed May 27, 1925; issued Mar. 19, 1929) for photovoltaic target; C. A. Sabbah, U.S. Pat. no. 1,694,982 (filed May 27, 1925; issued Dec. 11, 1929) for photoelectric target; and C. A. Sabbah, U.S. Pat. no. 1,747,968 (filed May 27, 1925; issued Feb. 18, 1930) for photoconductive target. In spite of the importance of these patents, Sabbah lost three patent interferences to Zworykin of Westinghouse, see VKZ Pat. File, Pat. Int. nos. 54,922, 54,923, and 55,448. Although General Electric and Westinghouse owned RCA jointly, their rivalry was deep and often bitter. They did not trust each other (or RCA) in spite of the fact that they were supposed to cooperate in the development of new products. Alexanderson was especially wary of sharing secrets with Westinghouse.

Dr. Benford's system was still being improved upon in June 1926. Motion picture film scanned by a Benford crystal moved slowly forward in front of a stationary photoelectric cell. At the receiver, a similar ten-sided glass crystal in front of an oscillograph (light control) focused a beam of light into a camera whose film was moved slowly forward. The Benford crystal turned out to be impractical and was abandoned. So Alexanderson turned to the old reliable Nipkow disc as the easiest way to generate a television signal. Alexanderson felt that the most profitable use of television in the near future would be the transfer of moving picture films over long distances by radio (Alexanderson File, Alexanderson to Lunt, June 2, 1926.)

24. On August Karolus (1893–1972), assistant professor at the Physical Institute of the Leipzig University, see Gerhart Goebel, "Das Fernsehen in Deutschland bis zum Jahre 1945," *Archv für das Post- und Fernmeldewesen* no. 5 (Aug. 1953), 278. See also A. Karolus, Br. Pat. no. 235,857, (filed June 20, 1924; issued Aug. 12, 1925).

Manfred von Ardenne claims his work was stimulated in 1924 with reports of Baird's experiments ("Evolution of the Cathode-Ray Tube," *Wireless World* 66, no. 4 (Jan. 1960), 28–30, 31.

25. Kenjiro Takayanagi (1899–1990) was Japan's pioneer inventor. He claimed to have became interested in the problem of television during the period between 1922 and 1924. He acknowledged that Campbell Swinton's 1924 article on electrical television encouraged his work. See "Research and Development of All Electronic Television System, Part 1: Discovery of Television and Decision of a Research Theme" (date unknown). This paper was given to me by Professor Takayanagi when I interviewed him in July 1981. It will be referred to hereafter as the Takayanagi File. In it, Takayanagi claims to have built his first camera tube in 1924 but that "due to our vacuum technique" the tube did not work (p. 4). Unfortunately, there are no lab notes or papers to support this early claim. If true, this would give him priority over Zworykin, who built his first camera tube in 1925–26. This is only one of many challenges to Zworykin's claims of priority.

26. VKZ MS, p. 79.

27. VKZ Pat. File, Pat. Int. no. 69,135, Preliminary Statement of Vladimir K. Zworykin, Nov. 13, 1934.

28. This information came from the files of the Westinghouse Electric research laboratories in East Pittsburgh, which I was allowed to inspect on June 20, 1980. This material will be known as the Westinghouse File. Westinghouse *Research Report* R-429A, marked confidential and dated June 25, 1926, was entitled "Problems of Television: Closing Report" and was signed by Zworykin. (See fig. 11.) The report confirms that Zworykin had indeed built, and presumably operated, camera tubes before the 1926 date. Until I unearthed this memorandum, I had serious doubts about most of Zworykin's claims to have built and demonstrated a television system to management at Westinghouse, since he had made so many conflicting statements that it was almost impossible to ferret out the truth. For instance, the reference to a cloth of quartz or Pyrex certainly indicates that he knew that a solid target plate (as specified his 1923 patent application) presented many insurmountable difficulties. So he changed to an open mesh screen which gave him the results he desired. As we shall see later, all these improvements were later incorporated into a July 13, 1925, patent application (see n. 33).

There is absolutely no proof that anything was built and operated before December 1923. However, Zworykin claimed to have reduced his patent to practice between Jan. 1 and June 2, 1924. See VKZ Pat. File, Pat. Int. no. 64,027, Record for Vladimir K. Zworykin; see also Pat. Int. no. 67,440 (Zworykin v. Mathes v. Round), Preliminary Statement of Vladimir K. Zworykin, Jan. 8, 1934. Again, there are no independent records of this claim, although it is probable that this is when he started actual work on the project.

29. V. K. Zworykin, "The Early Days: Some Recollections," *Television Quarterly* 1 (Nov. 1962), 69–73. Were there any actual tubes built in 1924–25? The answer must be yes even though none of them seem to have survived. Fortunately for Zworykin, in his interference with Farnsworth, he did have several witnesses including Metcalf and Tykociner who testified under oath that they had witnessed demonstrations of an early device (see Chapter 8, p. 119). See also VKZ Pat. File, Pat. Int. no. 64,027, Final Hearing, Apr. 24, 1934, p. 46, and Brief for Farnsworth, pp. 2–3, 5.

Zworykin always insisted that he built and operated an iconoscope in 1923, and unfortunately this has been accepted by most television historians as fact, although there is absolutely no proof that he even invented the iconoscope, much less built it, by 1923. For example, in July 1936, Zworykin stated, "Although the first of this type of tube was built as far back as 1923, many years of research and development had to be under taken before it was perfected sufficiently to meet the requirements of a satisfactory television system" (V. K. Zworykin, "Iconoscopes and Kinescopes in Television," *RCA Review* 1 [July 1936], 60). And in his manuscript he claims that he built an iconoscope for his demonstration to management at Westinghouse which included "a photo-electric mosaic in which every photo-electric element was combined with an individual condenser. The condenser charged continuously while the corresponding photo-electric element received light from a particular point of the transmitted picture." He even mentions that "it lasted for 1/30th of a second." While he gives no date, it is "within a few months"

224 / Note to Page 50

after he returned to Westinghouse, which would have to have been some time in June or July of 1923. (See VKZ Bio, pp. 78, 86–87.) Later, in October 1937, an article in *Newsweek* claimed that "In 1924, while working for the Westinghouse Electric and Manufacturing Co. in Pittsburgh, he demonstrated his achievements to company officials. They were enthusiastic. Then he [Zworykin], unwise in the ways of American selling, scotched his own case. His summation of difficulties that lay ahead was so convincing, officials ordered him back to practical research" ("Vladimir K. Zworykin Leaves Future of Television with the Engineers," *Newsweek* 10 [Oct. 25, 1937], 40). This is interesting for two reasons—it gives a 1924 (not 1923) date and, for the first time on record, states that the Westinghouse officials were "enthusiastic." In all other discussions of this episode, Zworykin goes to great lengths to claim that they were quite disappointed in the demonstration. However, he then contradicts himself again when he claims that "after we now had a satisfactory electronic receiver, a complete system of electronic television needed the Iconoscope" (VKZ MS, p. 82). See also Robert C. Bitting, who claims to have interviewed Zworykin on Jan. 17, 1963, and relates the usual story about Zworykin's work with Rozing. He refers to Rozing's tubes as "not of the sealed off variety: i.e. considerable pumping was required." Here Zworykin is confusing Rozing with Fernand Holweck, whom he met in 1928 (Bitting, "Creating an Industry," *Journal of the Society of Motion Picture and Television Engineers* 74 [Nov. 1965], 1015–23). See Chapter 6, pp. 71–72, for details of Zworykin's meeting with Holweck. According to Bitting, Zworykin also claimed that "by November 1923, he had assembled a working version of the iconoscope and had applied it in the world's first all-electronic pickup and display television system." He continued, "the system was capable of transmitting simple geometric figures, albeit at low brightness and contrast, and with a 60 line structure" (p. 1019). This is the first time I have come across this figure, and in fact Zworykin did not achieve sixty-line television until 1929–30. Likewise, Zworykin did not devise the picture tube (the kinescope) until December 1928, nor did he build the first one until April 1929 (see Chapter 6, p. 80). Once again, we see Zworykin manipulating the facts to agree with some imaginary timetable.

Still, this myth continued to grow. Most of this story seems to have come from Elmer Bucher, who says, "during April 1923, Zworykin furnished the Westinghouse Patent Department with complete information on his electronic television system and later that year [1923] he made a rough demonstration of the method before executives and patent officials of the Westinghouse Company" ("Television and Sarnoff," chap. 2, p. 25). Even the venerable Erik Barnouw states, "Late in December 1923, Vladimir Zworykin demonstrated for Westinghouse executives a crude yet practical form of a partly electronic television system" (*Tower of Babel: A History of Broadcasting in the United States* [New York: Oxford University Press, 1966], vol. 1, p. 154). Barnouw here relies on Orrin E. Dunlap, Jr., who states that Zworykin "actually had a complete television system operating on 60 cycles and showed a rough pattern on the face of the cathode-ray tube" and "also showed them [his Westinghouse employers] the kinescope, picture tube of the system" (*Radio and Television Almanac* [New York: Harper, 1951], pp. 75–76). Hofer also accepts the story that Zworykin had exhibited an electric television system to a group of Westinghouse executives "in the winter of 1923–24" ("Philo Farnsworth: Quiet Contributor," p. 51) This of course would destroy any chance of Farnsworth being first in this regard. Hofer relies here on Joseph Binns, "Vladimir Kosma Zworykin," in the National Geographic Society's *Those Inventive Americans* (Washington, D.C.: National Geographic Society, 1971). On the other hand, Asa Briggs states that "Zworykin had begun to develop electronic methods of scanning as early as 1925" (*The Golden Age of Wireless*, vol. 2 of *The History of Broadcasting in the United Kingdom* [London: Oxford University Press, 1965], p. 566). He is quoting from Zworykin, "Television with Cathode-Ray Tubes," *Journal of the Institute of Electrical Engineers* 73, (1933), 437. Here, however, Zworykin states only that "transmitting tubes of this type were actually built a few years ago, and they proved the soundness of the basic idea" (p. 38). Now a "few years ago" is quite vague, and agrees better with late 1929 or early 1930, when Zworykin did build his first operating cameras tubes, than 1923. But, as we will see in Chapter 7, problems with these early camera tubes led to a completely new design which finally yielded a practical camera tube with great potential.

Zworykin always called this early tube an "iconoscope" which was also the name of the camera tube he devised in May 1931. I have his original patent application in my files and it is obvious that there are four major differences between the tube described in his 1923 patent application and the tube built in 1931: (1) the original tube was gas filled, (2) it used gas focus, (3) it had a two-sided target, which means that the beam struck the rear of the target while the photoelectric material was on the front, and (4) it specified a continuous layer of photoelectric material. The later iconoscope was (1) gas free, (2) used electrostatic focus, (3) had a single-sided target, which means that the light and the electron beam impinged on the same side, and (4) and had a special mosaic target made of millions of separate globules. All of these changes were made over a period of years, with amendments to the original patent application to conform with the tube that was later built and operated. Many of these amendments were filed in 1931, after the new tube was built and operated. See VKZ Pat. File, Dec. 29, 1923, patent application (see n. 14 above).

30. VKZ Pat. File, 1923. Carr's letter was a reply to a letter of Aug. 19, 1925. Compare Zworykin's reply to Garret and Mumford, "History of Television" (n. 14 above). Zworykin also gave them the first picture of his November 1931 iconoscope. The 1925 date is also given in Novakosky, "100th Birthday of V. K. Zworykin," p. 65.

The Western Electric oscillograph tube used was described in J. B. Johnson, "Some Uses of the Cathode-Ray Oscillograph," *Bell Laboratories Record* 11 (Apr. 1926), 57–59. According to Zworykin, the rest of the apparatus was made by himself "practically alone" (VKZ MS, p. 78). This was a Herculean feat.

31. VKZ MS, pp. 78–79. Zworykin describes his demonstration to management at Westinghouse, but his acknowledgment that he received a Ph.D. in 1926, "after two years of evening courses and my work in the Westinghouse laboratory on photo-electric cells," clearly dates his work on television to no earlier than 1924, which probably was when the project was first started. Zworykin also cites this date (1924) in *Current Biography* (New York: H. W. Wilson Co, 1949), p. 654, as well as repeating his remark about "scotching his own case." In fact, this story has been told many times with minor variations, basically following what he described in his biography. For example, see Binns, "Vladimir Kosma Zworykin," pp. 192–93; and Ralph Dobriner, "Vladimir Zworykin: The Man Who Was Sure That Television Would Work," *Electronic Design* 18 (Sept. 1, 1977), 113. According to Sobel, "The apparatus was primitive, the picture occasionally unstable" (*RCA*, p. 121).

32. As noted earlier, Zworykin always specified alternating current (rotating) generators. In his 1923 patent application, he also shows them in his drawing (sheet 1) as items 22 and 23 but mentions that "triodes connected in oscillating circuits may be used in place of the alternating-current generators 22 and 23 (lines 21–23)." At the time, rotating generators were the easiest way to produce the deflection currents needed. Also by gearing two generators together at different speeds, it was possible to create a full raster on the screen that was basically locked together. See also Zworykin's July 13, 1925, patent application. While several other major advances were included, this use of rotating generators remained unchanged (see n. 34 below).

In spite of Zworykin's many claims that his 1925 system was "all-electronic" I have not found any evidence that he had ever built an electric (tube type) sine-wave or sawtooth scanning generator before 1929–30. If he had, of course, it would have given him priority over Philo Farnsworth on this most crucial point of invention. Instead, he seems to have used motor generators, which were readily available, easy to use, and could furnish the necessary power.

Two of the earliest patents for sawtooth or linear electric time base generators are N. Kipping, U.S. Pat. no. 1,892,274 (filed Aug. 23, 1923; issued July 13, 1926) and E. V. Appleton, J. F. Herd, and R. A. Watson Watt, Br. Pat. no. 235,254 (filed Feb. 11, 1924; issued Aug. 6, 1925). The first documented use of electric scanning generators for television was in Farnsworth's laboratories June or July 1929. See VKZ Pat. File, Pat. Int. no. 72,203 (Iams v. Ballard v. Farnsworth), Testimony of Harry Lubcke, p. 9. This interference was won by Farnsworth. At this time Lubcke was associate director of research at the Crocker laboratories in San Francisco. See "Notes," *Proceedings of the Institute of Radio Engineers* 17 (Sept. 1929), 1477; and Hofer, "Philo Farnsworth: Quiet Contributor," p. 58.

Oddly enough, Mrs. Farnsworth relates that the earliest sawtooth generators used a motor-driven variable resistor which delivered the sawtooth pulse. This of course is not all-electronic scanning. See Elma G. Farnsworth, *Distant Vision: Romance and Discovery on an Invisible Frontier* (Salt Lake City: Pemberly Kent, 1990), p. 113 (note).

33. V. K. Zworykin, U.S. Pat. no. 1,691,324 (filed July 13, 1925; issued Nov. 13, 1928), Br. Pat. no. 255,057 (conv. July 13, 1925, filed July 3, 1926; issued Mar. 31, 1927). This patent came to the attention of Campbell Swinton, who was so pleased with it that he wrote a letter to Alexanderson of GE expressing great interest in it (Alexanderson File, A. A. Campbell Swinton to Alexanderson, June 1, 1927).

Before this patent was applied for, Zworykin and Westinghouse signed an agreement on Mar. 27, 1925, to give Westinghouse full control of the patent at a later date, which turned out to be July 10, 1932. See "Westinghouse Electric and Mfg. Co Files Suit against RCA," *New York Times,* July 10, 1936, p. 6:7. Actually this is a stronger, technically superior patent than Zworykin's Dec. 1923 application, but Zworykin and the Westinghouse patent department wanted the earlier filing date for priority on his camera tube.

34. VKZ Pat. File, 1923. Patent examiner to O. E. Eschholz (Westinghouse patent dept.), June 18, 1936.

35. This was confirmed by Alexanderson; see Alexanderson File, Alexanderson to S. H. Blake (radio research committee), Apr. 8, 1926. Alexanderson told Blake that "Conrad of Westinghouse is doing research on picture transmission and television at their own expense. They have developed a new photo-cell and a new system of light control." On the mercury-arc device, see V. K. Zworykin and Dayton T. Ulrey, U.S. Pat. no. 1,856,087 (filed Sept. 28, 1923; issued May 3, 1932); also V. K. Zworykin, U.S. Pat. no. 1,709,647 (filed Mar. 17, 1924; issued Apr. 16, 1929); and V. K. Zworykin, U.S. Pat. no. 1,696,023 (filed Nov. 21, 1925; issued Dec. 18, 1928). By this date in 1926, Zworykin's direct involvement with any television project at Westinghouse was now terminated, though the patent department vigorously prosecuted various patent interferences on his behalf. For confirmation that Zworykin did no more work on television at Westinghouse after 1926, see Bucher, "Television and Sarnoff," chap. 2, p. 26: "Dr. Zworykin's research and development work at Westinghouse during 1923–1926 was devoted to further studies of electronic scanning." This agrees with what is known about his work on other projects until his trip to Paris in November 1928.

Regarding the Karolus Kerr cell, see Alexanderson File, Alexanderson to Adams (RCA), Dec. 18, 1925.

36. VKZ MS, p. 80. In 1925 Zworykin publicly revealed that Westinghouse Electric was working on television; see articles in the *New York Times,* Oct. 21, 1925, quotation p. 25:8; and Dec. 27, 1925, sec. 13, p. 12:2. See also V. K. Zworykin, U.S. Pat. no. 1,677,316 (filed July 15, 1925; issued July 17, 1928); and V. K. Zworykin, U.S. Pat. no. 1,709,763 (filed Oct. 15, 1926; issued Apr. 16, 1929). For details about Westinghouse, see Alexanderson File, Alexanderson to Blake, Apr. 8, 1926.

An impressive article about Zworykin, Jenkins, and Baird appeared in the *Scientific American* early in 1926 (Orrin E. Dunlap, Jr., "Seeing around the World by Radio," *Scientific American* 134 (Mar. 1926), 162–63. Here Zworykin is shown with his combined amplifier/photocell, which is claimed to be "the basis of television." Jenkins is shown with his early receiver, and there is a photograph of an early Baird sixteen-hole interlaced scanning disc.

37. Records from the Bell Telephone Laboratories Archives at Short Hills, N.J., which I was allowed to inspect, will be referred to hereafter as Bell Labs 33089. The earliest record that I could find pertaining to Bell's television research was a memo from H. E. Ives to H. D. Arnold, dated Jan. 1925. See also a memorandum from J. G. Roberts (patent attorney), Bell Labs 33089, May 1925.

38. Max Dieckmann and Rudolph Hell, DRP Pat. no. 450,187 (filed Apr. 5, 1925; issued Oct. 3, 1927). Hell later observed that none of their tubes had actually worked (see Goebel, "Das Fernsehen in Deutschland," p. 279).

39. "Jenkins Shows Far Off Objects in Motion," *New York Times*, June 14, 1925, p. 1:4; "First Motion Pictures Transmitted by Radio Are Shown in Capital," *Washington Post*, June 14, 1925, p. 1:3–4; Jenkins, *Radiomovies*, pp. 46–48. The members of Jenkins's staff are listed in his *Vision by Radio*, p. 3. I assume that having a prosaic staff had much to with Jenkins's lack of success. None of these people ever appeared in the literature after they left Jenkins.

40. Goebel, "Das Fernsehen in Deutschland," p. 279, citing an article in *Münchener Neueste Nachrichten* 21 no. 230 (Aug. 1925), 5, 21. See also Gustav Eichhorn, *Wetterfunk, Bildfunk, Television* (Leipzig: B. G. Teubner, 1926), which shows photos of Dieckmann's transmitter and receiver.

41. This information comes from Boris Rozing, "La participation des savants Russes au développement de la télévision électrique," *Revue Générale de l'Électricité*, Apr. 6, 1932, 507–18, a French translation of an article written in Russian by Rozing which appeared in a special number of *Electritchestvo* in May 1930. This translation is my main source of the history of Russian patents. The Russian patents were A. A. Tschernischeff, Russ. Pat. no. 3499 (filed Nov. 28, 1925; issued Aug. 31, 1927), which was for the photoconductive Vidicon camera tube, according to the *Soviet Encyclopedia* (vol. 29, p. 126); and B. P. Grabovsky, F. E. Popoff, and N. G. Piskounoff, Russ. Pat. no. 5592 (filed Nov. 9, 1926; issued June 30, 1928).

42. "Sound Recording and Reproduction," *Bell Laboratories Record* 1 (Nov. 1925), 95–101. See also Roland Gelatt, *The Fabulous Phonograph: 1877–1977* (New York: Macmillan, 1977), pp. 220–25. An excellent account of the growth of the phonograph and the introduction of the new Western Electric sound system is given in Oliver Read and Walter L. Welch, *From Tin Foil to Stereo: Evolution of the Phonograph* (Indianapolis: Howard W. Sams, 1959), pp. 243–54.

43. J. P. Maxfield, "The Vitaphone—An Audible Motion Picture," *Bell Laboratories Record* 2, no. 5 (July 1926), 200–204. An excellent account of the new Vitaphone process is detailed in Read and Welch, *From Tin Foil to Stereo*, pp. 286–88. For an overall review of the sound process in motion pictures see Edward W. Kellogg, "History of Sound Motion Pictures," *Journal of the Society of Motion Picture and Television Engineers* 64 (June 1955), 291–302; (July 1955), 356–74; and (Aug. 1955), 422–37.

44. See William Peck Banning, *Commercial Broadcasting Pioneer: The WEAF Experiment, 1922–1926* (Cambridge: Harvard University Press, 1946). The same story is told in Archer, *History of Radio,* pp. 267–69; and Archer, *Big Business and Radio,* pp. 285–88.

45. The story of the founding of NBC is also told in Archer, *Big Business and Radio,* pp. 278–99.

46. See VKZ MS, pp. 78–79; V. K. Zworykin, "A Study of Photocells and Their Improvement," *University of Pittsburgh Bulletin* 22, no. 31 (1926); and V. K. Zworykin, "Electrolytic Conduction of Potassium through Glass," *Physical Review* 27 (1926), 813.

47. "Television Perfected Asserts London Paper," *New York Times,* Jan. 23, 1926, p. 10:5; and "The 'Televisor,'" *Times,* Jan. 28, 1926, p. 9c. See also Sydney A. Moseley and H. J. Barton Chapple, *Television To-day and To-morrow* (London: Sir Isaac Pitman & Sons, 1930), p. 2; and Alfred Dinsdale, *Television* (London: Television Press, 1928), pp. 70–73. The same version was related sixty years later by R. W. Burns, *British Television: The Formative Years* (London: Peter Peregrinus, 1986), p. 47.

Baird applied for and received a patent on this device (see J. L. Baird, Br. Pat. no. 269,658 [filed Jan. 20, 1926; issued Apr. 20, 1927]). As noted, this process had been patented much earlier, see Georges-Pierre-Édouard Rignoux, Fr. Pat. no. 390,435 (filed May 20, 1908; issued Oct. 5, 1908); A. Ekstrom, Swed. Pat. no. 32,220 (filed Jan. 24, 1910; issued Feb. 3, 1912); and John Hays Hammond, Jr., U.S. Pat. no. 1,725,710 (filed Aug. 15, 1923; issued Aug. 20, 1929).

48. The "flying spot" method was really a breakthrough of the first magnitude, as Baird and his associates certainly realized. The big problem was that Baird had to hide the secret of his success. Having discovered the flying spot or, as the British called it, spot-light scanning process which made his demonstration a success, they could not reveal how it had been done. As a result, there was considerable subterfuge among his backers as they tried to protect the discovery. On Baird's fear of industrial espionage, see M. Exwood, *John Logie Baird: 50 Years of Television,* I.E.R.E. History of Technology Monograph (London: Institute of Electrical and Radio Engineers, 1976), p. 14. On the secrecy surrounding the Baird machines, see George Shiers, "Television 50 Years Ago," *Journal of Broadcasting* 19, no. 4 (Fall 1975), 394. According to Shiers, no one really saw Baird's machine in action because it was always "covered by screens of one sort or another." Still, Baird's secret was revealed by Dinsdale in 1928: "it is interesting to note that Mr. J. L. Baird, the British inventor, used this light-spot method in the course of some of his original experiments, the method being described by him in his British Patent No. 269,658 of Jan. 27, 1926, nearly eighteen months previous to the A. T. & T. demonstration" (*Television,* p. 100).

The great French historian Alexandre Dauvillier, author of one of the earliest and best histories of television to 1928, also commented on the Baird system. Dauvillier is quoted in Moseley and Barton Chapple as saying, "finally the Bell Telephone Company recently (April 7, 1927) succeeded in transmitting to a considerable distance, the human face, using (without saying so!) the Baird system" (*Television To-day and To-morrow*, p. 7). Dauvillier's excellent history appeared in two parts, see Alexandre Dauvillier, "La télévision électrique," part 1, "Étude des divers procédés projets ou réalisés," *Revue Générale de l'Électricité* 23 (Jan. 7, 1928), 5–23; part 2, "Téléphote et radiophote," ibid. (Jan. 14, 1928), 61–73, cont'd (Jan. 21, 1928), 117–28.

That Baird used the flying spot or spot-light system was confirmed in 1935 by John C. Wilson, who had worked for Baird: "Baird was undoubtedly the first, in 1925, to show natural pictures in half-tone. A great advance resulted in his invention in 1926 of the inverse method of scanning applied to three-dimensional views, and he followed up his early success by developing the Baird light-spot system" ("Twenty Five Years Change in Television," *Journal of the Television Society* [London] 2 [1935], 88). Ironically, Baird's patent was issued some thirteen days after the Bell lab's 1927 demonstration, so all of his attempts at secrecy were in vain.

49. "Moving Images Sent by Wire or Wireless by Professor Belin before Paris Experts," *New York Times*, July 29, 1926, p. 1:4; L. Fournier, "The Latest Advance toward Television," *Radio News* 8 (July 1926), 36–37, 84; P. C. [Pierre Chevallier?], "La télévision, par le procédé Édouard Belin," *Le Génie Civil Revue Générale de Techniques* 89, no. 25 (Dec. 18, 1926), 549–52; L. Fournier, "New Television Apparatus," *Radio News* 8 (Dec. 1926), 626–27; L. Lumière, "Belin et Holweck," *Comptes Rendus des Séances de l'Académie*, Feb. 28, 1927, 518–20; Auffray, *Édouard Belin*, pp. 98–102; and "Belin and Holweck," *Bulletin no. 243 de la Société Française de Physique* 8 (Mar. 1927), 35.

Belin's brilliant partner, Fernand Holweck (1890–1941), was born in Paris and was chief of staff of the Marie Curie Radium Institute from 1922 until his untimely death. He was famous for his vacuum pumps and high-vacuum electronic tubes. He joined forces with Belin sometime in 1923 and together they produced the first modern cathode-ray system in 1926. This collective effort, joined by Pierre Chevallier, also produced the first electrostatically focused cathode-ray tube.

50. A. Dauvillier, "Sur le téléphote, appareil de télévision par tubes à vide: Résultats expérimentaux préliminaires," *Comptes Rendus des Séances de l'Académie* 183 (Aug. 2, 1926), 352–54. See also Lucien Fournier, "Television by New French System," *Science and Invention* 15 (Mar. 1927), 988, 1066. The demonstration was described in "L'Exposition annuelle de la Société française de Physique," *Revue Générale de l'Électricité* 21, no. 26 (June 25, 1927), 1010.

Alexandre Dauvillier (1892–1979) was born in St.-Lubin-des-Joncherets, France. He studied at the Lycée Saint Louis et Faculté des sciences de Paris, from which he received his Doctor of Physical Sciences degree. Between 1944 and 1962 he was a professor at the Collège de France. Before that, from 1916 to 1930, he did extensive research into X-ray phenomena at the Louis de Broglie laboratories, and he studied various aspects of cosmology until his death. He made many valuable contributions to television, especially in cathode-ray tube technology.

51. Bell Labs 33089, memo from Gray to Ives, Nov. 16, 1926; see also memorandum from F. Gray, "Use of Cathode-Ray Tube Television Reception," May 1, 1927.

52. E. F. W. Alexanderson, "Radio Photography and Television," paper presented Dec. 15, 1926, and printed in *General Electric Review* 30, no. 2 (Feb. 1927), 78–84. The same paper was reprinted in *Radio News* 8 (Feb. 1927), 944–45, 1030–33. Alexanderson's multispot device became U.S. Pat. no. 1,694,301 (filed Oct. 19, 1926; issued Dec. 4, 1928). I could not find a single instance in the Alexanderson File of a successful demonstration of this device.

53. Sarnoff, *Looking Ahead*, pp. 199–204.

54. "Far Off Speakers Seen as Well as Heard Here in a Test of Television," *New York Times*, Apr. 8, 1927, p. 1:1. See also "Television: Development and Accomplishment," *Bell Laboratories Record* 4 (May 1927), 297–325; A. Dinsdale, "Television Demonstration in America," *Wireless World* 21 (June 1, 1927), 680–86; and H. Winfield Secor, "Radio Vision Demonstrated in America," *Radio News* 8 (June 1927), 1424–26.

55. Bell Labs 33089, memo from Gray to Ives, June 26, 1925; memorandum by Ives, July 10, 1925; F. Gray, "Engineering Notes," Feb. 10, 1926; Gray, "Progress Report," July 28, 1925; memorandum from Gray to Ives, Nov. 30, 1925. See also letter from Gray to Ives, "New Method of Television," May 2, 1926; and Frank Gray, "Use of a Moving Beam of Light to Scan a Scene for Television," *Journal of the Optical Society of America* 16 (Mar. 1928), 177–90, from a paper presented Dec. 5, 1927. See also H. E. Ives, "Television," *Bell System Technical Journal* 6 (Oct. 1927), 551–59. Since Baird had not yet revealed his use of the flying spot scanner, there was no reason for Gray to have suspected that Baird was also using this principle. But whereas Baird tried to keep it a secret, Gray revealed it in the Bell Telephone lab's Apr. 7, 1927, demonstration.

56. Alexanderson File, Apr. 8, 1926, Alexanderson wrote to S. H. Blake (radio research committee) that Conrad of Westinghouse was doing research on picture transmission and television at Westinghouse's expense.

57. Alexanderson File, Alexanderson to Barden (Victor Corp.), "R. D. Kell is in charge of television work," Oct. 2, 1929.

58. Bitting, "Creating an Industry," the section on Van Cortlandt Park appears on pp. 1016–17 and 1019–20. See also A. F. Van Dyck, "The Early Days of Television," *Radio Age* 15 (Apr. 1956), 10–12.

59. Takayanagi File, K. Takayanagi, "Experiments on Television," *Journal of the Institute of Electrical Engineers* (Japan) no. 482 (Sept. 1928), 108–13. It appears that Zworykin knew of Takayangi's work from this article. By 1928 Takayanagi was able to transmit human faces in half-tones, and he seems to be the first to have done this. An April 1990 documentary about Takayanagi first revealed that part of the secret of his success was that his tubes were viewed from the phosphor side.

Kenjiro Takayanagi is deservedly revered in Japan as the founder of the Japanese television industry, but there is a bit of a problem with several of his claims. For instance, he claimed to be "the first person in the world to succeed in receiving an image on an electron Braun tube in 1926." For this feat he received the Japanese Order of Cultural Merit on Oct. 20, 1981, and the Grand Cordon of the Order of the Sacred Treasure in 1989 (Takayanagi File). However, the record shows that this was not the first time anyone received an image on a cathode-ray tube; there were at least five prior successful efforts before Dec. 25, 1926: Dieckmann in Mar. 1909 (see p. 15), Rozing on May 9, 1911 (see p. 15), Zworykin on Oct. 2, 1925 (see p. 50), Belin and Holweck on July 26, 1926 (see p. 57), and Dauvillier on Aug. 2, 1926 (see p. 57). It is also interesting that Takayanagi claimed only to have transmitted a still figure, while both Belin and Holweck and Dauvillier had sent moving images of fingers, and so on, albeit without half-tones.

Takayanagi was one of several other pioneers who later were to challenge Zworykin's priority on both the iconoscope and kinescope. Without wishing to provoke controversy, I note that it seems to be a universal characteristic of inventors to skew the facts slightly in order to make a point. Zworykin, of course, was as guilty of this as anyone. But this does not lessen their real accomplishments, even if they are not "first."

60. Lucien Fournier, "New European Television Scheme," *Science and Invention* 15 (July 1927), 204–5; Georges Valensi, "L'état actuel du problème de la télévision," *Annales des Postes Télégraphes & Téléphone* 6 (Nov. 1927), 1047–67. This not only describes Valensi's work on cathode-ray television but gives a review of the work of Belin and Holweck and Dauvillier.

Chapter 6: The Kinescope

1. Most of this information came from several interviews with Harley Iams, between Jan. 16, 1976, and July 1980. Iams would not allow me to tape our sessions, so I had to sit down later and make sense of my notes. When he died, he left me a treasure trove of artifacts from his days at Westinghouse and RCA, including rare camera tubes, letters, and the mirror galvanometer from the original 1929 film projector that made television history. This material will be referred to hereafter as the Iams File.

Harley Iams (1905–84) was born on March 13, 1905, at Lorentz, W.Va. In 1927 he graduated from Stanford University and went to work for Westinghouse as a student engineer on Aug. 3, 1927. He told me that when he heard of Zworykin's work, he asked to be assigned to him. This started a friendship that lasted all of his life. He left Westinghouse in Dec. 1929 for a short while due to illness and returned to San Diego. He then came back to RCA in Apr. 1931. He was responsible for much of the early work on the iconoscope, the image iconoscope, and later, with Albert Rose, the orthicon. He left RCA in 1948 to join Hughes Aircraft.

Iams received two patents on facsimile; see H. Iams, U.S. Pat. no. 1,715,732 (filed Sept. 23, 1927; issued June 4, 1929) and V. K. Zworykin and H. Iams, U.S. Pat. no. 1,872,381 (filed Sept. 13, 1928; issued Aug. 16, 1932). See also VKZ Pat. File, Pat. Int. no. 73,203.

2. V. K. Zworykin, "Facsimile Picture Transmission," *Proceedings of the Institute of Radio Engineers* 17 (Mar. 1929), 536–50 (paper presented Jan. 2, 1929). This does not agree with Zworykin's account in his manuscript, in which he states that it used special paper, "with no photographic development" (VKZ MS, p. 80). See also V. K. Zworykin, U.S. Pat. no. 1,715,732 (filed Sept. 23, 1927; issued June 4, 1929).

3. See VKZ MS, p. 80, and V. K. Zworykin, U.S. Pat. no. 1,837,746 (filed Mar. 3, 1928; issued Dec. 22, 1931). It was described in V. K. Zworykin and E. D. Wilson, *Photocells and Their Application* (New York: John Wiley and Sons, 1930), pp. 53–54. See also V. K. Zworykin and E. D. Wilson, "The Caesium-Magnesium Photocell," *Journal of the Optical Society of America* 19 (Aug. 1929), 81–89.

Moreover, there was a more sensitive photocell, developed by Lewis R. Koller of GE, in existence at that time. Koller was credited with inventing the cesium-oxide-silver cell sometime in 1927. The only patent I could find was Lewis R. Koller, U.S. Pat. no. 1,725,651 (filed Oct. 14, 1927; issued Aug. 20, 1929). See Lewis R. Koller, "Characteristics of Photo-Electric Cells," *Transactions of the Society of Motion Picture Engineers* 12, no. 36 (Oct. 1928), 921–39; Koller, "Thermionic and Photoelectric Emission from Caesium at Low Temperatures," *Physical Review*, 2d series, 33 (Jan.–June 1929), 1082; Koller, "Some Characteristics of Photoelectric Tubes," *Journal of the Optical Society of America* 19 (1929), 135–45; and Koller, "Photo Electric Emission from Thin Films of Caesium," *Physical Review* 36 (Dec. 1, 1930), 1639. See also Alexanderson File, Alexanderson to Kintner, Jan. 31, 1928.

4. See Kellogg, "History of Sound Motion Pictures," pp. 294–98; Read and Welch, *From Tin Foil to Stereo*, pp. 253–83; and Hans Chr. Wohlrab, "Highlights of the History of Sound Recording on Film in Europe," *Journal of the Society of Motion Picture and Television Engineers* 85 (July 1976), 531–33.

5. See Archer, *Big Business and Radio*, pp. 329–34; Dreher, *Sarnoff*, pp. 103–10; Bilby, *The General*, pp. 93–95; Sobel, *RCA*, pp. 79–81; and Lyons, *David Sarnoff*, pp. 142–44.

6. V. K. Zworykin, L. B. Lynn, and C. R. Hanna, "Kerr Cell Method of Recording Sound," *Transactions of the Society of Motion Picture Engineers* 12, no. 35 (Sept. 1928), 748–59; and V. K. Zworykin, U.S. Pat. no. 1,802,747 (filed Apr. 12, 1927; issued Apr. 28, 1931). This patent was divided and became U.S. Pat. no. 1,939,532 (issued Dec. 12, 1933) as well.

7. VKZ MS, p. 80.

8. VKZ Pat. File, see Pat. Int. nos. 54,922 and 55,448, which he won on July 11, 1928. The quote, from a 1944 interview with Zworykin, is found in W. Rupert Maclaurin, *Invention and Innovation in the Radio Industry* (New York: Macmillan, 1949), p. 204.

9. George Everson, *The Story of Television* (New York: W. W. Norton, 1949); also in Hofer, "Philo Farnsworth: Quiet Contributor," pp. 9–42. See also Elma Farnsworth, *Distant Vision: Romance and Discovery on an Invisible Frontier* (Salt Lake City: Pemberly Kent Publishers, 1990). Philo Taylor Farnsworth (1906–71) was born near Beaver City, Utah, and raised on a farm in

Idaho. Though he lacked a formal education, he had a great imagination and an excellent mind. By the time he was in high school he was fascinated with the problem of television and made it his life's work. Allegedly he had worked out his system of television by 1922, though he started to work on electric television in 1926. By 1927 he had built a basic working system.

10. Philo T. Farnsworth, U.S. Pat. no. 1,773,980 (filed Jan. 7, 1927; issued Aug. 26, 1930); divided on Nov. 7, 1927, to become U.S. Pat. no. 1,806,935 (issued May 26, 1931); Hofer, "Philo Farnsworth: Quiet Contributor," p. 66; Everson, *Story of Television,* p. 52.

11. VKZ Pat. File, Pat. Int. no. 73,203 (Iams v. Ballard v. Farnsworth), citing Farnsworth's own notebooks. See pp. 11–12 of Farnsworth's notebook (reproduced in part in the patent interference) for the Sept. 7, 1927, demonstration and p. 72 for the Mar. 1, 1928, demonstration to GE. See also Hofer, "Philo Farnsworth: Quiet Contributor," p. 62.

12. Alexanderson File, letter from Bishop (Farnsworth financial banker) to C. E. Tuller (patents), May 22, 1928, quoted in A. G. Davis to J. A. Cranston, May 31, 1928; see also Hofer, "Philo Farnsworth: Quiet Contributor," p. 63.

13. "Radio Television to Home Receivers," *New York Times,* Jan. 14, 1928, p. 1:6; Bennett De Lacy, "Now You Can Be a 'Looker-In'!" *Popular Radio* 13 (Feb. 1928), 122–24, 162–64.

14. "Broadcasts Pictures," *New York Times,* May 6, 1928, p. 3:2; "RadioMovies and Television for the Home," *Radio News* 10, (Aug. 1928), 116–18, 173; "The Jenkins 'Radio-Movie' Reception Method," *Radio News* 10 (Nov. 1928), 420, 492–93; C. Francis Jenkins, "The Drum Scanner in RadioMovies Receivers," *Proceedings of the Institute of Radio Engineers* 17 (Sept. 1929), 1576–83. The drum scanner used as a receiver was U.S. Pat. no. 1,683,137 (filed June 2, 1926; issued Sept. 4, 1928).

15. Takayanagi File, "Research and Development of All Electronic Television System," publicity release, 1981, p. 7. Also see Takayanagi, "Experiments on Television," pp. 108–13.

16. "Mihaly System at German Broadcast Exhibition," *The Times* (London), Aug. 28, 1928, p. 11e (3d and 4th eds.); *The Times* (London), Sept. 1, 1928, p. 11b; "Shown at Berlin Radio Exhibition," *New York Times,* Sept. 1, 1928, p. 6:3; Goebel, "Das Fernsehen in Deutschland," pp. 281–82; "Television Method," *New York Times,* June 18, 1928, p. 32:2; "Successful Television Accomplished on Broadcast Band," *Radio News* 10 (Sept. 1928), 219–20, 277.

17. "Television Shows Panoramic Scene Carried by Sunlight," *New York Times,* July 13, 1928, p. 4:1; "Progress of Television," *The Times* (London), July 14, 1928, p. 13b; F. Gray and H. E. Ives, "Direct Scanning in Television," *Journal of the Optical Society of America* 17 (Dec. 1928), 428–34; Alexanderson File, Alexanderson to Dunham, Aug. 2, 1928.

18. "Radio Movies Demonstrated," *Science and Invention* 16 (Nov. 1928), 622–23, 666; "Radio 'Movies' from KDKA," *Radio News* 10 (Nov. 1928), 416–17; "Motion Picture Transmitter," *New York Times,* Aug. 19, 1928, sec. 8, p. 17:3. For the mercury arc device, see V. K. Zworykin, U.S. Pat. no. 1,709,647

(filed Mar. 17, 1924; issued Apr. 16, 1929). For the cesium magnesium photoelectric cell, see V. K. Zworykin, U.S. Pat. no. 1,837,746 (filed Mar. 3, 1928; issued Dec. 22, 1931). See Alexanderson File, Alexanderson to S. H. Blake, Apr. 8, 1926, confirming this.

19. "Smith Rehearses for Camera Men," *New York Times,* Aug. 22, 1928, p. 3:4. A picture of the camera is shown in *New York Times,* Aug 26, 1928, p. 13. This information came from interviews I had with Ray Kell in Princeton, in March 1977 and March 1978. Referred to hereafter as the Kell File.

Raymond Davis Kell (1904–86) was born in Kell, Illinois, on June 7, 1904. He graduated from the University of Illinois in 1926, and did graduate work until April 1927, when he joined General Electric. He went from the test section to the television section under Alexanderson in May 1927, and was soon directing the project under manager William A. "Doc" Tolson. He was an excellent engineer and designed and built most of the early GE television equipment.

Kell was an early believer in cathode-ray television at General Electric. After the separation of RCA from GE and Westinghouse in 1930, he was only too pleased to be transferred to RCA at Camden. He told me that he did not think highly of Alexanderson. At RCA, he and a cadre of ex-GE engineers, including Alda Bedford and Merrill Trainer, were set up in a special engineering section under Arthur Tolson. This group was later taken over by Elmer Engstrom.

Kell was a leader in the development of practical working television systems and spent his whole life at RCA. He was involved in every major project there, including during the war years and the great color battle with CBS. In 1946 he was in charge of television research at the RCA laboratories division with George Beers. Upon retirement, he moved to Arizona.

Alda V. Bedford was born on Jan. 6, 1904, in Winters, Texas. He graduated from the University of Texas in 1925. In 1926 he went to work for General Electric, first in general engineering and then in the test and research laboratory. There he worked on sound recording, audio amplifiers, loud speakers, and sound printers for film and finally television. He moved to RCA in September 1930 and became part of the group working under Tolson and then Elmer Engstrom, when he took over television research at RCA. After the war Bedford made invaluable contributions to the RCA color system, especially the concept of "mixed highs." I interviewed Bedford in March 1977, in Princeton. Cited hereafter as the Bedford File.

20. Details are found in Bell Labs 33089, Farnsworth to Dr. E. B. Craft, June 25, 1928; M. Fleager to E. B. Craft, Sept. 4, 1928.

21. "S. F. Man's Invention to Revolutionize Television," *San Francisco Chronicle,* Sept. 3, 1928, p. 1:3; "New Television System," *New York Times,* Sept. 4, 1928, p. 20:1.

22. "Play Is Broadcast by Voice and Acting in Radio-Television," *New York Times,* Sept. 12, 1928, p. 1:3; Robert Hertzberg, "Television Makes the Radio Drama Possible," *Radio News* 10 (Dec. 1928), 524–27, 587–90; "Drama via Television," *Science and Invention* 16 (Dec. 1928), 694–762.

23. Alexanderson File, Alexanderson to C. J. Young, Oct. 12, 1928; Kell to J. Huff, Dec. 14, 1928.

24. This information came from Pat. Int no. 69,135 (Henroteau v. Zworykin v. Farnsworth), statement of Henroteau, Oct. 1, 1934, pp. 5–6; and from questions asked about Koller's paper. See Koller, "Characteristics of Photo-Electric Cells," pp. 934–36. See also Zworykin, "Facsimile Picture Transmission," pp. 748–59.

25. V. K. Zworykin, U.S. Pat. no. 1,691,324 (filed July 13, 1925; issued Nov. 13, 1928); Alexanderson File, A. A. Campbell Swinton to Alexanderson, June 1, 1927. Many television historians have erroneously used this 1928 issue date to mark the birth of Zworykin's camera tube.

26. "Forging an Electric Eye to Scan the World," *New York Times,* Nov. 18, 1928, sec. 10, p. 3:1; David Sarnoff, "The New Age of Radio," paper read before the Economic Club of New York and published thereafter in pamphlet form (Nov. 27, 1928; pp. 10–14). Sarnoff's reference to an "electric eye" at this time intrigued me. It certainly is descriptive, but I was never quite sure what he was referring to. If Sarnoff saw something that interested him in Paris that summer, he gives no hint of it in the article. However, the fact that he sent Zworykin to Paris in November was no mere coincidence. It is known that neither Westinghouse, General Electric, or RCA had any business connections with Belin at the time. I can only surmise that what caught Sarnoff's attention must have been the Belin (and Holweck and Chevallier) cathode-ray tube. It is quite possible that someone from the Belin laboratories contacted Sarnoff during his visit and told him of their new advanced cathode-ray tube. Sarnoff, of course, could not turn down such an offer, and this has to be the reason why he sent Zworykin, his expert, to Paris in November 1928 to evaluate this new device.

Zworykin does not mention going to any other laboratory in Paris on this trip. He did not visit Dauvillier or Valensi, who were also working with cathode-ray tubes. As events turned out, Sarnoff's instincts were right; Zworykin's trip to Paris changed the course of television history.

27. VKZ MS, p. 84. Here I must apologize for the wrong dates given in my 1987 *History of Television* (pp. 122–23) when I said that Zworykin had gone to Europe in the summer of 1928 since he was not part of the August 1928 Westinghouse television demonstration. This information was given to me by Harley Iams, who told me that he was working on the deflection circuits and the design for the new picture tube in November *after* Zworykin returned from Europe. This information is also repeated in Pat. Int. no. 73,203 (Iams v. Ballard v. Farnsworth). This made sense as it gave Zworykin about a month to build his first tube (a rebuilt Western Electric oscillograph or a De War flask) and then go to Sarnoff in January 1929 and tell him that he had a "basic instrument" working in the laboratory.

I knew that Zworykin was in the U.S. in September 1928, as he attended a meeting in Lake Placid, N.Y. So I assumed that he had just returned home after being in Europe in the summer of 1928. I was quite amazed when I received FOIA/State Dept. PO34 (a passport application for Zworykin, dated Nov. 1, 1928), which revealed that Zworykin had not even applied for a passport until Nov. 1, 1928, for a voyage to start on Nov. 17, 1928. Zworykin stated that he was going to England, Germany, and France for one and a half months. He also stated that this was a "business trip." He returned on Dec. 24, 1928, and I could not figure out how he built a tube between then and the Jan. 3, 1929, meeting with Sarnoff in New York. This, of course, was during the Christmas holidays, and activities at the Westinghouse laboratory were probably at a standstill.

Another question is how he managed to order glass bulbs from Corning Glass at this time (Dec. 1928) and have them arrive by February 1? Clearly, he must have ordered the bulbs before he went to Paris. However, both Iams and Arthur Vance told me that Zworykin did have an operational tube before he visited Sarnoff. Of course he did have a tube—he had the Holweck/Chevallier tube, which was working quite well!

Perhaps this whole story is to cover up the fact that the Holweck/Chevallier tube would be used until Zworykin was able to build one of his own. This would explain why Iams was designing deflection circuits in November for a tube to be brought back from Europe in December. Iams certainly was relating the story as he remembered it some forty-eight years earlier. Even though Iams admired Zworykin, he was an honorable man and I don't believe he would make up such a story. On the other hand, Arthur Vance, who did not like Zworykin at all, tells a similar story but without specific dates. (Iams was at Westinghouse at the time, while Vance did not arrive until May 1929.) This is another of the mysteries of Zworykin's life that may never be solved.

28. Iams File, Harley Iams to Edward W. Herold, Feb. 3, 1984, p. 3. I have several of these postcards in my possession.

29. Albert Neuberger, "The Karolus System of Television," *Television* (London) 1, no. 8 (Oct. 1928), 35–37.

30. M. Du Mont, "Television on the Continent," *Television* (London) 1, no. 1 (Mar. 1928), 36–37. Zworykin relates his meeting with Holweck (VKZ MS, p. 84), but there is no mention of Belin, Ogloblinsky, or Chevallier. For details of the Belin facsimile machine see Lucien Fournier, "New Belin Photo-Transmitter," *Science and Invention* 16 (June 1928), 119, 174; G. Ogloblinsky, "Derniers progrès de la transmission Belinographique en France," *L'Onde Électrique* 7 (Oct. 1928), 446–55.

31. There is a cathode-ray tube on display at the Conservatoire National des Arts et Métiers, Musée National des Techniques in Paris. It has a placard claiming that "In 1928, Belin, Chevallier and Holweck tried with this apparatus to scan an image by means of two oscillating mirrors, one rapid to scan each line, the other slow for vertical deflection. This apparatus failed for a long time to display the ends of the lines, but such experiments were meanwhile considered as a great step to television, because these utilized for reception a portable cathode ray tube similar to that which is presented here." (See fig. 25.) For a description of the tube, see p. 72 above. I visited the museum on July 27, 1984, in search of some documentation on this

important artifact. I was shown only a sheet of paper stating that the tube had been "received in 1966," without any mention of who the donor was or where it had come from. All attempts to get more information from the museum curator were in vain. The tube on display is in two pieces, all glass, while all my other information indicates that the tube that Zworykin saw and brought home with him was metallic. According to Chevallier's patent even the glass tube should have been metallized in order to function. This tube has no metallic coating at all.

32. L. T. Jones and H. G. Tasker, "A Thermionic Braun Tube with Electrostatic Focusing," *Journal of the Optical Society of America* 9 (1924), 471–79. Their tube used a small amount of mercury vapor.

33. J. T. MacGregor-Morris and R. Mine, "Measurements in Electrical Engineering by Means of Cathode Rays," *Journal of the Institution of Electrical and Electronics Engineers* (London) 63 (Nov. 1925), 1056; O. S. Puckle, "History of Cathode Ray Tubes," *Wireless World* 16 (Mar. 23, 1937), 24–27, 78; Manfred von Ardenne, *Cathode-Ray Tubes* (London: Sir Isaac Pitman & Sons, 1939).

34. W. Rogowski and W. Grosser, DRP Pat. no. 431,220 (filed Dec. 21, 1923; issued Feb. 4, 1927); U.S. Pat. no. 1,605,781 (convention date [hereafter conv. date] Nov. 7, 1925; issued Nov. 2, 1926); W. Rogowski and W. Grosser, "Kathode-Ray Oscillograph," *Archiv für Elektrotechnik* 15 (Dec. 5, 1925), 377–84. For a history of "electron optics" see the following articles: Louis de Broglie, "A Tentative Theory of Light Quanta," *Philosophical Magazine* 47 (Feb. 1924), 446–58 (paper presented Oct. 1, 1923); Hans Busch, "Calculation of the Path of an Electron Beam in an Axial Symmetrical Electromagnetic Field," *Annalen der Physik* 81 (1926), 974–93 (paper presented Oct. 18, 1926); and J. Thibaud, "Effet magnétique longitudinal sur les faisceaux d'électron lents," *Journal de Physique* 10, no. 4 (1928), 161–76 (received Feb. 27, 1929). See also the following books: L. M. Myers, *Television Optics: An Introduction* (London: Sir Isaac Pitman & Sons, 1936); I. G. Maloff and D. W. Epstein, *Electron Optics in Television* (New York: McGraw-Hill, 1938); and Paul Harschek, *Electron-Optics* (Boston: American Photographic Publishing Co., 1944). Zworykin's contribution is related in L. Jacob, *An Introduction to Electron Optics* (London: Metheun & Co., 1951). According to Jacob, "the first high-vacuum oscillograph appeared through the labours of Zworykin in 1929" (p. 2). See also Myers, who gives Zworykin credit for a "more complicated electron lens system possessing more than one anode. One of the most important in this respect is due to Zworykin, who first described [such a lens system]" (*Television Optics,* p. 290); and V. E. Cosslett, who states that, although this type of tube was used for many years in experimental work, "otherwise, the tube remained substantially unchanged until Zworykin's work in 1928–9" (*Introduction to Electron Optics* [London: Clarendon Press, 1946], p. 4).

35. A. Dauvillier, Fr. Pat. no. 592,162, first addition 30,642 (filed Feb. 11, 1925; issued July 20, 1926). In the French patent system, it is possible to update an existing patent with an addition. Dauvillier stated that he had tried

acceleration of the beam after deviation and it did not work for him ("L'accélération tentée après déviation n'a pas donné de bons résultats"). See Dauvillier, "Sur le téléphote," p. 354.

36. See É. Belin, Fr. Pat. no. 638,661 (filed Oct. 28, 1926); first addition 33,669 (filed Mar. 4, 1927; issued Nov. 29, 1928). See also É. Belin and F. Holweck, "Présentation d'une expérience de télévision," *Bulletin de la Société Française de Physique* no. 243 (Mar. 4, 1927), 35–36. This issue also contained an article by Dauvillier (pp. 38–39) in which he was quite critical of the apparatus shown by Belin and Holweck. There was no love lost between Dauvillier and Holweck.

For information about how the new tube worked, see P. E. L Chevallier, Fr. Pat. no. 699,478 (filed Oct. 25, 1929; accepted Dec. 9, 1930; issued Feb. 16, 1931).

37. P. Selenyi, "Use of Negative Charges of Kathode Rays as a Means of Marking in the Kathode-Ray Oscillograph," *Zeitschrift für Physik* 47 (1928), 895–97 (received Feb. 6, 1928). See also P. Selenyi, "Electrical Charging of Glass by Kathode Rays and Its Practical Application," *Zeitschrift für technische Physik* 9, no. 11 (1928), 451–54; P. Selenyi, U.S. Pat. no. 1,818,760 (conv. date Feb. 1, 1928; filed Jan. 18, 1929; issued Aug. 11, 1931); also Br. Pat. no. 305,168 (conv. date Feb. 1, 1928; issued Mar. 27, 1929).

38. VKZ MS, p. 83; Bucher, "Television and Sarnoff," p. 44; FOIA/State Dept. PO34, passport application dated Nov. 1, 1928; and Westinghouse File. Purchase of the Holweck demountable cathode-ray oscillograph is confirmed in Westinghouse Electric and Manufacturing Research Department Report R-6706 A, V. K. Zworykin, "Cathode Ray Television Transmitters," May 28, 1930. See n. 31 above for a discussion of the differences between the tube that Zworykin was shown and brought back with him and the tube displayed in the Conservatoire Nationale des Arts et Métiers.

There is a strange parallel between Zworykin's trip to France and that of Thomas Edison. While Edison was in Paris for the 1889 exhibition, he visited the French physiologist Jules Étienne Marey. Marey had built a motion picture "camera" using a band of sensitized paper which took a series of impressions of birds in flight and animals. According to Gordon Hendricks, "Out of that brief visit with Marey came the famous kinetoscope" (*The Edison Motion Picture Myth* [Berkeley: University of California Press, 1961], p. 172). When Edison arrived home, he immediately filed for his fourth caveat for the motion picture. Even the names are similar, kinetoscope (Edison) and kinescope (Zworykin).

Likewise, there is no question that Zworykin brought back the idea for his new tube from Paris, and it is certain that this was the Holweck/Chevallier tube. As practical men, both Edison and Zworykin put the finishing touches on some else's great idea. Of more importance, they reduced the idea to practice. Unfortunately something later went wrong with the business relations between Westinghouse and the Belin/Holweck/Chevallier group.

39. Bucher, "Television and Sarnoff," p. 44; RCA Stockholders' Report, 1929, p. 3, David Sarnoff Research Center, Princeton; Archer, *Big Business*

and Radio, pp. 339–43; Sobel, *RCA,* p. 85; Lyons, *David Sarnoff,* pp. 156–58; and VKZ MS, p. 83.

40. "German Is Honored by Radio Institute," *New York Times,* Jan. 3, 1929, p. 29:6.

41. Lyons, *David Sarnoff,* p. 209.

42. Zworykin was very general in his manuscript about this meeting with David Sarnoff, noting that he had "eagerly accepted" Kintner's advice to go directly to Sarnoff as head of RCA and that Sarnoff had publicized the results of their meeting (VKZ MS, pp. 83–84). Other reports are diverse and a source of much misinformation. Bucher, for example, simply describes Zworykin as "recognizing the General [Sarnoff] as the top operating executive in 1929, trusting that the General would allow him the time to give full exposition to his electronic television system and what it had in store, should the necessary facilities and financial support be made available." Bucher continues, "Zworykin states that the General listened patiently for a half hour to his story on television and what could be accomplished through the further perfection of his iconoscope" ("Television and Sarnoff," pp. 46–47). We now know that it was really the picture tube (the kinescope) that Zworykin was describing, but for reasons which will become apparent, the iconoscope had far more publicity value to RCA than the kinescope and so the story was changed. This same story is told in a variety of ways. Dunlap writes, "On November 18, 1929, at Rochester, N.Y., he [Zworykin] showed the experts a noiseless television receiver. [No demonstration was given.] It used no motor or moving parts. Scanning was done electronically. . . . Zworykin, confident that he had developed the electronic 'eye' to do these various tricks, tried to convince industrialists that he had the right approach. The story is told how someone suggested that David Sarnoff, President of RCA, was the man who would appreciate the idea. For half an hour Mr. Sarnoff listened intently to Zworykin's tale of magic. 'It's too good to be true,' exclaimed Sarnoff. 'What will it cost to develop the idea?' 'Maybe about $100,000,' answered Zworykin" (O. E. Dunlap, Jr., *The Future of Television* [New York: Harper & Bros., 1947], pp. 156–57). According to W. R. MacLaurin, "The importance of the 'iconoscope' [which didn't exist at the time] was recognized almost immediately. Mr. Sarnoff as vice president and general manager of RCA, got in touch with Dr. Zworykin personally and asked him about his progress and the prospects for commercializing electronic television. Sarnoff, who had keen imaginative insight, was greatly impressed, and shortly thereafter Zworykin was told he could have additional assistance . . . , Zworykin dates the end of his struggles for recognition from this discussion with Mr. Sarnoff" (*Invention and Innovation in the Radio Industry* [New York: Macmillan, 1949], p. 203). See also J. Jewkes, D. Sawyers, and R. Stillerman, who claim that "In 1928 he [Zworykin] filed for his patent application on the revolutionary iconoscope, the device that transmits television images quickly and effectively: he thus removed a formidable obstacle from the development of commercial television. David Sarnoff, then the vice-president of R.C.A., became interested in Zworykin's progress in televi-

sion and assigned him four or five men to assist him, and steps were taken to improve his patent position" (*The Sources of Invention* [London: Macmillan, 1958], pp. 385–86); and Bitting, who says, "in 1927 or 1928 Sarnoff met Zworykin and made arrangements with Westinghouse for Zworykin to proceed with his television research and development" ("Creating an Industry," p. 1019). In his biography of Sarnoff, Lyons tells a similar story: "Early in 1929, accordingly, he [Zworykin] sought a private conference with the RCA vice-president and a meeting in New York was arranged. . . . The inventor laid his cards on the table. First he explained the principles on which his 'electric eye' was based. Then he demonstrated that basically the instrument was already operative, but it still required intensive and expensive development" (*David Sarnoff*, p. 209). Likewise Sobel: "In 1929, after the iconoscope-kinescope demonstration, Sarnoff sought out Zworykin to learn more of his work" (*RCA*, p. 124). Bilby, speaking of another nonexistent demonstration (in 1923), says, "Zworykin's superiors, dedicated refrigerator merchants, refused to commit the substantial development funds they judged would be needed." He continued with the meeting with Sarnoff, "In 1929, he [Sarnoff] was visited by a young Westinghouse scientist named Vladimir T. [*sic*] Zworykin. He, in his heavily Russian-accented English, outlined to an intent Sarnoff his concept of an electronic camera eye that would replace the spinning disc. He conceded the primitive nature of his laboratory apparatus and the need for extensive developmental work" (*The General*, pp. 120–21).

That Sarnoff and Zworykin had met previously is confirmed by Sarnoff's speech at Zworykin's retirement party in 1954. Sarnoff was quoted as saying that he had first met Zworykin "twenty-seven or twenty-eight years ago" ("Thirty Years of Progress in Science and Technology: Proceedings," seminar, Princeton University, Sept. 18, 1954, p. 7). Strangely enough, though both Zworykin and Sarnoff were present at many media events, there are no pictures available of them together (Phyllis Smith, archivist of the David Sarnoff Research Center, to author, phone call, May 13, 1993).

43. "Television Is Still in Laboratory Stage," *New York Times*, Jan. 27, 1929, p. 18:1.

44. Westinghouse File, Research Department Report R-6705 A, V. Zworykin, "Cathode Ray Television Receivers," closing report, May 28, 1930. See also VKZ MS, p. 81; Iams File, interview (1976) and Pat. Int. no. 73,203 (Iams v. Ballard v. Farnsworth), Testimony for Iams. Here the November 1928 date is given for work on deflection circuits and design of the cathode-ray tube. In 1976 I interviewed Arthur Vance at Laguna, Calif. Material from this interview will be known as the Vance File.

45. See M. Knoll, L. Schiff, and C. Stoerk, U.S. Pat. no. 2,036,532 (filed Dec. 12, 1928; conv. date Nov. 18, 1929; issued Apr. 7, 1936); R. H. George, "A New Type of Hot Cathode Oscillograph," *Journal of the American Institute of Electrical Engineers* 48 (July 1929), 884–90; R. H. George, "An Improved Form of Cathode-Ray Oscillograph," *Physical Review* 31 (Feb. 1928), 303 (paper presented Dec. 1927); VKZ Pat. File, Application of Roscoe Henry George, Affidavit of Roscoe Henry George, Mar. 25, 1933.

46. R. H. George, U.S. Pat. no. 2,086,546 (filed Sept. 14, 1929; issued July 13, 1937). "Variable velocity modulation," first invented by Boris Rozing in 1911 (see Br. Pat. no. 5,486 [filed Mar. 4, 1911; issued Feb. 29, 1912]), is a method of modulating the brightness of the electron beam by changing the horizontal velocity of the beam. As the beam slows down, it impinges on the phosphor for a longer period of time, thus brightening the spot. As the beam speeds up, it impinges on the phosphor for a shorter period of time, thus lessening the brightness of the spot. This was an alternate method used in the early days of cathode-ray television when electronic modulation of the beam was not possible. It was also an answer to the problem of stabilizing the beam when it was modulated which normally changed both its speed and brightness. Zworykin's new tube solved this problem for all time. Outside of a feeble effort by L. H. Bedford and O. S. Puckle in 1934 to promote this method, it was never used in commercial tubes ("A Velocity Modulation Television System," *Journal of the Institution of Electrical Engineers* [London] 71 [Jan. 1952], 63–85).

For details of the Purdue University television project see C. F. Harding, R. H. George, and H. J. Heim, "The Purdue University Experimental Television Station," *Purdue Engineering Bulletin,* Research Series no. 65, 23, no. 2 (Mar. 1939), 5–8, 114.

47. Westinghouse File, Research Memo 6-6705-1, Zworykin to Tolson, "Development of Television," p. 4.

48. VKZ MS, p. 81; Westinghouse File, Research Memo 6-6705-1, V. K. Zworykin to W. A. Tolson (General Electric, Schenectady), Jan. 18, 1930, "Development of Television in Westinghouse. Research Laboratory during February 1, 1929 to Jan. 1, 1930." See VKZ MS, p. 78, for the choice of the name Kinescope. See also W. D. Wright, "Picture Quality—The Continuing Challenge towards Visual Perfection," *Journal of the Royal Television Society* 16 (Jan.–Feb. 1977), 6. Here he tells of his early role in this project.

The members of Zworykin's original group were:

Arthur Vance (1904–80) graduated from Kansas State College in 1928 with a B.S. degree in electrical engineering. He started as a student engineer at Westinghouse in 1928 and in May 1929 was assigned to Zworykin's television group. He made the move to RCA in 1930 and stayed with Zworykin as engineer in charge of circuit research. He told me that he had left RCA under a cloud, and he was clearly very bitter about Elmer Engstrom, for reasons he would not tell me. He also said that he had no respect for Zworykin, and that Zworykin didn't even know the difference between inductive and capacitive reactance. Vance claimed that he stayed with Zworykin because he was able to work only three and four days a week and still get his quota of work done. (I heard from others who worked with Vance that he was an absolute wizard with electronic circuitry. His work on the iconoscope project was acknowledged by Zworykin in his 1933 paper to the Institute of Radio Engineers.) Vance was also quite bitter about the treatment given Farnsworth by the RCA patent department. He complained that much of his time was spent in designing devices and circuits to frustrate Farnsworth in the Patent Office. His last project for RCA was to design the power supplies for the Zworykin/Hillier electron microscope.

Randall C. Ballard (1902–?) was born in Chicago, and received a B.S. degree in electrical engineering from the University of Illinois in 1928. He was working on the television project at Westinghouse until he was transferred to Zworykin's group in September 1929. He went with the group to RCA and in 1932 invented the odd-line interlacing method that became standard in television around the world. In 1932 he left RCA and went to work for the U.S. Radio and Television Corporation in Marion, Indiana. He returned to RCA in 1935 and was with the RCA television research department until 1941, when he was transferred to radar work at the Princeton labs. After the war he returned to work on the RCA color system. By the time I caught up with him in 1977, I was told that he was suffering from Alzheimers disease and could not be interviewed.

Gregory Nathan Ogloblinsky (1901–34) is the unknown factor in this story. For his early years, the only information I could find comes from his application for membership in the Institution of Electrical and Electronic Engineers (a copy of that application was sent to me on Dec. 3, 1990). According to that, he was born on July 27, 1901, but there is no information about where he was born or about his education. He is recorded as working at the Belin laboratories in 1925, as a trained physicist. He worked with Chevallier and Holweck on all of their facsimile and television projects. He met Zworykin in Paris in December 1928 and was hired by Westinghouse to work in East Pittsburgh in July 1929. He moved with Zworykin to RCA in Camden in 1930 and worked there until his untimely death in an automobile accident in late 1934. Those I spoke to who had worked with him (Iams, Vance, Flory, W. D. Wright) were most impressed with him. More than one person told me that he solved most of the hard technical problems on Zworykin's iconoscope. I was quite shocked to discover the role that Ogloblinsky had played in the Chevallier/Zworykin picture tube controversy. It now appears that if he had not interceded on Chevallier's behalf the story would have been quite different. For Ogloblinsky's opinions on his co-workers, see Gregory Ogloblinsky to P. Chevallier, Sept. 11, 1929. This correspondence came from Jean-Jacques Ledos, who kindly furnished me with invaluable material (letters, pictures, and so on) from this era; Ledos to author, Apr. 19, 1993.

Dr. W. D. Wright is a brilliant optical engineer. He was with Zworykin when the first kinescopes were built and he designed the optical system of the first Zworykin oscillating mirror film projector. He then went back to England and in 1932 was working for EMI part-time on their television project. He remained with them until his retirement in 1947. Material from my interview with him (England, Nov. 17, 1978) will be known as the Wright File.

I have no information on John Batchelor at this time.

A few words about the caliber of Zworykin's staff are in order. There is no doubt that Zworykin's great success was due to their tremendous talent. Zworykin himself noted in his manuscript that he was unusually lucky in his associates. Although there was no long-range recruitment plan for the laboratory, the majority of the members of the laboratory were extremely competent and many of them became prominent in the electronic field (VKZ MS, p. 84).

Years later, James Hillier said of Zworykin that "In the laboratory he had the special ability to select or develop first rate research workers. I was never sure whether he was good at selecting or both. It does not really matter. The fact is that there is a substantial roster of people from the Doctor's activities who have developed international reputations" ("Dr. Zworykin's Memorial Service," Aug. 3, 1982, sent to me by Hillier in Sept. 1993).

Zworykin apparently was blessed by his ability to get the most out of his fellow workers and still retain their loyalty. His greatest fault seems to be that in his later years he insisted on putting his name on every laboratory project or paper simply because he was the department head. In his defense (which he really does not need), he was the "spark-plug" who made a project possible simply by just being there.

49. Iams File, interview (1976), about the sale of the receiving sets; the demonstration to Sarnoff was related to me by Vance in 1976.

50. For the formation of RCA Victor, see Archer, *Big Business and Radio*, pp. 346–49; Lyons, *David Sarnoff*, pp. 155–58; Bilby, *The General*, pp. 99–101; and Sobel, *RCA*, pp. 85–88.

51. J. Weinberger, T. A. Smith, and G. Rodwin, "The Selection of Standards for Commercial Radio Television," *Proceedings of the Institute of Radio Engineers* 17 (Sept. 1929), 1584–94 (paper presented May 14, 1929).

52. "Television Placed on Daily Schedule," *New York Times*, Mar. 23, 1929, p. 20:2; "Television Emerging from the Laboratory," *Radio News* 10 (June 1929), 1121; "Look In! RCA Is Televising," *Science and Invention* 17 (Dec. 1929), 725.

53. VKZ File, Ogloblinsky to Chevallier, Sept. 11, 1929.

54. Westinghouse File, Research Department Report R-6706 A.

55. Westinghouse File, Research Department Report R-6706 A, Zworykin, "Cathode Ray Television Transmitters," closing report. Details of the tube are found in V. K. Zworykin, U.S. Pat. no. 2,246,283 (filed May 1, 1930; issued June 17, 1941); also Fr. Pat. no. 715,912 (conv. date May 1, 1930; filed Apr. 23, 1931; published Dec. 11, 1931). Information on the first tube is in Zworykin's May 1, 1930, patent application, serial number 448,834. See also Harold J. McCreary, U.S. Pat. no. 2,013,162 (filed Apr. 10, 1924; issued Sept. 3, 1935) and Pat. Int. no. 54,922 (McCreary v. Zworykin), Feb. 8, 1932. McCreary lost to Zworykin and Westinghouse, apparently because the wording of the claims worked against McCreary's application.

56. VKZ MS, pp. 81–82.

57. Westinghouse File, Research Memo 6-6705-1, Zworykin to Tolson, "Development of Television," p. 3; and Research Department Report R-6705 A, Zworykin, "Cathode Ray Television Receivers," p. 3.

58. VKZ MS, p. 81–82.

59. V. K. Zworykin, U.S. Pat. no. 1,786,812 (filed Mar. 26, 1929; issued

Dec. 30, 1930); and reissue Pat. no. 19,314 (issued Sept. 11, 1934); see also V. K. Zworykin, U.S. Pat. no. 2,361,255 (filed July 5, 1929; renewed Aug. 13, 1931; issued Oct. 24, 1944). For the important patent on Zworykin's picture tube, see V. K. Zworykin, U.S. Pat. no. 2,109,245 (filed Nov. 16, 1929; issued Feb. 22, 1938); he also applied for a French patent (Fr. Pat. no. 705,523 [conv. date Nov. 16, 1929; filed Nov. 10, 1930; issued June 9, 1931]). I have the complete Patent Office file on the Nov. 16, 1929, patent application. P. Chevallier, Fr. Pat. no. 699,478 (filed Oct. 25, 1929; issued Feb. 16, 1931), U.S. Pat. no. 2,021,252 (conv. date Oct. 25, 1929; filed Oct. 20, 1930; issued Nov. 19, 1935).

60. V. Zworykin, "Television with Cathode Ray Tube for Receiver," Westinghouse Research Laboratory, Scientific Paper no. 393, Oct. 21, 1929. This is an early draft of the paper that Zworykin presented at the Institute of Radio Engineers on Nov. 18, 1929. It is complete in all details. It is interesting that Zworykin lists Rozing's 1907 patent (see Chap. 2, n. 6 above), Belin and Holweck's March 1927 paper (see Chap. 5, n. 49 above), Dauvillier's 1928 history (Chap. 5, n. 48 above), and Takayanagi's Sept. 1928 paper (Chap. 5, n. 59 above) as references. This paper includes the first use of the word *kinescope* for the tube (p. 5).

61. VKZ MS, p. 84; V. K. Zworykin, "Television with Cathode-Ray Tube for Receiver," *Radio Engineering* 9 (Dec. 1929), 38–41. See also V. Zworykin, "The Cathode-Ray Television Receiver," *Radio-Craft* 1 (Feb. 1930), 384–85; W. G. W. Mitchell, "The Cathode-Ray in Practical Television (Part 3)," *Television* (London) 2 (Feb. 1930), 590–93; A. Neuberger, "Das 'Kineskop,' ein neuer Fernseher," *Fernsehen* 4 (1930), 175–79; "Cathode-Ray Television Receiver Developed," *Scientific American* 142 (Feb. 1930), 147; and V. K. Zworykin, "Television through a Crystal Globe," *Radio News* 11 (Apr. 1930), 905, 949, 954. This last article states that it was "reprinted by courtesy of the Institute of Radio Engineers."

62. "V. Zworykin Demonstrates Non-Mechanical Receiver, Special Cathode Ray Tube Called 'Kinescope,'" *New York Times,* Nov. 19, 1929, p. 32:3. Bucher states that "In 1929, he [Zworykin] demonstrated an improved camera tube [the RCA Iconoscope] and on November 18, 1929, he demonstrated his electronic television system before the Convention of the Institute of Radio Engineers at Rochester, New York. This was the first public demonstration of the use of the cathode ray tube in television reception and antedated all other claimants to priority" ("Television and Sarnoff," p. 27). Neither of these statements are true. There was no iconoscope in existence in 1929 and while the kinescope was operating quite well in the Westinghouse laboratory, it was not demonstrated at this convention.

63. "Pictures Painted on Screen," *New York Times,* Nov. 24, 1929, sec. 11, p. 12:1. Over a year later, in January 1931, for those who had not seen the kinescope in operation, it seemed to offer little hope. For instance see A. Dinsdale, "De-Bunking Television," *Radio News* 12 (Jan. 1931), 594. Ironically this article features a picture of Zworykin holding the kinescope, which had solved all of the problems discussed in the article.

64. Bell Labs 33089, Ives to Charlesworth, Dec. 16, 1929.

65. Bell Labs 33089, memo from F. Gray dated Mar. 26, 1930. What is also ironic is that at this time the Bell labs had on its staff C. J. Davisson (1881–1958), who was later to win a Nobel Prize (1937) for his works on "electronic diffraction." See C. J. Davisson and C. J. Calbick, "Electron Lenses," *American Physical Society* 38 (1931), 585. There seemed to be no coordination between him and Gray, who was working on the cathode-ray tube problem. Later, an electrostatic focused tube (Western Electric 326A) was built and operated in the Bell labs for litigation purposes. See F. Smits, ed., *A History of Engineering and Science in the Bell System* (n.p.: AT & T Bell Laboratories, 1985), p. 144.

In 1937 the Bell labs were testing their new coaxial transmission system. In order to get a tube as perfect as possible, Davisson built a monstrous five-foot tube that varied the strength of the electron beam by deflecting it with two plates in the same manner used by Boris Rozing in 1907. See M. E. Streiby, "Coaxial Cable System for Television Transmission," *Bell System Technical Journal* 17 (July 1938), 447.

See Takayanagi File, "Research and Development of All Electronic Television System," pp. 9–10 (see Japanese Pat. no. 90,592 [filed Mar. 28, 1930; issued Mar. 5, 1931]).

66. "Television in Color Shown for First Time by Bell Telephone Laboratory," *New York Times,* June 28, 1929, sec. 10, p. 25:1; "Details of Color Television Receiving Apparatus," *New York Times,* July 7, 1929, p. 15:5; H. E. Ives, "Television in Color," *Science and Invention* 20 (Jan. 1930), 400–401, 474; and H. E. Ives and A. L. Johnsrud, "Television in Colors by a Beam Scanning Method," *Journal of the Optical Society of America* 20 (Jan. 1930), 11–22 (paper presented July 20, 1929).

67. See also H. E. Ives. "Two Way Television," *Bell Laboratories Record* 18 (May 1930), 399–404.

68. VKZ MS, p. 84. With the departure of Ogloblinsky, all television work by Belin and Holweck seems to have ceased. Dauvillier wrote me that he was unable to continue his research due to the Depression. The only television research in France was then being done by René Barthélemy, Henri de France, and Marc Chauvierre. In England, the Baird Television Company was broadcasting a thirty-line picture to a limited audience in London, but it too was undergoing hard times and was about to go into bankruptcy. In Germany there was work being done by Telefunken (in conjunction with General Electric), Fernseh (in conjunction with Baird Television), and independent research by Manfred von Ardenne and Kurt Schlesinger. Kenjiro Takayanagi was also making good progress in Japan. However, the most significant research was being done in the United States: by Zworykin in Camden and Farnsworth in San Francisco.

Chapter 7: The Iconoscope

1.　　　　. File, Pat. Int. no. 73,203 (Iams v. Ballard v. Farnsworth); Westinghouse File, Research Memo 6-6705-1, p. 4.

2. Bucher, "Television and Sarnoff," p. 45; Bitting, "Creating an Industry," p. 1017; Sobel, *RCA*, p. 124; Bilby, *The General*, p. 121; "W. R. G. Baker to Be Vice President in Charge of Manufacturing," *New York Times*, Nov. 19, 1930, p. 31:1; on Albert F. Murray's promotion to division engineer of research and development at RCA Victor, see "Institute Notes and Radio News," *Proceedings of the Institute of Radio Engineers* 18 (Mar. 1930), 569; interview with Murray at his home in Washington, D.C., on Apr. 18, 1978; see also Albert F. Murray to author, June 14, 1978. Engstrom's move was also noted in "Institute Notes and Radio News," *Proceedings of the Institute of Radio Engineers* 18 (May 1930), 912; and "Engstrom Replaces Retiring Burns at RCA," *Broadcasting*, Dec. 4, 1961, 76–77. I have not been able to find a contemporary reference about Engstrom being head of his own department in 1930. He did not become director of general research for the RCA manufacturing division of RCA Victor until 1937. See "Institute News and Radio Notes," *Proceedings of the Institute of Radio Engineers* 25 (Dec. 1937), 1511. Ulrey's position was confirmed in an interview with Iams.

3. VKZ MS, p. 85; VKZ Pat. File, Pat. Int. no. 73,203.

4. Bucher, "Television and Sarnoff," pp. 47–48; Iams File, Zworykin to Iams, Sept. 12, 1930; Westinghouse File, Research Department Report R-7295-2A, V. K. Zworykin, "Development of Television Transmitter," closing report, Nov. 17, 1930.

5. RCA Stockholders' Annual Report, 1929, pp. 3–4. This was also true for the Japanese Victor Company. It appears that although the telephone group was at odds with the radio group, they were still exchanging technical information because of cross-licensing agreements.

6. I had the good fortune to be the first television historian to visit EMI's archives in Hayes, Middlesex. This took place on Sept. 18–19, 1978. For some reason they had been closed to all visitors before this time. However, I wrote them and they replied that they would be pleased to have me go through their television files. I wish to thank Brian Samain and Leonard Petit for their cooperation. This will be known as the EMI File, Television (see esp. p. 74).

7. VKZ Pat. File, Pat. Int. no. 62,047, pp. 193–95. See also Pat. Int. no. 62,043.

8. See Everson, *Story of Television*, pp. 125–26; Hofer says that Zworykin's visit occurred between Apr. 10 and 13, 1930 (Thursday to Sunday) ("Philo Farnsworth: Quiet Contributor," p. 69). However Pat. Int. no. 64,027, Brief on Behalf of Philo T. Farnsworth, Testimony of D. K. Lippincott, says "April 16" (p. 270). For details of Farnsworth's magnetic focused picture tube, called the "Thermionic Oscillograph," see P. T. Farnsworth, U.S. Pat. no. 2,099,846 (filed June 14, 1930; issued Nov. 23, 1937). Basically the picture tube used the same focusing coil extending the entire length of the tube as did the dissector tube. This was an old idea and while it presented a practical method for focusing the electron beam, it prevented the tube from being modulated at low velocity and accelerated to high brightness by a second anode (as Zworykin had done in the kinescope). This resulted in a small, dim picture. This contradicts the claims of Farnsworth's supporters that the

present-day magnetically focused tube was invented by Farnsworth. No one else used Farnsworth's method and it soon became obsolete.

9. Knox McIlwain, "Survey of Television Pickup Devices," *Journal of Applied Physics* 10 (July 1939), 440; VKZ Pat. File, Pat. Int. no. 62,047, Testimony of Lubcke, p. 9; and Pat. Int. no. 73,203. For Farnsworth's "slope wave generator," see P. T. Farnsworth and H. G. Lubcke, U.S. Pat. no. 2,059,219 (filed May 5, 1930; issued Nov. 3, 1936). See also P. T. Farnsworth, U.S. Pat. no. 2,246,625 (filed May 5, 1930; issued June 24, 1941). This was for a scanning system.

10. VKZ Pat. File, Pat. Int. no. 64,027, p. 64; Brief on behalf of Farnsworth, pp. 70–71; Record for V. K. Zworykin, p. 198; see also Pat. Int. no. 73,203.

11. Alexanderson File, Alexanderson to H. E. Dunham (pat. dept.), June 4, 1930.

12. VKZ Pat. File, Pat. Int. no. 64,027, Brief on behalf of Farnsworth, p. 71.

13. V. K. Zworykin, U.S. Pat. no. 2,246,283 (filed May 1, 1930; issued June 17, 1941). This is the famous serial no. 448,834 application that was part of so many patent interferences. Several new claims were made after Nov. 23, 1938.

14. Work on these tubes was confirmed in interviews with Albert Murray, Arthur Vance, and Harley Iams. Both Murray and Wright confirmed in their interviews that the first tubes built were of the simpler linear version. For the patent history, see C. F. Jenkins, U.S. Pat. no. 1,756,291 (filed July 16, 1928; issued Apr. 29, 1930), and H. J. Round, U.S. Pat. no. 1,759,594 (conv. date May 21, 1926; filed May 11, 1927; issued May 20, 1930). The story of "charge storage" in television is quite complicated and would require a book of its own. But in going through the various patents files and interferences, I have come across several interesting items. The first reference is a proposal by R. C. Mathes of Bell labs to store the charges in a series of columns in order to save bandwidth (Bell Labs 33089, memorandum from R. C. Mathes, May 8, 1926). Next was the May 21, 1926, English patent application of H. J. Round, followed by a patent application by Mathes for a means of increasing the brightness of the Bell labs' large screen device (R. C. Mathes, U.S. Pat. no. 2,058,898 [filed Nov. 12, 1927; issued Oct. 27, 1936]). Then there is the July 16, 1928, application of C. F. Jenkins. Even Frank Gray sent out a memorandum on several cathode-ray transmitters using "charge storage"; see Bell Labs 33089, F. Gray, "Summary of Memorandum on Proposed Television Transmitters," May 20, 1930. Zworykin did not get around to filing for his patent until May 1, 1930.

Zworykin's patent lawyers included "a bank of condenser elements" on Sept. 30, 1932, as claim 32, which ultimately became claim 13 (VKZ Pat. File, 1923 application). Unless one knows the background to this patent, it appears that Zworykin did apply for a patent with "charge storage" in December 1923, when in fact this was added to the claims in 1932.

15. K. Takayanagi, Japanese Pat. no. 93,465 (filed Dec. 27, 1930; issued

Nov. 4, 1931); A. N. Konstantinov, Russ. Pat. no. 39,380 (filed Dec. 28, 1930; issued Nov. 30, 1934).

16. VKZ Pat. File, Pat. Int. no. 62,721, which Jenkins lost on Oct. 17, 1938. See also Zworykin, Fr. Pat. no 715,912 (conv. date Apr. 23, 1931; issued Dec. 11, 1931).

17. VKZ Pat. File, Pat. Int. no. 62,721; Westinghouse v. RCA, U.S. District Court for the District of Delaware, no. 1183 in Equity.

18. See Jenkins, *Radiomovies*, pp. 85–87. For pictures of Jenkins's apparatus, see "Outdoor Radiovisor," *Science and Invention* 17 (Jan. 1929), 840; and "Transmitting Board," *Science and Invention* 17 (June 1929), 151.

19. This was confirmed in interviews with Les Flory, Arthur Vance, and Harley Iams.

In 1978, I visited Les Flory at his home in Princeton. During this interview, he brought out a mosaic of one of these tubes which he had saved (see fig. 37), which closely resembles the tube described in Zworykin's patent applications of May 1 and especially July 17, 1930 (see Chap. 7, n. 30 below). For details of the difficulties involved in building such two-sided mosaics, see V. K. Zworykin and G. A. Morton, *Television: The Electronics of Image Transmission* (New York: John Wiley & Sons, 1940), p. 304.

Les Flory (born Mar. 17, 1907) received his B.S. degree in electrical engineering from the University of Kansas in 1930. He joined the research division of RCA Victor under Zworykin in 1931. He claims that his first assignment was the evaluation of the Farnsworth image dissector. He was engaged in research on television tubes and related electronic problems and was part of the original group that developed the iconoscope. In 1942, he was transferred to the RCA laboratories division in Princeton. From 1949 to 1954, he was in charge of work on storage tubes, and since 1949, he has been in charge of work on industrial television. More recently he has supervised the work of the General Research Laboratory on Medical Electronics, Astronomical Television and Highway Vehicle Control. He is a member of Sigma Xi and is a Fellow of the Institute of Radio Engineers. Material from my interviews with him will be referred to as the Flory File.

20. Iams File, Zworykin to Iams, June 20, 1930.

21. Iams File, Zworykin to Iams, June 20, 1930.

22. Zworykin and Wilson, *Photocells and Their Application*.

23. Details of this demonstration were explained in several memos in the Alexanderson file, starting with "Minutes of the Technical Committee Meeting of April 1, 1930." Alexanderson then wrote to Sarnoff, "I believe that a public showing of the apparatus as now set up should first be made in Schenectady and then the apparatus should be transferred to New York where the entertainment can be properly staged" (Alexanderson to Sarnoff, Apr. 2, 1930). This proposed New York demonstration never happened.

This solves a problem that had long plagued me. According to Barton Kreuser, "RCA gave a large-screen demonstration at the RKO 58th St Theatre on January 16, 1930" ("Progress Report—Theatre Television," *Journal of the Society of Motion Picture Engineers* 53 [Aug. 1949], 128–36). Although I have searched carefully, I could not find a single reference to this in either Bucher, "Television and Sarnoff," or the *New York Times* or in any other contemporary source. Apparently the demonstration was planned, but it never occurred. See also Van Dyck, "Early Days of Television," pp. 10–12.

24. Alexanderson was quite concerned that General Electric get credit for the first large-screen television program just as KDKA got credit for radio (Alexanderson File, memo from Alexanderson to M. P. Rice [GE publicity], May 8, 1930). See also Alexanderson File, "Engineering Report for 1930," Radio Consulting Dept., undated; "RCA Has Affiliated with Radio-Keith-Orpheum to Install Television Sets in Theatres across the Country," *New York Times*, June 14, 1930, p. 1:5; "Television on Stage," *The Times* (London), May 23, 1930, p. 14b; "Television Advances from Peephole to Screen," *Radio News* 12 (Sept. 1930), 228–30, 268–69.

25. Vance File; Vance said that he was in charge of the Zworykin/Westinghouse demonstration. See also Alexanderson File, B. R. Cummings (RCA) to Alexanderson, June 16, 1930. This spoke of having R. Kell bring a forty-eight-line television receiver and pickup equipment to Camden to be in working order by July 15. See also Bucher, "Television and Sarnoff," pp. 47–48; see also Kell's note to W. Rider of GE, "The W.E. case 14718 (the Kinescope) of Dr. Zworykin is a very important invention in the television art. It covers the cathode ray tube as now used for television reception" (Kell to Rider, July 11, 1930). See also VKZ Pat. File, Pat. Int. no. 73,203.

26. Bucher, "Television and Sarnoff," pp. 47–48.

27. Iams File, Zworykin to Iams, Sept. 12, 1930.

William Arthur Tolson (b. 1896) was born at San Angelo, Texas. He received his B.S. degree in electrical engineering from Texas A. and M. in 1923. He was with General Electric from 1923 until 1930 and joined RCA Victor in Camden in 1930. He was head of "Development Television Test" until late 1931, when he was transferred to lesser duties on the Zworykin television project.

Merrill A. Trainer was born (b. 1905) in Philadelphia and received his B.S. degree from Drexel Institute in 1927. He was an assistant in Alexanderson's radio consulting laboratory, part of General Electric, from 1927 to 1930. He joined RCA Victor in Camden in 1930.

28. For the complete story see Archer, *Big Business and Radio*, pp. 351–86. Archer stated, "Indeed, when one observes the whole picture, before, during and after separation, one cannot escape the conclusion that RCA was the company that gained the greatest advantages in this stoutly contested and long drawn controversy" (p. 386). See also *New York Times*, Nov. 27, 1932, sec. 4, p. 1:8; RCA Stockholders' Report, Mar. 14, 1932, pp. 5–6 and Nov. 21, 1932, pp. 1–3; also in RCA Stockholders' Report, Mar. 1, 1933, pp. 5–6; and in Lyons, *David Sarnoff*, p. 166.

29. Bucher, "Television and Sarnoff," p. 56; "Institute Notes and Radio News," *Proceedings of the Institute of Radio Engineers* 18 (Aug. 1930), 1478; "In Memoriam: Robert E. Shelby," *RCA Review* 18 (Mar. 1956), i.

30. V. K. Zworykin, U.S. Pat. no. 2,157,048 (filed July 17, 1930; issued May 2, 1939); V. K. Zworykin, U.S. Pat. no. 1,955,899 (filed Sept. 25, 1930;

issued Apr. 24, 1934); and V. K. Zworykin, U.S. Pat. no. 2,084,364 (filed Dec. 24, 1930; issued June 22, 1937).

31. Iams File, Zworykin to Iams, Sept. 12, 1930. In an interview with me in July 1981, Iams confirmed the fact that Ogloblinsky was actually running the camera tube project and John Batchelor was responsible for most of the work on the picture tube.

32. VKZ File, Ogloblinsky to Chevallier, Sept. 11, 1929. VKZ Pat. File, the Oct. 9 date is from the Chevallier patent file.

33. VKZ File, Ogloblinsky to Chevallier, June 18, 1930 (italics in original).

34. VKZ File, Ogloblinsky to Chevallier, Aug. 14, 1930.

35. VKZ File, Ogloblinsky to Chevallier, Aug. 22, 1930.

36. VKZ File, Chevallier to Zworykin, Nov. [date has been erased], 1930. On the international convention, see Howard I. Forman, ed., *Patents, Research and Management* (New York: Central Book Co., 1961), pp. 258–59. P. E. L. Chevallier, U.S. Pat. no. 2,021,252 (conv. date, Oct. 20, 1930; filed Oct. 25, 1929; issued Nov. 19, 1935). Chevallier's changes to the American patent include an additional element d in fig. 1 and an additional J and d in fig. 2. This is explained as follows, "In both forms of apparatus cylinders d, d together form concentration element P." In fig. 2 of the original French patent, there are two connections to the glass tube, whereas the American patent only shows one at T, the rear of the tube. Also claim 8 of the American patent states for the first time, "accelerating said electron beam *after* deflection." In the British patent (no. 360,654) there appears a new fig. 3, which shows an enlarged detail of the concentration electrode. This clearly came from Chevallier's second American patent (no. 2,021,253), which was divided out of U.S. Pat. no. 2,021,252. In addition, the British patent shows for the first time a high-voltage connection, similar to Zworykin's, so that Chevallier could claim the advantages of modulation and deflection at low voltage and high brightness. Since none of this is described in the original French patent there may be questions as to whether these items should have been included, but they were accepted.

In any case, outside of the fact of electrostatic focus, there were enough differences between Chevallier's original French application and Zworykin's Westinghouse application to have given Zworykin's improvements legitimacy. The biggest difference, of course, was the position of the deflection units in Zworykin's patent between the first and second anodes. Chevallier's patent showed the deflection unit part of the second anode section. Why the RCA patent office panicked when they first saw the Chevallier American patent application we shall never know. But they quickly decided not to challenge him and ultimately gave him everything he demanded.

37. VKZ File, Chevallier to Zworykin, Dec. 2, 1930; Zworykin to Chevallier, Dec. 18, 1930. Zworykin's reaction to Chevallier's patent was quite correct. It was merely an update of the French 1927 patent. However, the RCA patent department took a rather dim view of it.

38. P. E. L. Chevallier, Fr. Pat. no. 699,478 (filed Oct. 25, 1929; accepted Dec. 9, 1930; published Feb. 16, 1931).

39. VKZ Pat. File, the Kinescope patent file. On July 19, 1932, Grover

became Zworykin's patent attorney at RCA. So he was handling both his and Chevallier's patents at the same time. See Westinghouse (Zworykin), Fr. Pat. no. 705,523 (conv. date Nov. 10, 1930; filed Nov. 16, 1929; issued June 9, 1931); P. E. L. Chevallier, U.S. Pat. no. 2,021,253 (filed Oct. 20, 1930; divided Aug. 25, 1931; issued Nov. 19, 1935).

I was first told of the Chevallier/Zworykin Kinescope story by Dr. James D. McGee, whom I interviewed in 1978, when he was Emeritus Professor of Applied Physics at the University of London and Senior Research Fellow at the Imperial College. He let me know from the start that he had no appreciation for Zworykin, and the first thing he told me was the Zworykin had *not* invented the kinescope. When I pressed him for details, he suggested that when I got home to look up the patents of Pierre Chevallier. He also suggested that Zworykin did not invent the iconoscope but had copied it from Campbell Swinton. McGee then went on to tell me that all of his work on camera tubes at EMI was based solely on Campbell Swinton's theories. In his written work, as in this interview, the theme was always that neither EMI nor the British owed anything to either RCA or Zworykin. When I left, he gave me an unpublished paper entitled "The Early Development of the Television," which he had written in 1976–77, on his work at EMI from 1932 on. All references to it and to this interview will be called the McGee File.

In an article on Campbell Swinton, McGee had this to say about Zworykin's 1923 patent application: "Unfortunately the structure of the mosaic is such that on scanning each mosaic element the integrated charge simply neutralizes the electrostatically held charge on the signal plate and so produces no signal. Thus the device cannot operate as claimed and hence the patent must be considered invalid" ("The Contribution of A. A. Campbell Swinton F.R.S. to Television," *Royal Society of London, Notes and Records* 32 [July 1977], 91–105, quotation on p. 97). However, this was not true, as Zworykin indicated that while no storage was involved, the mosaic did give off a pulse that produced a signal ("Early Days," p. 71).

James Dwyer McGee (1903–87) was born in Canberra. Following graduation from St. Patrick's College, Goulburn, he entered Sydney University and received a B.S. degree in 1927 and an M.S. in 1928. He went to Clare College, Cambridge, where he worked in the Cavendish Laboratory as a research student for three years under Lord Rutherford and Sir James Chadwick, receiving his doctorate in 1931. In January 1932 he joined the EMI research laboratories under G. E. Condliffe, working on development of television devices. During the war he developed various electronic infrared image converters and the instruments in which they were used. Upon retiring from teaching at Imperial College, London, he returned to his native Australia.

40. George, "New Type of Hot Cathode Oscillograph."

41. R. H. George, U.S. Pat. no. 2,086,546 (filed Sept. 14, 1929; issued July 13, 1937). RCA took over processing of this patent on Sept. 7, 1934.

42. Marc Chauvierre, "Qui a inventé la télévision?" *La Liason* nos. 127 and 128 (1940?), 17.

43. Spencer's visit is related in VKZ Pat. File, Pat. App. no. 468,610 (U.S.

Pat. no. 2,157,048) of July 17, 1930; Spencer to Goldsborough, Jan. 30, 1931 ("applicant is requested to file affidavits showing that the device is operative," p. 2); Goldsborough to Spencer, July 1931, p. 3. See for instance Pat. Int. no. 54,922 (McCreary v. Zworykin), Feb. 8, 1932, p. 1.

44. Sobel, *RCA*, p. 105; Bilby, *The General*, p. 122.

45. "Sarnoff Predicts Era of Television," *New York Times*, May 6, 1931, p. 27:1; "Radio-Vision Era Is Dawning," *New York Times*, May 31, 1931, sec. 9, p. 9:1, featuring a picture of Zworykin with his kinescope. See also Bucher, "Television and Sarnoff," p. 56.

46. Bucher, "Television and Sarnoff," pp. 62–63; E. W. Engstrom, "A Study of Television Image Characteristics," *Proceedings of the Institute of Radio Engineers* 21 (Dec. 1933), 1631–51. See "Engstrom Replaces Retiring Burns at RCA," *Broadcasting* 61 (Dec. 4, 1961), 76–78; "Dr. E. W. Engstrom Elected President of RCA," *Broadcast News* 112 (Dec. 1961), 5. For the radio transmitters, see L. F. Jones, "A Study of the Propagation of Wave-Lengths Between Three and Eight Meters," *Proceedings of the Institute of Radio Engineers* 21 (Mar. 1933), 349–85.

47. Philo T. Farnsworth, "An Electrical Scanning System for Television," *Radio Industries* 5 (Nov. 1930), 386–89, 401–3; "Visual Broadcasting Still an Experiment," *Radio News* 12 (Feb. 1931), 761; A. Dinsdale, "Television by Cathode Ray," *Wireless World* 28 (Mar. 18, 1931), 286–88; P. T. Farnsworth, "Scanning with an Electric Pencil," *Television News* 1 (Mar.–Apr. 1931), 48–51; A. Dinsdale, "Television Takes the Next Step," *Science and Invention* 19 (May 1931), 46–47; A. H. Halloran, "Scanning without a Disc," *Radio News* 12 (May 1931), 998–99, 1015; W. G. W. Mitchell, "Developments in Television," *Journal of the Royal Society of Arts*, May 22, 1931, pp. 616–42. According to Mitchell, "Farnsworth is using the cathode-ray tube for reception, and in addition, he is about the only important worker making use of strictly electrical methods for transmission" (pp. 636–37). For the Farnsworth demonstration, see "Institute News and Radio Notes," *Proceedings of the Institute of Radio Engineers* 19 (Mar. 1931), 331.

48. See Everson, *Story of Television*, pp. 132–36, and Maclaurin, *Invention and Innovation*, pp. 208, 214.

49. The details of Sarnoff's visit are given in Schatzkin and Kiger, "Philo T. Farnsworth," pp. 18–19. See also Everson, *Story of Television*, p. 199; Sobel, *RCA*, p. 125. The date is hard to pin down, but Sarnoff was in California during the week of May 18, 1931, so I presume that that is when he visited Farnsworth's laboratory. See Bucher, "Television and Sarnoff," p. 54. In "Radio-Vision Era Is Dawning," Sarnoff discloses plans and discusses probable influence on home, stage, and screen. This seems to have been the last of Sarnoff's optimistic statements about television for a few years. Above all, Sarnoff was a realist; he knew that a few more years in the laboratory could only improve the RCA television system.

50. Zworykin and Morton, *Television*, p. 304.

51. Iams File, Iams to author, July 9, 1981. Here Iams states that he, Flory, Essig, Ogloblinsky, and Zworykin worked on the new tube. (For some

reason he left out Arthur Vance). In a letter of July 25, 1983, he related his role in the new iconoscope. He enclosed a copy of his U.S. Pat. no. 2,141,789 (filed Dec. 12, 1930; issued Dec. 27, 1938), which seems to prove that such an arrangement would work and would result in increased electrical output. He was right of course; this was proven by the success of the new single-sided tube. On the importance of Farnsworth's work, Les Flory told me, "Farnsworth was the first person anywhere to demonstrate a workable electronic television system. The fact that other later developments resulted in better performance is beside the point" (Flory File).

52. See S. F. Essig, U.S. Pat. no. 2,065,570 (filed Feb. 24, 1931; issued Dec. 29, 1936) and U.S. Pat. no. 2,020,305 (filed July 30, 1931; issued Nov. 12, 1935). The story of Essig's accident was told to me by Iams on Aug. 2, 1981 (Iams File).

53. This was mentioned to me by Iams on Aug. 2, 1981 (Iams File). Compare VKZ MS, pp. 81, 82–83, and Zworykin's engineering notebook, which I found at the David Sarnoff Research Center in Princeton. An entry dated Oct. 23, 1931, mentions this event. See also Garret and Mumford, "Zworykin successfully tested a tube which had been built in his laboratory in November 1931" ("History of Television," p. 35). They include a picture of this iconoscope with the caption "Fig. 14.—An early iconoscope—made in Zworykin's laboratory on the 9th November 1931, and successfully tested by him on the following day" (p. 37).

54. V. K. Zworykin, U.S. Pat. no. 2,021,907 (filed Nov. 13, 1931; issued Nov. 26, 1935); Kolomon Tihanyi, Br. Pat. no. 313,456 (conv. date June 11, 1928; filed June 11, 1929; void; but it was published Feb. 3, 1931); also Br. Pat. no. 315,362 (filed July 12, 1928; also void); K. Tihanyi, Fr. Pat. no. 676,546 (conv. date June 11, 1928; filed June 11, 1929; issued Feb. 24, 1930); K. Tihanyi, U.S. Pat. no. 2,158,259 (filed June 10, 1929; divided Mar. 8, 1936; issued May 16, 1939). (I was able to locate and interview Tihanyi's daughter, Katrina Glass, in Los Angeles in 1977. She repeated the claim that her father had invented the iconoscope. She also told me that her father worked for Siemens in Berlin the late 1920s.) Francois C. P. Henroteau, U.S. Pat. no. 1,903,112 (filed May 29, 1929; renewed Aug. 13, 1931; issued Mar. 28, 1933); also U.S. Pat. no. 1,903,113 (filed May 29, 1929; divided and refiled Sept. 8, 1930; issued Mar. 28, 1933). There is also a Br. Pat. no. 335,958 (filed June 4, 1929; issued Oct. 6, 1930). Henroteau claims that he first applied for a U.S. patent on Aug. 10, 1928 (serial no. 298,809). He sent a copy to Zworykin at Westinghouse on Oct. 1, 1928 (received Oct. 3, 1928). It was allowed on Aug. 18, 1930, but for some reason Henroteau's patent attorney decided to withdraw but not to abandon it and stopped the application on Feb. 3, 1931. It was then renewed on Aug. 13, 1931. It is almost certain that RCA's lawyers contacted Henroteau because of his British patent. RCA was negotiating with him in March 1935. See Pat. Int. no. 69,135 (Henroteau v. Zworykin), which he won on Jan. 4, 1936. This was one of the few interferences that Zworykin and RCA lost. See also George J. Blake and Henry D. Spooner, Br. Pat. no. 234,882 (filed Feb. 28, 1924; is-

sued May 28, 1925); Riccardo Bruni, Br. Pat. no. 310,424 (filed Apr. 25, 1928; issued Apr. 3, 1930).

55. There have been other claims as to who invented the iconoscope. For instance the *Soviet Encyclopedia* says, "S. I. Kataev invented the iconoscope in 1931" (vol. 11, p. 483). However, this claim is modified in the section on electronics: "S. I. Kataev and V. K. Zworykin, working independently of each other, developed iconoscopes in 1931 and *1931*" (vol. 30, p. 145); and, in the entry on Zworykin himself, "in 1931, Zworykin built the first iconoscope" (vol. 9, p. 702). In spite of all the claims for both Kataev and Konstantinov, the first Russian iconoscope was the work of Krueser, who actually built and operated the first iconoscope in 1935. See Paul Shmakov, "The Development of Television in the USSR," *Journal of the Television Society* (London) 2 (Jan. 1935–Dec. 1938), 100–101. See also Novakosky, "100th Birthday of V. K. Zworykin," p. 66. For Takayanagi, see Japanese Pat. no. 100,037 (filed June 13, 1932; issued Mar. 9, 1933). Takayanagi had previously filed for a patent covering "charge storage"; see Japanese Pat. no. 93,465 (filed Dec. 27, 1930; issued Nov. 4, 1931). Takayanagi stated in a meeting with Zworykin in Camden in 1934 that Zworykin had applied for a patent but had been rejected by the Japanese patent office on the grounds that his (Takayanagi's) principle had already been established as a Japanese patent. Takayanagi continued, "Zworykin had applied for a Japanese patent for his iconoscope apparatus and method, thinking that I was unaware of them. Contrary to his expectations, my principle on the iconoscope had been patented in 1931" (Takayanagi File, "Research and Development of All Electronic Television System," p. 5). Zworykin's 1930 patent application certainly was sent to the Victor Company of Japan as they were receiving all of the RCA licensee bulletins as well as patent applications, just as EMI, Marconi, and Telefunken were. Whether these were shown to Takayanagi is not known. However, if he was watching the patent situation, he certainly knew of the Jenkins patent on "charge accumulation," which was issued Apr. 29, 1930. This patent, as I indicated before, had a tremendous influence on inventors all over the world. Zworykin's 1930 camera tube patent application was issued in France on Dec. 11, 1931.

Of the various contenders, three (Round, Tihany, and Henroteau) preceded Zworykin; and four (Konstantinov, Takayanagi, Kataev, and McGee) were after him. However the point to be repeated is that only Zworykin designed, built and operated a working tube. So his priority is beyond question.

56. See Knox McIlwain, "Survey of Television Pickup Devices," *Journal of Applied Physics* 10 (July 1939), 440.

57. See V. K. Zworykin, U.S. Pat. no. 2,021,907 (filed Nov. 13, 1931; issued Nov. 26, 1935). This early explanation came from Zworykin and Morton, *Television,* pp. 233–34.

58. The success of the RCA image dissector was related to me by Les Flory in several interviews. He furnished me with the picture of himself and Ballard on the roof at Camden testing out an RCA dissector camera (see fig.

39). The dissector's performance was also confirmed by Iams; see Iams File, interview of Aug. 2, 1981.

59. VKZ Pat. File, 1923 patent application, p. 75.

60. See RCA Stockholders' Report, Mar. 14, 1932, which says that in 1931 "the Gramophone Company Ltd., in which RCA held a substantial interest, was unified with the Columbia Graphaphone Co. Ltd. These two British companies were joined under a new holding company called Electric and Musical Industries, Ltd. As a result of this unification [RCA] now owns approximately 27 per cent interest in EMI" (p. 4). See also Briggs, *Golden Age of Wireless*, p. 76; Burns, *British Television*, pp. 187–88; EMI File, Television, p. 74, "EMI Directors Report," Oct. 24, 1932.

61. "New Strides Made by His Master's Voice Gramophone Co.," *New York Times*, Jan. 8, 1931, p. 12:2; "A New Television System," *Wireless World* 28 (Jan. 14, 1931), 38–39; W. G. W. Mitchell, "London Looks in at New Television Departure," *Science and Invention* 19 (May 1931), 21, 78–79; C. O. Browne, "Multi-Channel Television," *Journal of the Institution of Electrical Engineers* (London) 70 (Mar. 1932), 340–53. See also EMI File, Television, p. 144.

62. The fact that the Gramophone Company was working with Zworykin's cathode-ray tube was revealed in C. O. Browne, "Technical Problems in Connection with Television," *Journal of the Institution of Electrical Engineers* (London) 69 (Oct. 1931), 1232–34 (paper presented May 6, 1931). The personnel listing comes from EMI File, Television, p. 74. However, according to S. G. Sturmey, "the tubes which the EMI workers obtained from America were soft [gas filled], unstable and unable to be focused. . . . The first successful hard tube *ever* produced commercially were results of this [EMI] work." None of this is true of course. See Sturmey, *The Economic Development of Radio* (London: Gerald Duckworth & Co., 1958), p. 201.

63. See Browne, "Technical Problems," p. 1233. The information about the receiver came from my interview with Wright on Dec. 18, 1978. He stated that in the summer of 1931 RCA sent one of Zworykin's kinescopes to EMI. He fitted it up with scanning circuits and tried to receive one of Baird's television transmissions (Wright File).

64. Manfred von Ardenne, "New Television Transmitters and Receivers Using Cathode Ray Tubes" *Fernsehen* 2 (Apr. 1931), 65–80 (in which von Ardenne discusses the use of the cathode-ray tube as the solution of presenting pictures of eight to ten thousand elements); "Illustration of Baron M. von Ardenne's Cathode-Ray Television Sender," *New York Times*, Aug. 16, 1931, sec. 9, p. 8:4; E. H. Traub, "Television at the 1931 Berlin Radio Exhibition," *Journal of the Television Society* 1 (Dec. 1931), 100–103. See M. von Ardenne, Br. Pat. no. 387,536 (conv. date Mar. 27, 1931; filed Mar. 29, 1932; issued Feb. 9, 1933. Von Ardenne had been on the RCA list of those receiving licensee bulletins, patent applications, and so on since 1926. He also confirmed that he knew of Zworykin's early work on camera tubes in an interview I had with him in 1984. Von Ardenne's cathode-ray tube was quite popular with the British in their early work on radar. This dependence on a foreign supplier, especially a German source, gave the British the impetus

to start producing cathode-ray tubes themselves. Von Ardenne continued to work on television until the German television industry was put under the control of the German Air Ministry in August 1935. He then turned to work on the electron microscope and was about five years ahead of everyone else in this field. He, like Zworykin, turned to medical electronics in 1964 and operates a clinic for cancer research at his home laboratory.

Manfred von Ardenne was born in Hamburg on Jan. 20, 1907. He graduated from the gymnasium at sixteen and attended the University of Berlin for two years (1925 and 1926). By 1928, he had established a laboratory in the basement of his home in Lichterfelde, and had already received several patents and published his first book. His experiments with cathode-ray tubes led him to the all-electric cathode-ray system that he demonstrated in 1931. He worked on television until 1935, when all research was taken over by the Luftwaffe. He then experimented with image tubes and was one of the earliest inventors of the electron microscope, including the scanning electron microscope and the X-ray projection microscope.

During World War II he made significant contributions to the German atomic bomb project. In 1945 he voluntarily went to the USSR to work on their atomic bomb project. He returned to Dresden in 1955. In 1964 he set up a lavish castle on the Elbe with a medical laboratory for cancer research. I interviewed von Ardenne in his home on Aug. 3, 1984. Before I left he gave me most of his papers relating to his work on television. This material will be referred to as the Ardenne File.

65. EMI File, Television, pp. 144, 194; Bucher, "Television and Sarnoff," p. 56.

66. J. D. McGee, "The Life and Work of Sir Isaac Shoenberg," *Journal of the Royal Television Society* 13 (May–June 1971), 210.

Isaac Shoenberg (1880–1963) was born in Pinsk, a small city in northwestern Russia. He had wanted to be a mathematician but, because of his Jewish heritage, he had to settle for a degree in electrical engineering from the Polytechnical Institute of Kiev University. He first worked at a chemical engineering company, but very soon joined the Russian Marconi Company. There he was responsible for the research, design, and installation of the earliest wireless radio stations in Russia.

For EMI, he led the great engineering team that finally produced the first high-definition television system, which was adopted by the BBC in 1937. This system lasted until the start of World War II and returned to broadcasting in 1945. It remained in use until 1986. He was knighted for his services.

67. In spite of all the rhetoric from certain British historians claiming that EMI never received any technical information from RCA, this exchange was only the start of a fruitful friendship between the two companies. The official EMI line was expressed by S. J. Preston, who says that "it has often been suggested that British television is merely a copy of American television . . . for example, the Emitron is nothing more than the Iconoscope and the CPS Emitron is only the Orthicon." He denied this by stating, "Throughout the

formative period of television, RCA and EMI had certain commercial relations but these did not involve ownership of EMI or control of EMI policy by RCA." This of course was not true. David Sarnoff, president of RCA, sat on the board of directors of EMI and wielded enormous influence even though he only controlled 27 percent of the stock. He continued, "Nor was there was any exchange of research or manufacturing information between the two Companies. It is true that technical publications and copies of patent applications were exchanged, but 'know how' was not exchanged because RCA at the time only prepared to agree to such an exchange for a payment, which EMI was not prepared to make" ("Birth of a High Definition Television System," *Journal of the Television Society* [London] 7 [1953], 115–25, quotations p. 123). Preston's interpretation agrees with (and probably came from) McGee's assessment of the same situation. McGee said, "Now I can state categorically that there was no exchange of 'know-how' between the two companies during the crucial period of 1931 to 1935." However, he continues, "We did, of course, receive a steady stream of literature including patent specifications from many sources and we had a preview of patent applications from several companies such as R.C.A. and Telefunken, A.G. Our difficulty was to decide what was meaningful. The key word is 'know how.'" As far as I can determine EMI was given patent specifications and other important information that included "know how." What apparently happened was that while EMI was only too pleased to receive all of this information and help in their formative years, it was later decided that EMI could have done all of this on their own with no help from RCA. Certainly RCA did not have to hold EMI's hand and lead them through every step of the way. See McGee File, unpublished paper, "Early Development of Television," p. 53. I hope that the vital assistance given to EMI by RCA during the period of 1931 to 1933 will finally be appreciated, without lessening the great accomplishments made in Shoenberg's laboratory some three thousand miles away. From 1934 on, EMI owed RCA nothing and gave the world the first high-definition television service.

68. Additional details of Zworykin's new picture tube were given in V. K. Zworykin, "Improvements in Cathode-Ray Tube Design," *Electronics* 2 (Nov. 1931), 188–90. Strangely, there is not a word of this tube's application to television. It was to be used for the observation and study of repeating phenomena, observation and study of continuous phenomena, and observation of transient phenomena. What is of interest here is that though both Westinghouse and General Electric were forbidden to do any more television work, they both brought out advanced versions of Zworykin's original kinescope for oscillographic purposes. This emphasizes that there was no love lost between them after the consent decree. On Westinghouse, see W. O. Osbon, "A New Cathode-Ray Oscilloscope," *Electric Journal* 28 (May 1931), 322–24; on General Electric, G. F. Metcalf, "A New Cathode-Ray Oscillograph Tube," *Electronics* 3 (May 1932), 158–59.

According to the *Soviet Encyclopedia,* S. I. Kataev "developed the first high-vacuum receiving television tube in the USSR in 1931–1932" (vol. 11, p. 483). The period of 1931–32 also saw the entry of Allan B. Du Mont, who had been chief engineer of DeForest Radio Corporation, into the production of cathode-ray tubes. After DeForest went bankrupt, Du Mont, carefully observing the state of the art of television devices, started to build cathode-ray tubes for oscillographic purposes. Late in 1931, he set up a laboratory in his garage and started to build electrostatic focused, dual accelerating anode tubes similar to Zworykin's kinescopes. See Allan B. Du Mont, "An Investigation of Various Electrode Structures of Cathode Ray Tubes Suitable for Television Reception," *Proceedings of the Institute of Radio Engineers* 20 (Dec. 1932), 1863–77 (paper presented May 4, 1932). He later claimed that the Zworykin/RCA picture tube "was an old art and that he was not infringing on RCA's patent position" (Maclaurin, *Invention and Innovation*, p. 219). This marked the beginning of Du Mont's rise to importance in the television industry. See Electric & Musical Industries (R.C.A. Victor Co. Inc.) Br. Pat. no. 391,887 (filed Nov. 19, 1931; issued May 1, 1933).

69. The annual RCA Stockholders' Report for 1933 said of RCA's interest in EMI that "the latest reports are encouraging." But the report also noted that while EMI was engaged in the "manufacture of radio sets, electrical and mechanical phonographs, records, etc.[, i]t is not engaged in the field of communications."

70. See MacGregor Morris, "Cathode Ray Television," *Journal of the Television Society* (London) 1 (1931–34), 69–70. The other two papers, one by von Ardenne entitled "The Cathode-Ray Tube Method of Television" and the other by a student member, Coryton E. C. Roberts, entitled "A Vacuum Photo-Cell Type of Transmitter," were both published in the same issue of the journal (pp. 71–74 and 82–83). Roberts's image dissector tube never worked, nor did Dieckmann's. So while Farnsworth may not have really invented the image dissector, if we go by application date, he was the first and only one to make it work. See C. E. C. Roberts, Br. Pat. no. 318,331 (filed June 22, 1928; issued Sept. 5, 1929). It is interesting that this patent was filed after Farnsworth's but granted before his. Many historians go by issue date and thus put Dieckmann first, Roberts second, and Farnsworth third. Both the British and French patent offices were very quick to issue new patents because they were not as thorough as the U.S. Patent Office.

71. This was related to me in interviews by all of the Zworykin television group. See "R.C.A.-Victor to Reopen Plant Expecting Trade Gain," *New York Times*, Jan. 4, 1931, p. 1:2. For an overall assessment, see Archer, *Big Business and Radio*, p. 362.

Chapter 8: "A Century of Progress"

1. "Test Television Progress: Images Sent from Top of Empire State Building in Experiment," *New York Times*, May 18, 1932, p. 24:3; Bucher, "Television and Sarnoff," p. 64.

2. "News from Abroad (From Our Own Correspondent)," *Television* 1 (July 1932), 174; Bucher, "Television and Sarnoff," pp. 64–65.

3. "Television Images Are Leaping from a Skyscraper Pinnacle," *New York*

Times, May 22, 1932, sec. 8, p. 10:1; Bucher, "Television and Sarnoff," p. 71. Bucher, quoting from the 1933 stockholders' report, stated, "It is impossible to anticipate the exact time when this development can be introduced on an industrial basis."

4. VKZ MS p. 85. Here Zworykin confirms that in Camden he was joined by a group of General Electric engineers, headed by Elmer Engstrom, which had worked on television problems under Alexanderson. He gives no specific date. I had written RCA about this and was answered by Al Pinsky, who said that "Engstrom took charge of the television project in 1932"; Pinsky to author, May 20, 1983. There is some confusion about who was actually running the Empire State Building television project in 1931–32. According to "In Memoriam: Robert E. Shelby," "He [Shelby] was placed in charge of NBC's first experimental television installation atop the Empire State Building." This probably means that Shelby took over from T. A. Smith, who had been in charge of studio operations. Thus Shelby was now in charge of its cameras (both live and film) and monitors for NBC while Engstrom of RCA was in charge of its radio transmitter and antennas (construction and technical) operations. RCA often did this with experimental projects, turning them over to NBC only when operational.

See also "Dr. E. W. Engstrom Elected President of RCA," *Broadcast News* 112 (Dec. 1961), 4–5; "Who's Who at 'Radio Headquarters,'" *Broadcast News* 10 (Feb. 1934), 25; and Bitting, "Creating an Industry," p. 1020.

Elmer W. Engstrom (1901–85) was born in Minneapolis and received a B.S. degree from the University of Minnesota in 1923. He was with the radio engineering department of General Electric from 1923 to 1930, when he was transferred to the engineering department at RCA. He took over the television research department early in 1932. By 1937 he was in charge of all research for the RCA Manufacturing Company. In 1942 Engstrom was appointed director of general research at Princeton. In 1949 he received an honorary doctorate from New York University. In 1951 he was made vice president of RCA laboratories. In 1961 he was made president of RCA, a position he kept until 1965 when Robert Sarnoff took over. See E. W. Engstrom, "A Study of Television Image Characteristics," part 1, *Proceedings of the Institute of Radio Engineers* 21 (Dec. 1933), 1631–51.

5. Flory File, Flory to author, Aug. 17, 1981.

6. See V. K. Zworykin, G. A. Morton, and L. E. Flory, "Theory and Performance of the Iconoscope," *Proceedings of the Institute of Radio Engineers* 25 (Aug. 1937), 1071–92. See also J. D. McGee and H. G. Lubszynski, "E.M.I. Cathode Ray Television Transmission Tubes," *Journal of the Institution of Electrical and Electronic Engineers* (London) 84 (Apr. 1939), 468–75 (paper presented July 20, 1938; rev. Oct. 1938).

7. See figs. 41–44. Here you will see at least four different types of iconoscopes that were tried. How many more were lost or destroyed may never be known.

Les Flory provided me with pictures and information that indicated that by August 1932 the iconoscope was working quite well. It is interesting that

Flory called it the "Ogloblinsky iconoscope." See picture of Zworykin, Richard Campbell, and Ogloblinsky on the roof at Camden, observing the eclipse of the sun (fig. 47).

8. V. K. Zworykin, "Description of an Experimental Television System and the Kinescope," *Proceedings of the Institute of Radio Engineers* 21 (Dec. 1933), 1655–73. This article gives a figure of 1.5 candle-power for a second anode voltage of 6000 volts.

9. All of these patents incorporated interlaced scan: J. L. Baird, Br. Pat. no. 253,957 (filed Jan. 1, 1925; issued Aug. 25, 1926); R. D. Kell, Br. Pat. no. 308,277 (conv. date Mar. 20, 1928; filed Mar. 20, 1929; issued May 15, 1929); J. L. Baird, Br. Pat. no. 321,389 (filed June 5, 1928; issued Dec. 27, 1929); U. A. Sanabria, U.S. Pat. no. 1,805,848 (filed June 7, 1929; issued May 19, 1931). This was for a forty-five-hole triple interlaced scanning system. T. A. Smith, Br. Pat. no. 349,773 (conv. date June 27, 1929; filed May 27, 1930); Fritz Shröter, DRP Pat. no. 574,085 (filed Sept. 27, 1930; issued Mar. 23, 1933); M. von Ardenne, Br. Pat. no. 387,087 (conv. date Dec. 22, 1930; filed Dec. 21, 1931).

10. Randall C. Ballard, U.S. Pat. no 2,093,395 (filed Jan. 6, 1932; issued Sept. 14, 1937); Br. Pat. no. 394,597 (issued June 29, 1933). This was the second RCA patent application (the first was Zworykin's) that showed the new configuration of the single-sided iconoscope.

11. Randall C. Ballard, U.S. Pat. no. 2,152,234 (filed July 19, 1932; issued Mar. 28, 1939); Br. Pat. no. 420,391 (issued Nov. 30, 1934). For some reason, Ballard left the Zworykin research laboratory in Apr. 1932 and joined the U.S. Radio and Television Corporation in Chicago. See "Institute News and Radio Notes," *Proceedings of the Institute of Radio Engineers* 20 (Apr. 1932), 606. He rejoined RCA in Sept. 1935. See "Institute News and Radio Notes," *Proceedings of the Institute of Radio Engineers* 23 (Sept. 1935), 977.

12. See E. W. Engstrom, "Study of Television Image Characteristics," part 1, and part 2: "Determination of Frames Frequency for Television in Terms of Flicker Characteristics," *Proceedings of the Institute of Radio Engineers* 23 (Apr. 1935), 295–310. Earlier papers were E. W. Engstrom, "An Experimental Television System," *Proceedings of the Institute of Radio Engineers* 21 (Dec. 1933), 1652–54; V. K. Zworykin, "Description of an Experimental Television System and the Kinescope," *Proceedings of the Institute of Radio Engineers* 21 (Dec. 1933), 1656–73. For a description of the apparatus used at the Empire State Building, see R. D. Kell, "Description of Experimental Television Transmitting Apparatus," *Proceedings of the Institute of Radio Engineers* 21 (Dec. 1933), 1674–91; G. L. Beers, "Description of Experimental Television Receivers," *Proceedings of the Institute of Radio Engineers* 21 (Dec. 1933), 1692–1706; all of these papers received Feb. 20, 1933.

13. Vladimir K. Zworykin and Gregory N. Ogloblinsky, U.S. Pat. no. 2,178,093 (filed Mar. 10, 1932; issued Oct. 31, 1939). This was the subject of Pat. Int. nos. 69,636 and 73,812, both won by Farnsworth on Apr. 26, 1938.

14. Hofer, "Philo Farnsworth: Quiet Contributor," pp. 75–76.

15. VKZ Pat. File, Pat. Int. no. 64,027, Brief for Philo Farnsworth, 1933, pp. 49–60. This statement best characterized Zworykin's actions for the rest of his life. In his perception, certain facts were true and nothing could change his mind. Whether this was due to coaching of the RCA patent department or was of his own volition is not known. See Pat. Int. no. 64,027, Brief for Farnsworth, pp. 2–3; Final Hearing, Apr. 24, 1934, pp. 5, 46, 48.

16. VKZ Pat. File, Pat. Int. no. 60,779 (Zworykin v. Jenkins v. Mathes v. Round), Sworn affidavit of Harley A. Iams, Oct. 11, 1934; Briefs on behalf of Applicant (Vladimir K. Zworykin), paper no. 94, p. 54. Iams built four tubes for this affidavit. In 1934, Zworykin stated that "Transmitting tubes of this type were actually built quite a few years ago and proved the soundness of the basic idea. During the succeeding years this development was carried on in the Research Laboratories of the Westinghouse Electric and Manufacturing Company in East Pittsburgh." See Zworykin, "The Iconoscope—A Modern Version of the Electric Eye," *Proceedings of the Institute of Radio Engineers* 22 (Jan. 1934), 16–31. This was late 1925. It is apparent from Zworykin's July 13, 1925, patent application and patent amendments that in 1924–25 he had built and experimented with actual tubes and experienced the problems associated with them. I feel reasonably sure that, while his dates may be off, Zworykin did indeed give a demonstration to Westinghouse management (albeit with dire results) of the first electronic camera tube in 1925. It would certainly explain why he was not working on any active television system for Westinghouse after 1926.

After establishing that Zworykin did build and operate a camera tube in 1924–25, I was allowed to examine this "first camera," which was on display at the David Sarnoff Research Laboratories in Princeton. The tube was certainly different from what had been described in the literature. For instance the target is made of mica rather than a thin film of aluminum oxide. The photoelectric substance was turning yellow, which means that it probably wasn't potassium hydride, and distilling all over the tube. There is no grid structure in front of the target. I have also found, by looking at various pictures, that there may be more than one tube in existence, as there are subtle differences in their shapes. This disturbed me greatly. I asked Harley Iams about these in an interview on Jan. 16, 1977. He startled me by stating that the so-called 1924–25 tubes are "phonies" which he had never seen before. I asked him if they were any of the tubes he had built in 1934 for Pat. Int. no. 60,779. He stated emphatically that they were not the tubes described in his affidavit.

These tubes were never publicly shown before Zworykin retired in 1954. The first picture that I can find of this "1924" pickup tube was shown in A. G. Jensen, "The Evolution of Modern Television," *Journal of the Society of Motion Picture and Television Engineers* 69 (Nov. 1954), 181 (paper presented May 7, 1954). This was just before Zworykin retired from RCA (see fig. 68). Where had it been kept for thirty years? In his office with his papers and other mementoes? Possibly, yet no one had ever reported Zworykin showing it to them in any of his many interviews. He spoke of it, but never showed it. Yet the tube displayed is broken, and matches very closely the tube shown in figs. 8 and 9. The mesh screen is barely visible and there is no enclosure between it and the target. On the other hand, RCA later sent out publicity pictures of an unbroken tube, with a conical enclosure between the target and the mesh screen. This tube appears to be quite new and bears the number 40596. It seems obvious that it was a reconstruction specially made for his retirement festivities.

Eventually I did find out why the tube on display is not the original "1923 Iconoscope" as claimed. This explanation came from a tube collector, Howard Shrader of Princeton. In 1977, he showed me a display of television tubes that he had gathered from the RCA plant in Camden. Shrader told me that in 1942 when the RCA labs moved from Camden to Princeton he heard that the RCA patent museum was going to be eliminated due to lack of space. Every item was to be thrown out. The word went out to all collectors in the area that on a certain day all these items would be available for free collection. All tubes that were not taken were to be immediately destroyed. He told me that he and four other collectors worked very hard to stay ahead of the men with the hammers. He gathered about a thousand tubes and took them back with him to Princeton. Later, he was quite generous with them and gave me a dozen of his choice tubes. At one time he offered me his whole collection, but I had no room for them and turned him down (I have regretted this decision ever since). He did give me a variety of tubes which included a rare 1931 sleeve type iconoscope and a 1932 miniature iconoscope. I have never received the rare 1931 RCA dissector which he promised to me. So it is possible that the original 1924–25 camera tubes just did not survive this wanton destruction. Then when Zworykin retired in 1954, RCA simply had some new "old" tubes built and used them for display. This is just one more mystery that may never be solved.

It is interesting that in 1975 Zworykin is shown at his desk holding what the caption calls an "early iconoscope of the type he invented in 1922." It is really a commercial type of tube that had long been retired from service. It would have made more sense historically to have had him at least hold the exhibition tube that had been made for him in 1954.

17. See Pat. Int. no. 64,027.

18. See "Television's 50th Anniversary Issue," *Journal of the Royal Television Society* (London) 16 (Nov.–Dec. 1977), 93. This is a letter from Zworykin to W. G. W. Mitchell, secretary of the Television Society of London, Sept. 20, 1932; P. Hémardinquer, "Notre enquête sur la télévision," *La Nature* 60 (Sept. 1, 1932), 242.

19. E. W. Engstrom, "An Experimental Television System," *Proceedings of the Institute of Radio Engineers* 22 (Nov. 1934), 1241–45. In 1934 the *Proceedings of the Institute of Radio Engineers* published a series on RCA's experiemental television system. See E. W. Engstrom, "Introduction: An Experimental Television System"; R. D. Kell, A. V. Bedford, and M. A. Trainer, "The Transmitter"; R. S. Holmes, W. L. Carlson, and W. A. Tolson, "The Receiver"; and Charles J. Young, "Radio Relay Link for Television Signals," *Proceedings of the*

Institute of Radio Engineers 22 (Nov. 1934), 1241–94. See also R. D. Kell, A. V. Bedford, and M. A. Trainer, "Scanning Sequence and Repetition Rate of Television Images," *Proceedings of the Institute of Radio Engineers* 24 (Apr. 1936), 559–76.

20. O. E. Dunlap, Jr., "Television Lurks behind the Economic Curtain—Secrecy Makes Research Mysterious," *New York Times,* Jan. 1, 1933, sec. 8, p. 5:1.

21. VKZ MS p. 96; V. K. Zworykin, "On Electron Optics," *Journal of the Franklin Institute* 215 (May 1933), 535–55 (paper presented Feb. 24, 1933). In this article Zworykin cites E. Bruch and H. Johanson, *Annalen der Physik* 5 (Nov. 1932), 129; E. Bruch, *Zeitschrift für Physik* 78 (Sept. 1932), 3–4, 179–98; and M. Knoll, F. Q. Houtermans, and W. Schultze, *Zeitschrift für Physik* 78 (Oct. 4, 1932), 340–62.

There were also many other articles written on this subject. See, for example, M. Knoll and E. Ruska, "Nachtrag zur Mitteilung: Die magnetische Sammelspule für schnelle Elektronstrahlen," *Zeitschrift für technische Physik* 12 (1931), 448; and M. Knoll and E. Ruska, "Beitrage zur geometrischen Elektronoptik," *Annalen der Physik* 12 (Feb. 1932), 607–40, 641–61; M. Knoll and E. Ruska, "Elektronenmikroskop," *Zeitschrift für Physik* 78 (Oct. 4, 1932), 5–6, 318–39. This described a cold-cathode magnetic electron microscope.

22. "New 'Electrical Lens' an Aid to Television," *New York Times,* Mar. 5, 1933, p. 10:3. The advantages of a "short coil" (thin lens) were described by L. M. Myers in *Electron Optics: Theoretical and Practical* (New York: D. Van Nostrand Co., 1939). He claimed that "magnetic focusing is by far the simplest, neatest and most effective method known in electron optics," and that "in television tubes the spot obtainable is invariably better, inasmuch as it is sharper, smaller and free of aberration, especially halo" (p. 76). While the use of a magnetic coil for focus was the oldest method, no one before had combined it with the accelerating second anode that Zworykin had introduced with his kinescope; see Myers, *Television Optics,* pp. 290–91. Since there was no way to protect this method in the Patent Office, it was soon being used by everyone, including Farnsworth, who switched from long focus coils (thick lens) to short ones. Zworykin's tube was the beginning of the modern cathode-ray tube that is in universal use today.

23. Murray File, Murray to author, July 14, 1978.

24. Everson, *Story of Television,* pp. 133–36; Maclaurin, *Invention and Innovation,* pp. 208, 214–15; and Hofer, "Philo Farnsworth: Quiet Contributor," pp. 71–72.

25. VKZ Pat. File, all patent interferences won on May 19, 1933; quotation from "D. Sarnoff," *Radio Engineering* 13 (May 1933), 2.

26. Zworykin's paper, "The Iconoscope—A New Version of the Electric Eye," was announced in "Institute News and Radio Notes," *Proceedings of the Institute of Radio Engineers* 21 (June 1933), 745.

27. O. E. Dunlap, Jr., "Novel Radio Optic 'Sees,'" *New York Times,* June 25, 1933, sec. 9, p. 7:1; Bucher, "Television and Sarnoff," pp. 68–73.

28. V. K. Zworykin, "The Iconoscope—A Modern Version of the Electric

Eye," *Proceedings of the Institute of Radio Engineers* 22 (Jan. 1934), 16–32 (paper received June 14, 1933); Bucher, "Television and Sarnoff," p. 71. Similar versions of the Zworykin paper appeared in the following: "The Iconoscope—A Modern Version of the Electric Eye," *Broadcast News* no. 8 (Aug. 1933), 6–13; "Television with Cathode-Ray Tubes," *Journal of the Institute of Electrical Engineers* (London) 73 (Oct. 1933), 437–51 (paper received July 17, 1933); "Système de télévision par tubes à rayons cathodiques," *L'Onde Électrique* 12 (Nov. 1933), 501–39 (paper received July 26, 1933); "Fernsehen mit Kathodenstrahlröhren," *Hochfrequenztechnik und Elektroakustik* 43 (Apr. 1934), 1–37 (paper presented Sept. 4, 1933); "Television," *Journal of the Franklin Institute* 217 (Jan. 1934), 1–37 (paper presented Oct. 18, 1933).

29. W. L. Laurence, "Human-like Eye Made by Engineers to Televise Images," *New York Times*, June 27, 1933, p. 1:1; O. E. Dunlap, Jr., "Outlook for Radio-Sight," *New York Times*, July 2, 1933, sec. 9, p. 6:1 (including the first published photograph of an iconoscope). See profile of Zworykin in *Newsweek*, July 8, 1933, p. 1:25. See also "The Iconoscope, America's Latest Television Favorite," *Wireless World* 33 (Sept. 1, 1933), 197; Merle Cummings, "Television Advances," *Radio News* 15 (Oct. 1933), 214–15, 245–46; and A. Church, "Recent Developments in Television," *Nature* 132 (Sept. 30, 1933), 502–5.

30. "Television Device Shown," *New York Times*, July 11, 1933, p. 6:3; "The 'Eye' Gains Prestige," *New York Times*, July 16, 1933, sec. 10, p. 7:5; and "Television Advances," *Radio News*, 215, 246–47.

31. Bell Labs 33089, Frank Gray, "Note on Zworykin's Iconoscope," July 6, 1933; memorandum to H. D. Arnold from O. E. Buckley and H. E. Ives, July 6, 1933. Their comments are quite interesting: "Since the apparatus was neither demonstrated nor shown at the Chicago meeting, we have no way of judging how well it works in practice. But we know of no reason why successful results could not be ultimately obtained." See also memorandum from H. E. Ives to O. E. Buckley, July 20, 1933; Frank Gray, "Suggested Outline for Development Work on a Cathode-Ray Transmitter," July 31, 1933.

32. FOIA/State Dept. PO33, Zworykin's passport application dated June 16, 1933; VKZ MS p. 94; VKZ Pat. File. Zworykin brought much material with him on his trip to London in July 1933 (Al Pinsky to author, Nov. 14, 1978). Pinsky states that "Dr. Zworykin says Sarnoff sent him [Zworykin] to England to convey a great deal of RCA know-how to EMI." This was never confirmed by Shoenberg in his testimony before the Selsdon committee. However, on p. 29 of the committee's report he did admit that "Zworykin was the first to give a method for preparing a surface consisting of a great number of photo-cells insulated from each other" which EMI soon began to use. This information came from a visit to the British Post Office Archives in Sept. 1979. I found the files to be in great disorder, with pages (sometimes whole sections), photographs, and diagrams missing. This will be referred to as the Selsdon Committee File. I have also relied on Briggs, *Golden Age of Wireless*, and Burns, *British Television*, for many details of the Selsdon committee report.

33. EMI File, Television, pp. 162 and 218.

34. EMI File, Television, p. 193. This stated that a "hard" crt was developed by mid-1934 with a life of 1700 hours. At that time most gas-filled tubes had a life of around 500 hours.

35. Schoenberg's opinions found in the Selsdon committee report, p. 32; RCA Stockholders' Report, 1933: "EMI's agreement with RCA forbade it from engaging in the field of communications." See also Burns, "The HMV philosophy was that they should aim to develop an effective cathode-ray tube picture receiver, but should leave any attempt to develop a television cameras to others" (British Television, p. 293).

36. For the story of the 1932 Tedham and McGee camera tube see McGee File, "Early Development of Television," pp. 17–30. McGee claims that "during the first half of 1932 Tedham and I had long discussions as to how a tube embodying the essentials of Campbell Swinton's suggestions could be constructed" (p. 20). He admits that "Tedham and I could not resist the temptation to disobey orders [it was against company policy to build such a tube] and give it a trial" (p. 27). He continued, "a working tube was built and only lasted a short while until the gas evolution of electron gun cathode was too much for the photo cathode" (p. 30). See W. F. Tedham and J. D. McGee, Br. Pat. no. 406,353 (filed Aug. 25, 1932; issued Feb. 26, 1934). This was really for a method of depositing a substance through a wire mesh in order to get a great number of insulated photoelectric elements. Since this patent was applied for only three days after Shoenberg became director of research at EMI, I am now certain that he was not only aware of the experiment but part of it. I have long considered the McGee tale as a cover-up for the fact that they had willfully violated the RCA/EMI agreement. Shoenberg certainly resented RCA's restraints on EMI. See EMI File, Television, p. 144.

There is another side to this story. According to Keith Geddes, "By virtue of EMI's link with the Radio Corporation of American (RCA) they had access to work that was being carried out in the RCA laboratories on a revolutionary new camera tube the 'iconoscope.' This proved to be a crucial factor" (Broadcasting in Britain—1922–1972 [London: Her Majesty's Stationery Office, 1972], p. 22). See also Sturmey, Economic Development of Radio. According to Sturmey, "in 1931 he [Zworykin] produced the iconoscope—the details of which were published in 1933. The iconoscope after considerable modification became the basis of the emitron camera developed in England by EMI. . . . The decision was made before Zworykin published his iconoscope, but EMI had access to Zworykin's early work and did not need to start from scratch in developing the Emitron" (pp. 200–201).

37. RCA Stockholders' Report, 1934, reported that EMI was "engaged in the development of television transmitting and receiving apparatus." According to Burns, "Late in Autumn of 1933 McGee was asked to lead the group charged with the development of a pickup tube" (British Television, p. 293).

38. The date of Baird's "retiring" to his home laboratory in No. 3, Crescent Road, Sydenham, varies. Margaret Baird says that "we settled there in

July 1933" (*Television Baird,* p. 128). This would be shortly after West took over as chief engineer. Hutchinson says Baird returned to his home in Sydenham in 1934, cutting himself off from his company (*Baird,* p. 25). This is also confirmed by J. D. Percy, who says that in 1934, "Having seen the shape of television to come, Baird retired to his home where in a specially built private laboratory, he concentrated entirely on the problems of large screen, colour and stereoscopic television" ("John L. Baird, The Founder of British Television," memorial issue of the Television Society of London, March 1950, *Journal of the Television Society* 6 [1950–52], 14). Burns quotes from West's interview with Noel Ashbridge on July 13, 1933, "he [West] added that Moseley was out of the concern altogether, and that Baird was in the process of going out, but would probably be retained in a kind of secondary capacity for detail research" (*British Television,* p. 270). See also "John Logie Baird," in *35 Years of World Television,* special ed. of *International Broadcast Engineer Supplement,* 1971, p. 13.

39. This information about RCA and Russia came from Loren Jones; Jones to author, July 5, 1993. At this late date in my research I came across a reference to Jones. I wrote him, and he has kindly sent me much material concerning his adventures in Russia in 1930 and later, in 1937–38. He was also a close friend of Zworykin and furnished me with much valuable information on him as a man. I also interviewed Jones at his home on Aug. 9, 1993. This material will be known as the Jones File.

40. VKZ MS pp. 87–94. See Novakosky, "100th Birthday of V. K. Zworykin," p. 67. This was confirmed in an article by Henrikas Yushkiavitshus, vice chairman of the USSR's State Committee for Radio and Television, in his Shoenberg Memorial Lecture for the Royal Television Society in 1989. According to Yushkiavitshus, Zworykin did indeed visit the Soviet Union at the invitation of the Soviet government to lecture about his inventions and was in the USSR in 1934 and on Feb. 2, 1935 ("The Past and Future of Soviet Television," *Journal of the Royal Television Society* 26, no. 6 [Nov.–Dec. 1989], 316–22). There is no other source for this information. In his manuscript, Zworykin has lumped all of his visits to the Soviet Union into one section. The fact that he made a trip in August 1933, late in 1934, and probably in Feb. 1935 is all made part of one trip. Only the Freedom of Information Act and inquiries to the Soviet Union have made it possible for me to know when he actually visited it. For reasons unknown, there has never been a description of these visits to the Soviet Union in any American, French, German, or British journal that I could find. None of the newspapers carried so much as a hint of his travels there, although all of his travels elsewhere seem to have been reported on.

However, the FBI files I was sent do not go back to 1933. Instead they start in July 1944. I have written the FBI asking for any information prior to 1944 and am awaiting an answer. They seem to be very selective as to what information they will furnish.

It appears from the State Department records that they welcomed his early visits; the ease and speed with which he received a passport is proof. The

Soviet Union was a police state which closely guarded its borders. While it is known that literally hundreds of spies were at work for Russia during this period, it is certain that the U.S. had very few of our own in Russia. Certainly the U.S. welcomed Zworykin's visits, as limited at they were. His reports would be of great value and much could be deduced from them.

But, as Dr. Zworykin relates, these visits were regarded with suspicion by members of the émigré Russian colony. Surely they were the source of many FBI reports. These people were not too happy with Zworykin and his rise to fame and power and probably they did much to create suspicion about him. Certainly they were not aware of what information he may have brought home with him. And the FBI is not careful about its sources of information.

See Press Release, Sat., Nov. 18, 1933, "Establishment of Relations with the Union of Soviet Socialist Republics," *U.S. State Department Bulletin*, Nov. 22, 1933.

41. VKZ MS p. 93. On Shröter's attitude toward the information Telefunken received from RCA, see E. H. Traub, "Television at the 1933 Berlin Radio Exhibition," *Journal of the Television Society* 1 (1931–34), 273–85; and Fritz Shröter, *Handbuch der Bildtelegraphie und des Fernsehens* (Berlin: Julius Springer, 1932), pp. 61–62. About a year later, Shröter applied for a camera tube patent (similar to Zworykin's) using a single row of separate elements; see Telefunken, Br. Pat. no. 431,904 (filed Dec. 12, 1933; issued July 7, 1935).

In 1978 I interviewed Hans Gerhart Lubszynski in Argyle, Scotland, about his work at Telefunken under Shröter. He had gone to work at Telefunken some time in 1931. He found Shröter to be quite stuffy, with very little imagination, and quite conservative. He ran the laboratory with an iron hand. In 1933, all non-Germans (Lubszynski was a Polish national) were thrown out of Telefunken and forced to look for work elsewhere. Lubszynski applied at EMI and was received warmly by Shoenberg. His work on the super emitron and on the orthicon tubes was classic. This material will be referred to as the Lubszynski File.

42. "Gain in Television Amazes Marconi," *New York Times*, Oct. 11, 1933, p. 25:5. The article states that Marconi was accompanied by Sarnoff as the guest of E. T. Cunningham, president of RCA Victor. There is no mention of Zworykin being present.

43. W. D. Wright, "Picture Quality—The Continuing Challenge towards Visual Perfection," *Journal of the Royal Television Society* 16 (Jan.–Feb. 1977), 6–10 (this was originally the Shoenberg Memorial Lecture of the Royal Television Society). On Dr. Leonard Broadway, see *Here's Looking at You: The Story of British Television 1908–39* (London: BBC, 1984), p. 105. Information on the Russian visitors was sent to me by T. A. Vasilieva, Lenin State Library; Vasilieva to author, Nov. 1, 1990.

44. "Super-Microscope Uses Television to Open Vast Ranges for Science," *New York Times*, Feb. 25, 1934, p. 1:2; "A New Tool for Microscopy," *New York Times*, Mar. 4, 1934, sec. 8, p. 8:1. Zworykin claimed that this device was an adaptation of his television apparatus. Since it used the new

iconoscope he claimed that it had an "electrical memory." It was described in V. K. Zworykin, "Electric Microscope," *Congresso Internazionale di Elettro-Radio Biologia* 1 (Sept. 1934), 672–86; and V. K. Zworykin, "Latent Image Mechanism," *Science* 79 (Mar. 2, 1934), 10. A picture and a description of the device appear in V. K. Zworykin and L. E. Flory, "Television in Medicine and Biology," *Electrical Engineer* 71 (Jan. 1952), 40–45. It is important to understand that this was not an electron microscope, which would depend on proper manipulation of the electron beam to enlarge the image by use of either electromagnetic or electrostatic fields. Zworykin's advantage was in his kinescope, which was the best display device available at the time. Zworykin worked on an electron microscope at a later date. See M. Knoll and E. Ruska, "Beitrag zur geometrischen Elektronoptik," 607; and M. Knoll and E. Ruska, "Elektronenmikroskop," 318–39, which described a cold-cathode magnetic electron microscope. See also "Magnification Is Trebled by Electron Microscope," *New York Times,* July 25, 1934, p. 2:4. Knoll and Ruska had patented this device; see Siemens-Schuckerwerke Akt-Ges., Br. Pat. no. 402,781 (conv. date May 30, 1931; filed May 30, 1932; issued Nov. 30, 1933). The Bell Telephone labs under C. J. Davisson were also conducting research into an electron microscope, which they claimed was crude and yielded minimal results. See "Electron Microscope," *New York Times,* Sept. 23, 1934, sec. 9, p. 7:1.

45. Briggs, *Golden Age of Wireless,* p. 563; Burns, *British Television,* pp. 227–28.

46. For the best description of the conflict between Baird Television and EMI during this period see Briggs, *Golden Age of Wireless,* pp. 566–82. It is also related in more detail in Burns, *British Television,* pp. 241–301.

47. For information on the intermediate film process, see "New Film Devised to Aid Television," *New York Times,* Aug. 19, 1932, p. 20:8; and Moseley and Chapple, *Television To-day and To-morrow,* pp. 201–5 (the Fernseh apparatus is shown on p. 200). See also J. C. Wilson, *Television Engineering* (London: Sir Isaac Pitman & Sons, 1937), pp. 404–7.

48. Briggs, quoting Baird: "It was of utmost importance that 'broadcasting in this country should not be under foreign influence—directly or indirectly'" (*Golden Age of Wireless,* p. 573). Briggs continues, "EMI insisted for its part that all of its equipment, including cathode ray tubes and photocells, everything down to the last screw, was home-manufactured and that most of its key patents were British" (p. 588). See also Burns: "For Baird could not acknowledge—for some considerable time—that EMI's television system was being developed by British workers in a British factory using British resources. For him the Radio Trust [*sic*] of America, through its associated companies in London, was the mainspring of EMI's progress" (*British Television,* p. 248). This was reaffirmed by the BBC. "It is not true to say that the EMI system was American, having regard to the fact 'that it had been entirely developed and wholly manufactured by an English staff in English workshops and that therefore it had already given employment to a very considerable number of workers in this country'" (Burns, p. 255). For Baird's letter to the king, see Burns, p. 262.

49. Briggs, *Golden Age of Wireless*, p. 583.

50. Selsdon Committee File. See also "Television Committee to Visit U.S. and Germany," *The Times* (London), Oct. 22, 1934, p. 19a; "Lord Selsdon Heads Brit. Govt Commission Studying American Progress," *New York Times*, Nov. 11, 1934, sec. 9, p. 15:3; "Federal Communications Commission Holds Dinner in Honor of Britons," *New York Times*, Nov. 18, 1934, sec. 9, p. 13:1.

51. On McGee's 1934 tubes, see EMI File, Television, p. 194. According to Garret and Mumford, EMI was still using a Nipkow-disc studio flying spot scanner in June 1934 ("History of Television," p. 35). On RCA's standard, see L. M. Clement and E. W. Engstrom, "RCA Television Field Tests," *Television* (RCA Institutes Technical Press) 2 (Oct. 1937), 33.

52. "Marconi Co. in Merger to Promote Television," *New York Times*, May 23, 1934, p. 10:4; "New Company, Marconi-EMI," *The Times* (London), May 23, 1934, p. 16g; and Burns, *British Television*, pp. 318–19. See also Briggs, *Golden Age of Wireless*, p. 568; the £100,000 figure is given on p. 578. This was an enormous figure. In 1994 dollars it amounts to well over five million dollars.

There may be one explanation for this. Many scientific advisors to the British government felt that a war with Germany was inevitable, and that expenditures in electronics, especially in the new field of radar, would give them a manufacturing base of technology, manpower, and facilities to cope with this coming crisis. By the time war broke out, Great Britain had a chain of highly developed radar stations along the English Channel that helped win the Battle of Britain. The radio transmitters were built by Metropolitan Vickers and the receivers by A. C. Cossor Ltd.

The argument against this of course, is that no British government has ever been accused of being so far-sighted. Still the facts speak for themselves. The money was made available, the plants were built and research and development was funded.

One of the weaknesses of the early British research on radar was the availability of a suitable cathode-ray tube. The most suitable for their purposes was the one designed by von Ardenne and built by Leybold in Cologne. See R. A. Watson Watt, *Cathode-Ray Oscillography. Radio Research* (London: His Majesty's Stationery Office, 1933), p. 4. See also von Ardenne, "Evolution of the Cathode-Ray Tube," p. 30.

53. H. G. Lubszynski and S. Rodda, Br. Pat. no. 442,666 (filed May 12, 1934; issued Feb. 12, 1936); also U.S. Pat. no. 2,244,466 (conv. date, May 4, 1935; issued June 3, 1941). VKZ Pat. File, Pat. Int. no. 74,655, decided in Bedford's favor on Aug. 24, 1937.

The image iconoscope or Super Emitron, as EMI called it, enjoyed great favor in Europe. It was in constant use until the middle 1950s. Only the improved image orthicon of RCA finally made it obsolete. Udelson completely misstates the patent status of the image iconoscope when he says, "And yet, RCA was unable to use its version of the improved pickup tube, the image Iconoscope. The radio giant was trapped in a patent dilemma" (*Great Television Race*, pp. 99–100). Udelson gives no source for this information, but it is incorrect—RCA not only could use this tube, it owned the rights to it. Still, RCA preferred not to use it because Iams and Rose were developing a new generation of camera tubes using low-velocity electron beam scanning in their laboratories at Harrison, N.J. RCA was about to lose control of the iconoscope which reverted back to Westinghouse in 1938 (see Chap. 10, n. 3). RCA did have a patent problem with Philo Farnsworth, when they found that he had applied for a patent in Apr. 1933 for the first low-velocity camera tube while working for Philco. Udelson makes no reference at all to Farnsworth's most important invention.

54. "Award Liebman Memorial Prize to Dr. V. K. Zworykin for Television Inventions," *New York Times,* June 3, 1934, sec. 9, p. 9:4. See also "Vladimir K. Zworykin, Morris Liebman Memorial Prize Recipient for 1934," *Proceedings of the Institute of Radio Engineers* 22 (July 1934), 938. The citation stated that "Zworykin was made director of the electronic research laboratory in 1934."

55. "Television Nears Technical Solution," *Electronics* 5 (June 1934), 172–73.

56. The story of Major Edwin Howard Armstrong is told in Lessing, *Man of High Fidelity.* Armstrong was one of America's greatest inventors. He was responsible for the invention of the regenerative or "feedback circuit" in 1914, the heterodyne in 1917–18, the superregenerative circuit in 1921, and the superheterodyne in 1922. He was also responsible for the first practical frequency modulated (FM) system of radio transmission. It seems that in 1922 a research engineer, John Carson of the American Telephone & Telegraph, "proved" that FM transmission was impractical. He stated that "this type of modulation inherently distorts without any compensating advantages whatever" ("Notes on the Theory of Modulation," *Proceedings of the Institute of Radio Engineers* 10 [Feb. 1922], 57–64). This inhibited most researchers from further work on it. Carson was working with a narrow-band system. Armstrong made it work by extending the necessary bandwidth. See E. H. Armstrong, U.S. Pat. no. 1,941,066 (filed July 30, 1930; issued Dec. 26, 1933); also U.S. Pat. nos. 1,941,067, 1,941,068, and 1,941,069 (all issued Dec. 26, 1933). See also Lyons, *David Sarnoff,* pp. 212–14; Bilby, *The General,* pp. 198–200; and Sobel, *RCA,* pp. 128–31.

57. Everson, *Story of Television,* pp. 164–65. As a result of this visit, Takayanagi sent Iams two volumes of the Japanese television journal, *Transactions of the Television Engineers of Japan,* for 1935 and 1936. I was given this material by Iams, along with a letter from Takayanagi to Iams, dated Jan. 25, 1935, thanking him for his hospitality.

58. K. Takayanagi, "Recent Development of Television Technic in Europe and America," *Journal of the Institution of Electrical Engineers* (Japan) 5, no. 3 (Mar. 1935), 172–80. For a definitive work on the early German efforts on radar, see David Pritchard, *The Radar War: Germany's Pioneering Achievement 1904–45* (Wellingborough, Eng.: Patrick Stephens, 1989). Pritchard claims that Germany started work on radar in May 1934. See also Philo T. Farnsworth, "Television by Electron Image Scanning," *Journal of the Franklin Insti-*

tute 218 (Oct. 1934), 411–44; A. H. Brolly, "Television by Electronic Methods," *Journal of the American Institute of Electrical Engineers* 53 (Aug. 1934), 1153–60.

I had the good fortune to interview Arch Brolly at his home in Saratoga, California, in July 1977. He was Farnsworth's chief engineer in both San Francisco and Philadelphia. In 1931 he was left behind with a skeleton staff to keep the labs in San Francisco operating while Farnsworth was in Philadelphia with Philco. He told me that they were first able to get an outdoor picture from a dissector camera in August 1933. (See "Outdoor Test of Television Proves Merit," *San Francisco Chronicle*, Sept. 14, 1933; "Television Declared Ready to Broadcast Starting Sectionally, with Relays Later," *New York Times*, Aug. 11, 1933, p. 18:2.) When I told him that RCA had been getting live pictures from a dissector since early in 1931, he was not the least surprised. He recognized that Farnsworth's limited resources just could not compete with the Zworykin laboratory in Camden. Although he admired Farnsworth, his one complaint was that Farnsworth's mind was so active that he was always coming up with new ideas on how to improve his system. As a result, he would order a new circuit or part and by the time it had arrived or was built, it was obsolete. He admitted that the lack of sensitivity of the dissector was its greatest handicap, but he claimed that they did get it to work under controlled circumstances and of course it was quite good as a film scanner. This material will be referred to as the Brolly File.

59. Details of new financing from Seymour Turner are found in Schatzkin and Kiger, "Philo T. Farnsworth," p. 20. "Tennis Stars Act in New Television," *New York Times*, Aug. 25, 1934, p. 14:4; Everson, *Story of Television*, pp. 142–45; "Television Progress," *Electronics* 7 (Sept. 1934), 272.

60. Everson, *Story of Television*, p. 147. According to Sir Harry Greer, chairman of Baird Television, it was in June 1934; see "Sir Harry H. Greer," *The Times* (London), June 21, 1935, p. 25a. "British Empire Gets Rights to Farnsworth Television Apparatus," *New York Times*, June 20, 1935, p. 21:1. However, Tony Bridgewater, the eminent British television historian who was working for the BBC at the time, told me that the Farnsworth equipment was never shown to the Selsdon committee.

61. See "Television: A Survey of Present-Day Systems," *Electronics* 7 (Oct. 1934), 300–305; Philco's iconoscope tube was described as a "cathode ray camera tube" on p. 305. For the first Russian iconoscope, see Shmakov, "Development of Television in the USSR," pp. 97–105. Shmakov credited Krusser with developing the Russian iconoscope in 1934. See also Novakosky, "100th Birthday of V. K. Zworykin," p. 67. Bell Labs 33089, John R. Hofele, "Television Transmission System Using Cathode-Ray tubes," memorandum, July 31, 1934. Telefunken's tube is described in Goebel, "Das Fernsehen in Deutschland," pp. 290–91.

62. A. V. Bedford, U.S. Pat. no. 2,166,712 (filed Oct. 26, 1934; issued July 18, 1939) and U.S. Pat. no. 2,293,147 (filed May 18, 1934; issued Mar. 26, 1935).

63. On RCA's standard, see Clement and Engstrom, "RCA Television Field

Tests," p. 33; on Marconi-EMI's, see the classic three-part article, "The Marconi-E.M.I. Television System," that appeared in *Journal of the Institution of Electrical Engineers* (London) 83 (Dec. 1938), 758–801 (papers presented Apr. 21, 1938)—part 1: "The Transmitted Wave-Form," by A. D. Blumlein; part 2: "The Vision Input Equipment," by C. O. Browne; and part 3: "The Radio Transmitter," by N. E. Davis and E. Green.

Chapter 9: A Return to Russia

1. Zworykin's manuscript states that his visit took place in "the fall of 1934" (VKZ MS, p. 97). It is described in VKZ MS, pp. 88–94; FOIA/State Dept. PO32 (Zworykin to Dept. of State, Sept. 27, 1935) gives dates of Sept. 19 and Nov. 4, 1934; Novakosky, "100th Birthday of V. K. Zworykin," pp. 68–67. This is confirmed by Yushkiavitshus, "Past and Future of Soviet Television," p. 316. See also Francis Wheen, *Television: A History* (London: Century Publishing, 1985), p. 32.

2. See VKZ MS, pp. 88–95. As I have noted earlier, in his memoir, Zworykin relates only one visit to the Soviet Union. I have taken the position that his description is a composite of all of his early trips to the USSR. Now I know that there were at least three separate visits, one in 1933 (confirmed in "The 'Eye' Gains Prestige," *New York Times,* July 16, 1933, sec. 10, p. 7:5), one in 1934 (see VKZ MS, p. 95), and one in 1935 (Yushkiavitshus, "Past and Future of Soviet Television"). On p. 95 of his memoir, Zworykin starts a new chapter entitled, "1934–1939 (1934 Return to U.S.A. from USSR)." This probably means that the 1934 visit was the most important. I believe that the 1933 visit was quite short and consisted mainly of the one lecture on the iconoscope in Moscow. The visit in 1935 must have also been short. One interesting note: Zworykin had always insisted that Professor Rozing died soon after the Russian Revolution. I even wrote him and told him that Rozing had died in 1933—he simply ignored me.

Zworykin's encounter with Beria is of interest. Beria went on to head the dreaded Soviet secret police, the NKVD, in 1938 and his cruelty apparently knew no limits. After Stalin's death, Beria became deputy premier. He was later charged with treason and executed in June 1952.

3. Iams File, interview, July 1981, Iams had the greatest respect for Albert Rose.

Dr. Albert Rose (1910–90) was born in New York City. He received a B.A. from Cornell University in 1931 and a Ph.D. in physics in 1935. In that year, he went to work for the RCA Manufacturing Company's research and engineering department in Harrison. Here he joined Harley Iams in developing a series of low-velocity electron beam scanning camera tubes starting with the Orthicon in 1937.

Rose transferred to the Princeton labs in 1942 and went on to invent the two-sided glass target that made the image orthicon possible. This radical tube was developed with Paul Weimer and Harold B. Law, and made the rapid expansion of television in the United States possible after the war. He then started research on tubes using photoconductive surfaces which led to the invention of the Vidicon. Rose spent two years in Zurich and finally retired from RCA in 1975. He then spent much time in travel and teaching at various universities. He wrote several books, and wrote the preface to my *History of Television, 1880 to 1941*, and received many honors for his work. I had the pleasure of interviewing him at his home in Princeton on Apr. 6, 1979.

Rose and Otto Schade set new standards for the comparisons of film, television, and the human eye. For example, see A. Rose, "The Relative Sensitivities of Television Pickup Tubes, Photographic Film, and the Human Eye," *Proceedings of the Institute of Radio Engineers* 30 (June 1942), 293–300; A. Rose, "A Unified Approach to the Performance of Photographic Film, Television Pickup Tubes, and the Human Eye," *Transactions of the Society of Motion Picture Engineers* 47 (Oct. 1945), 273–93; and O. H. Schade, "Electro-optical Characteristics of Television Systems," part 4: "Correlation and Evaluation of Electro-optical Characteristics of Imaging System," *RCA Review* 9 (Dec. 1948), 686. Rose and Schade proved that 525-line television had a resolution quite close to that of 35mm motion picture film. This of course was contrary to the Eastman Kodak Company's claim of film's presumed 2000- to 4000-line resolution. High-definition studies have proved that the 35mm film projected on a large screen has about 800 lines of resolution, which is equaled quite easily with a 1125-line system. See Louis L. Pourciao, "High Resolution TV for the Production of Motion Pictures," *Journal of the Society of Motion Picture and Television Engineers* 93 (Dec. 1984), 1119; Arthur Kaiser, Henry W. Mahler, and Renville H. McMann, "Resolution Requirements for HDTV Based upon the Performance of 35mm Motion-Picture Films for Theatrical Viewing," *Journal of the Society of Motion Picture and Television Engineers* 94 (June 1985), 654–59; K. Blair Benson and Donald G. Fink, *HDTV Advanced Television for the 1990's* (New York: McGraw Hill, 1991).

4. See Yushkiavitshus, "Past and Future of Soviet Television," p. 316; and Shmakov, "Development of Television in the USSR," p. 105

5. "Institute News and Radio Notes," *Proceedings of the Institute of Radio Engineers* 23 (Mar. 1935), 266; VKZ File, Zworykin's RCA Executive Bio., dated Sept. 1954. Zworykin is credited with several electron multiplier patents, U.S. Pat. nos. 2,144,239, 2,147,825, 2,150,573, 2,189,305, 2,205,055, 2,231,697, and 2,231,698.

6. See Selsdon Committee File, "Report of the Television Committee," presented by the Postmaster General to Parliament by command of His Majesty, Jan. 1935 (London: His Majesty's Stationery Office, 1935).

7. "Television Gains Reported by RCA," *New York Times*, Feb. 27, 1935, p. 21:4.

8. "Alexandra Palace Picked as British Television Site," *New York Times*, Apr. 4, 1935, p. 1:2; "London Television Station, Alexandra Palace Chosen," *The Times* (London), June 7, 1935, p. 13d; "Television Next Year: Public Service from London," *The Times* (London), Aug. 13, 1935, p. 12f.

9. "First Field Tests in Television Costing $1,000,000 to Begin Here," *New*

York Times, May 8, 1935, p. 1:2; "Television in U.S.A. $1,000,000 to Be Spent in Experiments," *The Times* (London), May 9, 1935, p. 18e; "Million Dollar Plan Brightens Television," *New York Times,* May 12, 1935, sec. 10, p. 11:1; "Television Progress in the USA, America to follow Britain's Lead," *Television & Short-Wave World* 8 (Aug. 1935), 436; Bucher, "Television and Sarnoff," pp. 77–82.

10. See Lessing, *Man of High Fidelity,* p. 183. On Clement, see "Institute News and Radio Notes," *Proceedings of the Institute of Radio Engineers* 18 (Nov. 1931), 1906; "Institute News and Radio Notes," *Proceedings of the Institute of Radio Engineers* 24 (Feb. 1936), 152–53. These same notes reported that Baker had left RCA to take charge of radio engineering and manufacturing for GE at Bridgeport. See also "Institute News and Radio Notes," *Proceedings of the Institute of Radio Engineers* 24 (Feb. 1936), 158.

11. "London Television Station, Alexandra Palace Chosen," *The Times* (London), June 7, 1935, p. 13d; "Marconi-E.M.I. Television," *Television & Short-Wave World* 8 (July 1935), 380; "Television Transmissions, Details of the Baird and Marconi-E.M.I. Systems," *Wireless World* 37 (Oct. 4, 1935), 373; "High-Definition Television from the Alexandra Palace," *Television & Short-Wave World* 8 (Nov. 1935), 631–34; "High Definition Television Service in England," *Journal of the Television Society* 2 (1934–35), 34–43; M. von Ardenne, "Interlacing and Definition," *Television & Short-Wave World* 8 ((Dec. 1935), 719–21.

12. "P. T. Farnsworth Demonstrates Gains at Television Laboratories Inc.," *New York Times,* July 31, 1935, p. 15:4; "Television Transmitters Planned," *Electronics* 8 (Sept. 1935), 294–95; "A Demonstration of the Farnsworth System," *Television & Short-Wave World* 8 (Nov. 1935), 628; Samuel Kaufman, "Demonstrated High-Definition Television," *Radio News* 17 (Nov. 1935), 265, 268; Samuel Kaufman, "Farnsworth Television," *Radio News* 17 (Dec. 1935), 330–331, 381.

13. Bell Labs 33089, Foster C. Nix, "Photoconducting TV Transmitter," Aug. 6, 1935. This was an amazing feat. No one else was able to produce a workable photoconducting camera tube until 1949–50 when the RCA Vidicon was introduced. I do not know why Bell has never publicized this achievement.

14. "Sarnoff off to Europe," *New York Times,* July 27, 1935, p. 16:5; "Sarnoff Denies Change in E.M.I. Setup," *Wall Street Journal,* Oct. 2, 1935, p. 84; Bucher, "Television and Sarnoff," p. 126; RCA Annual Stockholders' Report, 1935, p. 3. According to Sobol, "RCA sold its shares to investors in that country for $10.2 million, intending to plow the funds into domestic operations" (*RCA,* p. 101).

15. "Company Meetings," "Mr. Alfred Clark on Importance of Company's Research Work," *The Times* (London), Nov. 16, 1935, p. 20a, b, c. This report of the EMI stockholders' meeting held on Nov. 15, 1935, was quite interesting. Clark made two important points. The first was the value of its research engineers, and the second referred to the recent sale of shares by RCA. He stated that "You realize of course, that the company cannot acquire

its own shares . . . , and negotiations to that end are being pursued on the point of compensation. We fully approve of what is being done." See also Sobol, *RCA*, p. 101.

16. Bucher, "Television and Sarnoff," p. 126. Here Sarnoff is quoted as saying that "the system is known abroad as the Marconi-E.M.I. Television system which is fundamentally based on the RCA Television system developed in the RCA Laboratories in the United States." The truth of this is confirmed in Burns, *British Television*, where Baird's bitter statement also appears (p. 440). The BBC's last transmission using the Baird system was on Jan. 30, 1937.

17. "New Radio Device Spurs Television," *New York Times*, Oct. 24, 1935, p. 2:5; V. K. Zworykin, G. A. Morton, and L. Malter, "The Secondary Emission Multiplier," *Proceedings of the Institute of Radio Engineers* 24 (Mar. 1936), 351–75 (paper presented Oct. 23, 1935). Zworykin's presence at Rochester on Nov. 18, 1935, was confirmed in "Institute News and Radio Notes," *Proceedings of the Institute of Radio Engineers* 23 (Oct. 1935), 1269. See also the picture of Zworykin holding this multiplier tube on the front cover of *Radio Engineering* 16 (Apr. 1936). See also V. K. Zworykin and E. D. Massa, U.S. Pat. no. 2,078,304 (filed Oct. 30, 1935; issued Apr. 27, 1937). For information on Farnsworth's prior cold-cathode electron multiplier, see D. K. Lippincott and H. E. Metcalf, "Cold-Cathode Multiplier Tube," *Radio Engineering* 14 (Nov. 1934), 18–19. See also Everson, *Story of Television*, pp. 129–31. This simple device was built into Farnsworth's image dissector and increased its sensitivity to the point where it could be used in bright daylight for 240-line scanning. See Television Laboratories, Br. Pat. no. 450,138 (conv. date Oct. 7, 1933; issued July 6, 1936).

18. "$2,000,000 Soviet Order to RCA," *New York Times*, Dec. 16, 1935, p. 40:5. See also "Radio Pictorial," *Radio-Craft* 8 (Nov. 1936), 269. This article had a photo of the first Soviet public cathode-ray television using a transmitter (camera) similar to the Zworykin iconoscope. While Russian scientists were quite capable of building Zworkyin's iconoscope, they depended on the great facilities of RCA to build a complete television system, including the transmitter, control rooms, cameras, and receivers. On the Moscow television installation, see Frederick O. Barnum III, ed., *"His Master's Voice" in America* (Camden, N.J.: General Electric, 1991), p. 202. See also VKZ MS, p. 95.

19. "Third Eye Demonstrated," *The Times* (London), Jan. 4, 1936, p. 13g; "Zworkyin Shows New Electron Tube," *Radio-Craft* 8 (Jan. 1936), 391. There were at least two patents on the image tube; see Br. Pat no. 489,846 (conv. dates Nov. 30, 1935, Dec. 24, 1935, and Dec. 31, 1935), and Br. Pat. no. 488,920 (filed Nov. 30, 1935; accepted June 30, 1938). These were assigned to Marconi Wireless Telegraph Company.

20. V. K. Zworykin, "Electron Optics and Their Applications," *Journal of the Institution of Electrical Engineers* (London) 79 (July 1936), 1–10 (paper presented Feb. 5, 1936). See "Dr. Zworykin on the Electron Multiplier," *Television & Short-Wave World* 9 (Mar. 1936), 153–54, 191; and "The Elec-

tron Image Tube," *Radio-Craft* 8 (Apr. 1936), 594, 622. FOIA/State Dept. PO32 (Zworykin to Dept. of State, Sept. 27, 1935) shows that Zworykin applied for a passport on Sept. 27, 1935. In it he stated that he was to leave for Europe from New York on the SS *Bremen* on Oct. 26, 1935, for a stay of about two months. The passport was issued on Oct. 1 and was to expire on June 21, 1937. For some reason, this trip was postponed; he did not go until the middle of Jan. 1936. He was in Europe from Feb. to Apr. 1936, and returned to give a lecture in Washington on May 1, 1936. See "Institute News and Radio Notes," *Proceedings of the Institute of Radio Engineers* 24 (Apr. 1936), 544. See also V. K. Zworykin and G. Morton, "Applied Electron Optics," *Journal of the Optical Society of America* 26 (1936), 181–89.

21. VKZ MS, p. 97. Here Zworykin stated that the EMI system was 240 lines. It was 405. Baird was still on 240 lines. Zworykin is pictured in a British film on the construction of the Alexandra Palace.

22. V. K. Zworykin, "'L'Optique Électronique' et ses applications," *L'Onde Électrique* 15 (May 1936), 293–96 (paper presented Mar. 10, 1936). René Barthélemy (1889–1954) was a French physicist who made several minor advances in early French television. All of his work was done for the Compagnie pour la Fabrication des Compteurs et Matériel d'Usines à Gaz of Montrouge. It is claimed he gave his first television demonstration on Apr. 14, 1931.

23. "Dr. Zworykin Visits Berlin," *Television and Short-Wave World* 9 (Apr. 1936), 216. This is one of those very rare reports of Zworykin visiting Europe. See also VKZ MS, pp. 97 and 104. "Berlin Television Begins," *The Times* (London), Mar. 23, 1935, p. 11d; "Television for Air Defense," *The Times* (London), Aug. 10, 1935, p. 12d. For the definitive work on the early German efforts on radar, see Pritchard, *Radar War.*

24. Von Ardenne File, interview, Aug. 1984.

25. VKZ MS, p. 100. See P. Selenyi, "Electrography, A New Electrostatic Recording Process and Its Application to Picture Telegraphy," *Elektrotechnische Zeitschrift* 56 (Aug. 29, 1935), 961–63. See also P. Sélenyi, "Methoden, Ergebnisse und Aussichten des elektrostatischen Aufzeichnungverfahrens (Elektrographie)," *Zeitschrift für technische Physik* no. 12 (1935), 607–14 (paper presented Sept. 28, 1935). This shows photos of this work. See Br. Pat. no. 449,824 (filed Mar. 22, 1934; issued July 6, 1936). It is of interest that Chester Carlson, who is given credit for inventing this copy process, had a coworker, a refugee from Hungary, who helped him develop this invention. It is claimed that this was conceived in 1937 though the patent was not applied for until 1939. See C. Carlson, U.S. Pat. no. 2,297,691 (filed Apr. 4, 1939; issued Oct. 6, 1942); see also C. F. Carlson, U.S. Pat. no. 2,588,649 (filed Aug. 27, 1943; issued Mar. 11, 1952).

26. VKZ MS, pp. 99–100.

27. VKZ MS, p. 94; *Sov. Enc.*, vol. 11, p. 405. Kapitza's story is told in *History of Physics*, ed. Spencer R. Weart and Melba Phillips (New York: American Institute of Physics, 1985), p. 219.

28. See "Institute News and Radio Notes," *Proceedings of the Institute of Radio Engineeers* 24 (July 1936), 1060.

Chapter 10: "The World of Tomorrow"

1. Orrin Dunlap, Jr., "Outdoor Scene Is Broadcast in Successful Television Test," *New York Times,* Apr. 25, 1936, p. 1:4–5.

2. "Radio Corp. to Start Tests," *New York Times,* June 28, 1936, sec. 9, p. 10:1; "Description of Apparatus," *New York Times,* June 29, 1936, p. 17:3; "Experiments Held in Secret," *New York Times,* June 30, 1936, p. 21:6; "Television Stages First Real Show," *New York Times,* July 8, 1936, p. 21:3. Don Lee, "Television on the West Coast," *Radio-Craft* 8 (Aug. 1936), 76, 110; "Inauguration of Daily Television Broadcast Schedule," *Radio Engineering* 16 (June 1936), 24. See "Radio Progress during 1936," part 4, *Proceedings of the Institute of Radio Engineers* 25 (Feb. 1937), p. 204. "Philco Radio and Television Corp. Holds Demonstration," *New York Times,* Aug. 12, 1936, p. 21:8; "Philco Television," *Radio Engineering* 16 (Sept. 1936), 9–10. See also A. F. Murray, "The New Philco System of Television," *Radio-Craft* 8 (Nov. 1936), 270, 315; Television Reporter, "Television as Good as Home Movies," *Radio News* 17 (Nov. 1936), 265–66, 308.

3. "Patent Suit Filed on RCA Television," *New York Times,* July 10, 1936, p. 6:7; VKZ Pat. File, see Pat. Int. no. 62,721, paper no. 94, Briefs on behalf of Appellant Vladimir K. Zworykin.

4. "Television of the Games, Disappointing Results," *The Times* (London), Aug. 3, 1936, p. 12a; "Pictures of Olympic Events Are Broadcast to Berlin Theatres," *New York Times,* Aug. 6, 1936, p. 24:8; "Used to Show Olympic Relay," *New York Times,* Aug. 10, 1936, p. 12:6; Hubert Gibas, "Television in Germany," *Proceedings of the Institute of Radio Engineers* 24 (May 1936), 741–50.

The sale of television sets to the general public was predicted in July 1939; see "Television Trends Abroad," *New York Times,* July 19, 1936, sec. 10, p. 10:1. Although promised many times, this never happened in Germany. (Also, in spite of many rumors to the contrary, there is absolutely no evidence that Hitler ever appeared live on German television.) See "Fernsehen bei den Olympischen Spielen 1936," *Fernsehen und Tonfilm* 7 (Aug. 1936), 57–59; W. Federmann, "Fernsehen wahrend der Olympischen Spiele," *Telefunken* no. 75 (1937), 18–22; "La télévision aux jeux Olympiques et a l'exposition de T. S. F. de Berlin," *L'Onde Électrique* 10 (Nov. 1936), 729–49. See Goebel, "Das Fernsehen in Deutschland," pp. 356–57, for details of the Olympic programming. Frithjof Rudert, "50 Years of 'Fernseh,' 1929–1979," *Bosch Technische Berichte* 6 (May 1979), 35.

5. "B.B.C. and Television, Trial Programs for a Month," *The Times* (London), Sept. 15, 1936, p. 16e; "Television Programmes, Regular Service to Begin on Nov. 2," *The Times* (London), Oct. 10, 1936, p. 10a; "Broadcasting, Inauguration of Television," *The Times* (London), Nov. 2, 1936, p. 21b; "Television in London, Opening of Regular Service," *The Times* (London), Nov. 3, 1936, p. 9a; "B.B.C. Television Programme," *The Times* (London), Nov. 3, 1936, p. 9c; L. M. Gander, "The First of Many," fiftieth anniversary issue, *Journal of the Royal Television Society* 16 (Nov.–Dec. 1977), 42–43; "BBC

to Start Regular Broadcasts," *New York Times,* Nov. 1, 1936, sec. 10, p. 12:4; "*London Times* and *London Telegraph* Comment on Prospect; Comment on Dedication of London System," *New York Times,* Nov. 29, 1936, sec. 12, p. 12:7. On defects in the Baird system see "Television in the Home: Demonstration of Baird System," *The Times* (London), Aug. 27, 1936, p. 10b. The official version of why the EMI system was so superior to Baird's was related in G. Cock, "Report on Baird and Marconi-EMI systems at Alexandra Palace," TAC paper no. 33, Dec. 9, 1936, cited in Burns, *British Television,* pp. 424–28. See also "Britain Inaugurates Television for Public Use," *Radio News* 19 (Jan. 1937), 391–92, 446; M. P. Wilder, "Television in Europe," *Electronics* 10 (Sept. 1937), 13–15; H. M. Lewis and A. V. Loughren, "Television in Great Britain," *Electronics* 10 (Oct. 1937), 32–35, 60–62; Allan B. Du Mont, "Is Television in America Asleep?" *Radio-Craft* 8 (Nov. 1937), 268, 306. For technical details of this system, see Blumlein, Browne, Davis, and Green, "Marconi-E.M.I. Television System." See also G. R. M. Garret, *Television: An Account of the Development and General Principles of Television as Illustrated by a Special Exhibition Held at the Science Museum—June–September 1937* (London: His Majesty's Stationery Office, 1937), pp. 58–64; Burns, *British Television,* pp. 360–474; Norman, *Here's Looking at You,* pp. 118–80; and Briggs, *Golden Age of Wireless,* pp. 594–622. On the choice of EMI, see "London Television Service, E.M.I. Transmissions in Future," *The Times* (London), Feb. 5, 1937, p. 14c; "British Television Bars Baird System," *New York Times,* Feb. 5, 1937, p. 8:6.

6. "Report of the RMA Television Committee," *Radio Engineering* 16 (July 1936), 19–20; "Philco Television," *Radio Engineering* 16 (Sept. 1936), 9–10; "More Television Planned for Philadelphia Area," *New York Times,* Jan. 10, 1937, sec. 10, p. 12:2; "Philco Radio and Television Holds Test with New 441 Line Pictures," *New York Times,* Feb. 12, 1937, p. 25:8; "TV Show Reveals State of the Art," *New York Times,* Feb. 21, 1937, sec. 10, p. 12:1 (RCA also went to 441 lines; see "'Empire State' Television Shows Marked Advance," *Radio News* 18 [July 1937], 7–8, 60); "Pictures Found to Be Clearer by Suppression of One Side Band of Wave," *New York Times,* Nov. 29, 1937, sec. 12, p. 10:5. It had been found that it was not necessary to transmit both sides of the normal radio wave for television. (They were symmetrical.) By suppressing one side band, it was possible to transmit a wider radio signal, with sharper results, within the framework of the restricted television channel. See, for example, W. J. Poch and D. W. Epstein, "Partial Suppression of One Side Band in Television Reception," *Television* 2 (Oct. 1937), 134–50.

7. "New Tele-Lens Aids Big Screen," *New York Times,* May 9, 1937, sec. 9, p. 12:7; "New Lens Projector Flashes Television on a Screen," *New York Times,* May 16, 1937, sec. 10, p. 12:1; "The Projection Kinescope Makes Its Debut," *Radio-Craft* 9 (Aug. 1937), 83, 110; "Big Screen Television Pictures," *Radio News* 19 (Sept. 1937), 143, 173; V. K. Zworykin and W. H. Painter, "Development of the Projection Kinescope," *Proceedings of the Institute of Radio Engineers* 25 (Aug. 1937), 937–53 (paper presented May 12, 1937). This convention celebrated the I.R.E.'s silver anniversary. Zworykin credited his

original engineering team of J. C. Batchelor, G. N. Ogloblinsky, and Arthur Vance for their work on this project. See R. R. Law, "High Current Electron Gun for Projection Kinescopes," *Proceedings of the Institute of Radio Engineers* 25 (Aug. 1937), 954–76 (paper presented May 12, 1937).

8. See Barnum, *"His Master's Voice" in America*, pp. 202–3.

9. "Moscow TV Center to Use American Devices," *New York Times*, May 2, 1937, sec. 11, p. 12:2. It still was not operating in Aug. 1938. See "A Transmitter for Moscow," *Television & Short-Wave World* 10 (Aug. 1938), 469–70.

10. "License Agreement Signed by Farnsworth Television Inc. and the American Telegraph and Telephone Comp," *New York Times*, July 27, 1937, p. 8:4. See Everson, *Story of Television*, pp. 155–59.

11. "Vladimir K. Zworykin Leaves Future of Television with the Engineers," *Newsweek* 10 (Oct. 25, 1937), 40. The motion picture industry in Hollywood was always quite concerned about the role television would play in the future. They set out to study its status as early as July 1936. See "Television from the Standpoint of the Motion Picture Producing Industry," *Journal of the Society of Motion Picture Engineers* 26 (July 1936), 74, and (Aug. 1937), 144–48. In a 1977 interview, Les Flory complained of Zworykin's habit of putting his name on everything that came out of the laboratory; Flory File.

12. Harley Iams and Albert Rose, "Television Pickup Tubes with Cathode-Ray Beams," *Proceedings of the Institute of Radio Engineers* 25 (Aug. 1937), 1048–70 (paper presented May 12, 1937).

13. "Television Gains Told by Engineers," *New York Times*, June 18, 1938, 13:8; "I.R.E. Convention 1938," *Electronics* 11 (July 1938), 8–12.

14. See Harley Iams, G. A. Morton, and V. K. Zworykin, "The Image Iconoscope," *Proceedings of the Institute of Radio Engineers* 27 (Sept. 1939), 541–47 (paper presented June 17, 1938). They credited the late G. N. Ogloblinsky, Rose, H. B. Devore, Les Flory, and E. A. Massa for making their work possible. See Alda V. Bedford, U.S. Pat. no. 2,258,728 (filed Sept. 29, 1934; issued Oct. 14, 1941); H. G. Lubszynski and S. Rodda, Br. Pat. no. 442,666 (filed May 12, 1934; issued Feb. 12, 1936); VKZ Pat. File. The use of only one RCA image iconoscope camera was related to me in an interview with Robert B. Morris in July 1976. For his other activities, see "Robert Morris in Charge of NBC Television System," *Electronics* 11 (Jan. 1938), 10. See also "New Camera Is Promised, " *New York Times*, Sept. 5, 1937, sec. 10, p. 10:7; "A New Emitron Camera," *Television & Short-Wave World* 10 (Jan. 1938), 11–12; "The Latest Emitron Camera," *Television & Short-Wave World* 10 (July 1938), 397; J. D. McGee and H. G. Lubszynski, "E.M.I. Cathode-Ray Television Transmission Tubes," *Journal of the Institution of Electrical Engineers* (London) 84 (Apr. 1939), 468–75 (paper presented July 20, 1938).

15. "The Cenotaph Ceremony," *The Times* (London), Nov. 12, 1937, p. 11d; "Super Emitron Camera," *Wireless World* 41 (Nov. 18, 1937), 497–98; "The Derby Television Broadcast," *Television & Short-Wave World* 10 (July 1938), 409–10; "Enthusiastic Welcome for Mr. Chamberlain" (see photograph showing the EMI camera covering the event), *The Times* (London),

Oct. 1, 1938, p. 7e. Also in Gordon Ross, *Television Jubilee: The Story of 25 Years of BBC Television* (London: W. H. Allen, 1961), photo p. 168.

16. Bucher, "Television and Sarnoff," pp. 128–32; "RMA Announces Receiving Set Production Plans," *New York Times*, Oct. 21, 1938, p. 25:1.

17. "New Era for Radio Seen Opening Here," *New York Times*, Jan. 3, 1939, p. 31:3; "RCA Plans Television Set Marketing for April," *New York Times*, Jan. 20, 1939, p. 4:5; Lyons, *David Sarnoff*, pp. 215–16.

18. VKZ Pat. File, U.S. Pat. no. 2,141,059, application no. 683,337 (filed Dec. 29, 1929); see also Zworykin, Br. Pat. no. 255,057 (conv. date July 13, 1925; application date July 3, 1926; accepted Mar. 31, 1927). Here this process was described quite clearly, "the best results are obtained by stopping the operation before the bright blue color shows more than slight indications of turning violet. The process described appears to form a colloidal deposit of potassium hydride consisting of minute globules" (p. 2, ll. 46–50). "Wins Basic Patent in Television Field," *New York Times*, Dec. 22, 1938, p. 38:6; "Notes on Television," *New York Times*, Dec. 25, 1938, sec. 9, p. 12:7; "Basic Television Patent Issued," *Broadcasting* 16 (1939), 71.

19. "Atom and Gas Molecules Shown Million Times Enlarged on Screen," *New York Times*, Dec. 30, 1938, pp. 1:4, 9:3.

20. VKZ File, Zworykin's RCA Executive Bio., dated Sept. 1969, p. 4.

21. "The Proposed Television Standards," *Radio-Craft* 10 (Jan. 1939), 402–3, 432; "Television Transmitters," *Electronics* 12 (Mar. 1939), 26–28; "Television Receivers in Production," *Electronics* 12 (Mar. 1939), 23–25, 78–81; "Television Receivers," *Broadcasting* 16 (May 1, 1939), 22. RCA Victor sets were built by the RCA Manufacturing Company. See "RCA Manufacturing Co. Formed by Consolidation of RCA Victor and RCA Radiotron," *New York Times*, Dec. 27, 1934, p. 30:1.

Allan B. Du Mont started building television receivers in 1937 after a visit to London. He purchased a "Cossor Television" receiver and brought it back to the United States. He then converted it to 441-line images. A few months later he was building his own receivers. See Will Boltin, "Sight and Sound News," *Radio News* 25 (May 1939), 32. Crosley Radio was working in cooperation with Du Mont laboratories.

22. "Fair Radio Exhibit to Be in Huge 'Tube,'" *New York Times*, Feb. 7, 1938, p. 32:5; O. E. Dunlap, Jr., "Telecast to Homes Begins on April 30—World's Fair Will Be on Stage," *New York Times*, Mar. 19, 1939, sec. 11, p. 12:1; "Plans Completed for RCA's Video Exhibition at Fair," *Broadcasting* 16 (Apr. 1939), 31, 45; D. H. Castle, "A Television Demonstration for the New York World's Fair," *RCA Review* 4 (July 1939), 6–13. A photograph of the transparent receiver was shown in *Electronics* 12 (June 1939), 12. "CBS Takes Time in Preparation for Video Operations," *Broadcasting* 16 (June 1, 1939), p. 74.

23. Sarnoff, *Looking Ahead*, pp. 100–101; "Dedication of RCA Seen on Television," *New York Times*, Apr. 21, 1939, p. 16:1; "New York Display Dedicated," *Broadcasting* 16 (May 1, 1939), 21; "David Sarnoff Was Photographed . . . and Televised at New York World's Fair," *Radio-Craft* 11 (May 1,

1939), 21; "America Makes a Start," *Television & Short-Wave World* 10 (June 1939), 330; David Sarnoff, "The Birth of an Industry," *RCA Review* 4 (July 1939), 3–5; Bucher, "Television and Sarnoff," pp. 132–35.

24. O. E. Dunlap, Jr., "Today's Eye-Opener," *New York Times*, Apr. 30, 1939, sec. 11, p. 12:1; O. E. Dunlap, Jr., "Ceremony Is Carried by Television as Industry Makes Its Formal Bow," *New York Times*, May 1, 1939, p. 8:3; O. E. Dunlap, Jr., "Television's First Week at World's Fair Teaches No End of Lessons," *New York Times*, May 7, 1939, sec. 10, p. 12:1.

25. "Television Impresses the Public," *Broadcasting* 16 (May 15, 1939), 16, 65; "Notes on Television, Radio Today Survey of N. Y. C. Area Set Sales," *New York Times*, May 21, 1939, sec. 10, p. 8:5; "Television Ads Meager," *New York Times*, July 14, 1939, p. 22:4; "Inadequate Telecasts Blamed for Slow N.Y.C. Set Sales," *New York Times*, Aug. 27, 1939, sec. 9, p. 10:6; "FCC Hangs Up an Amber Light," *New York Times*, Nov. 9, 1939, sec. 10, p. 17:3.

26. "Notes on Television," *New York Times*, June 11, 1939, sec. 10, p. 8:5. This article mentions a new television eye called an "orthocon" described by Iams and Rose.

27. "The Orthicon," *Electronics* 12 (June 1939), 11–14, 58–59. Albert Rose and Harley Iams, "The Orthicon, A Television Pick-up Tube" *RCA Review* 4 (1939), 186–99. See Albert Rose and Harley Iams, "Television Pick-up Tubes Using Low-Velocity Electron Beam Scanning," *Proceedings of the Institute of Radio Engineers* 27 (Sept. 1939), 547–55 (paper presented June 7, 1939). For an early form of low-velocity scanning, see H. Iams, U.S. Pat. no. 2,213,548 (filed May 31, 1938; issued Sept. 3, 1940); also A. Rose, U.S. Pat. no. 2,213,174 (filed July 30, 1938; issued Aug. 27, 1940). This was the first patent covering what eventually became known as the orthicon.

On Farnsworth's patent application, filed while he was an employee of Philco, see P. T. Farnsworth, U.S. Pat. no. 2,087,683 (filed Apr. 26, 1933; issued July 29, 1937). This patent has often been overlooked because Farnsworth called it an "Image Dissector." A casual glance at the patent specification reveals what looks like a normal dissector with a finger (anode) at one end and a photoelectric screen at the other end. Yet the mode of operation was not the same as the image dissector's. It featured a low-velocity scanning beam that scanned a target made of a large number of insulated islands. The beam struck the target at cathode (zero) potential and discharged it, then the modulated beam returned along the same path. Here it entered the anode and produced the video signal. This was quite an advance. Also, to keep the beam under control, the tube was immersed in a magnetic field (as in the image dissector), which was Farnsworth's major contribution.

So Philo Farnsworth must be given credit for the idea of low-velocity scanning at cathode potential. However, he never claimed to have built such a tube. This was left to Iams and Rose. One vital factor that Farnsworth did not know of was the need for normal (perpendicular) landing of the beam on the target. This meant that the beam approached the target at a constant angle. This was later described by Lubszynski of EMI in 1936. See H. G. Lubszynski, Br. Pat. no. 468,965 (filed Jan. 15, 1936; issued July 15, 1937). The tube also required a two-sided target to make it practical.

While Farnsworth never did build a low-velocity electron beam camera tube, he did build countless varieties of image dissectors. In May 1938 he described a camera tube which used secondary emission before scanning; see "A New Farnsworth 'Pickup-Tube,' Amplification before Scanning," *Television & Short-Wave World* 9 (May 1938), 260. Another variation was described in November 1938. See "P. T. Farnsworth Describes New Vacuum Tube for News Transmission to I.R.E. and the RMA," *New York Times,* Nov. 15, 1938, p. 19:3.

28. "Institute News and Radio Notes," *Proceedings of the Institute of Radio Engineers* 27 (Jan. 1939), 74; "Institute News and Radio Notes," *Proceedings of the Institute of Radio Engineers* 27 (May 1939), 352, 354; "Institute News and Radio Notes," *Proceedings of the Institute of Radio Engineers* 27 (June 1939), 406.

29. The information about Taunton Lakes came from an interview with Loren Jones, who was quite friendly with Zworykin, on Aug. 3, 1993, and from James Hillier. It appears that Zworykin lived there by himself. According to State Department records, his main address was 4800 Walnut Street, Philadelphia; FOIA/State Dept. PO30, passport application, Aug. 12, 1937. See also the RCA publication *RCA Family* 2, no. 8 (Aug. 1939), 5.

30. VKZ MS, pp. 100–103. FOIA/State Dept. PO29 (Zworykin to Passport Division, July 12, 1939) states that Zworykin was going to Italy, Egypt, Palestine, Egypt, Assyria, Switzerland, Germany, France, Belgium, Holland, and England. The passport was issued on July 14, 1939. See also "Crisis Interrupts Science Discussion," *New York Times,* Sept. 1, 1939, p. 19:3.

31. VKZ MS, pp. 102–3; "RCA Plans World's Largest Radio Laboratory at Princeton," *Broadcasting* 19 (Mar. 10, 1941), 58.

32. Briggs, *Golden Age of Wireless,* p. 622; Burns, *British Television,* pp. 472–73. See also Ross, *Television Jubilee,* p. 64.

33. Burns, *British Television,* p. 472.

34. T. H. Hutchinson, "Programming the Television Mobile Unit," *RCA Review* 4 (Oct. 1939), 154–61; "Television in the Field," *Electronics* 12 (June 1939), 13; "Need for More than One Camera at Sports Telecasts Discussed," *New York Times,* Mar. 3, 1940, sec. 10, p. 10:3. On sales of sets, see Bilby, *The General,* p. 134.

35. "Agree on Patents," *New York Times,* Oct. 3, 1939, p. 32:3. On Sarnoff's statement in 1940, see U.S. Senate, Committee on Interstate Commerce, "Development of Television: Hearings Committee," Apr. 10, 11, 1940 (Washington, D.C.: GPO, 1940), p. 38; cited by Sol Taishoff, "Flexible Television Is Urged by President," *Broadcasting* 17 (Apr. 15, 1940), 18, and by Hofer, "Philo Farnsworth: Quiet Contributor," pp. 82–83. Everson devotes an entire chapter ("R.C.A. Licenses") to the subject (*Story of Television,* pp. 242–47).

36. "FM Gets Its Day in Court," *Electronics* 13 (Apr. 1940), 14–16, 74–77; "Green Light or Red Light," *Electronics* 13 (May 1940), 11–13, 79–81; Lessing, *Man of High Fidelity,* pp. 240–45; Lyons, *David Sarnoff,* p. 218; Bilby, *The General,* pp. 134–37.

37. See R. B. Janes, R. E. Johnson, and R. S. Moore, "Development and Performance of Television Camera Tubes," *RCA Review* 10 (June 1949), 191–223.

38. Albert Rose, U.S. Pat. no. 2,506,741 (filed Sept. 20, 1940; continued Nov. 28, 1945; issued May 9, 1950); also Br. Pat. no. 613,003 (filed on Sept. 20, 1940; issued on Nov. 22, 1948). It is of interest that there was a prior Russian patent concerning a glass target; see G. Braude, Russ. Pat. no. 55,712 (filed Feb. 3, 1938; issued Sept. 30, 1939). See also G. Braude, "A New Type of the Mosaic for the Television Pick Up Tube," *Journal of Physics* (Moscow) 9 (1945), 348–50; and a review of Braude's "Ein neus Fernsehsystem" (*Technical Physics* 4, no. 9 [n.d.], 671–706) that appeared in *Fernsehen und Tonfilm* 10 (1938), 79. Braude claimed that a tube of this type was realized and tested before the war, so he is given well-deserved priority. However, Braude's target lacked two features necessary for its success, low-velocity electron beam scanning and normal landing of the beam. Therefore, Rose must be given the credit for not only inventing the glass target but building it into a tube and making it work successfully. Rose's tube was the parent of the two-sided orthicons and image orthicons that made possible both postwar television and future work on tubes using photoconducting surfaces.

39. V. K. Zworykin and G. A. Morton, *Television: The Electronics of Image Transmission* (New York: John Wiley & Sons, 1940).

40. "RCA's Electron Microscope," *Broadcasting* 18 (May 1, 1940), 43. According to this article, work was done under Zworykin's supervision. L. Marton, M. C. Banca, and S. F. Bender, "A New Electron Microscope," *RCA Review* 5 (Oct. 1940), 232–43; J. Hillier and A. Vance, "Recent Developments in the RCA Electron Microscope," *Proceedings of the Institute of Radio Engineers* 29 (Apr. 1941), 167–76 (paper presented Jan. 9, 1941); V. K. Zworykin, "Electrons Extend the Range," *Electrical Engineering* 59 (Nov. 1940), 441–43; V. K. Zworykin, J. Hillier, and A. W. Vance, "An Electron Microscope for Practical Laboratory Service," *Transactions of the American Institute of Electrical Engineers* 60 (Apr. 1941), 157–62 (paper presented Jan. 27, 1941). Zworykin gave a speech, "A Three Hundred Kilovolt Electron Microscope," on Nov. 10, 1941, at the fall meeting of the I.R.E.; see "Program," *Communications* 21 (Nov. 1941), 9. Hillier told me he was hired in 1940; James Hillier to author, Dec. 22, 1988.

41. "Coverage of GOP Convention to Include Television Pickups," *Broadcasting* 18 (June 15, 1940), 22; "Republican Convention, Images Establish Distance Record, 1800 miles," *New York Times,* June 29, 1940, p. 5:6; O. B. Hanson, "RCA-NBC Television Presents a Political Convention as First Long-Distance Pickup," *RCA Review* 5 (Jan. 1941), 267.

42. "FCC Grants Limited Commercial Operations," *New York Times,* Mar. 1, 1940, p. 18:1; "RCA Cuts Prices on Receivers," *New York Times,* Mar. 13, 1940, p. 34:1; "Drive to Promote Video Sets Started by RCA," *New York Times,* Mar. 15, 1940, p. 8:6; "FCC Stays Start of Television, Rebukes R.C.A. for Sales Drive," *New York Times,* Mar. 25, 1940, p. 1:6; "FCC Pledges Help in Television Field," *New York Times,* Aug. 1, 1940, p. 23:5; "Ten Video Stations Granted by FCC as Interest Slackens," *Broadcasting* 18 (Aug. 1, 1940), 17; "Television Committee Organizes," *Electronics* 13 (Aug. 1940), 34; "National Television Systems Committee Meets to Develop Standards," *New York Times,*

Oct. 14, 1940, p. 20:2. The same basic story of this struggle between RCA and the FCC is told in Lyons, *David Sarnoff*, pp. 214–20; Sobel, *RCA*, pp. 132–35; and Bilby, *The General*, pp. 132–138.

43. "Color Television Achieves Realism," *New York Times*, Aug. 30, 1940, p. 21:3; "CBS Tests Dr. P. Goldmark's Color Transmitting Process," *New York Times*, Sept. 5, 1940, p. 18:6; "Color Television Demonstration by CBS Engineers," *Electronics* 13 (Oct. 1940), 32–34, 73–74; "Color Television," *Communications* 21 (Feb. 1941), 20.

44. "Industry Accord over Television Standards Is Seen," *Broadcasting* 18 (Oct. 15, 1940), 102; "Industry Drafting Television Report," *Broadcasting* 19 (Jan. 20, 1941), 54; "NTSC Television Standards," *Communications* 21 (Feb. 1941), 12–13; "National Television Systems Committee Representatives Reports Revisions in Recommendations," *New York Times*, Mar. 21, 1941, p. 22:3; "FCC Authorizes Commercial Broadcasting as of July 1st," *New York Times*, May 3, 1941, p. 17:2. For the full story of the committee's work, see Donald G. Fink, *Television Standards and Practice* (New York: McGraw-Hill Book Company, 1943).

45. "RCA Plans World's Largest Radio Laboratory," p. 58.

46. "RCA Sees Commercial Television Retarded by Defense Program," *New York Times*, May 7, 1941, p. 27:3; "RCA Studies Future of Video but Sees Rather Dim Future," *Broadcasting* 19 (May 12, 1941), 92. Lyons gives the date of May 27, 1940; see *David Sarnoff*, p. 220. This figure is given in Sobel, *RCA*, p. 135. See also Bilby for the number of sets actually sold (*The General*, p. 138). Where did all of these receivers disappear to? There are no more than a dozen or so in collectors' hands today.

It appears that the Office of Production Management (OPM) had given radio (and television) the next to the lowest classification for procurement of basic metals. According to an article that appeared in *Broadcasting*, "There is no doubt about a bogging down of interest, notably in visual radio. . . . Television is regarded in many quarters as a forgotten industry for the duration" ("Defense Measures May Disrupt Industry," *Broadcasting* 19 [May 5, 1941], 11, 51).

47. "Novel Commercials in Video Debut," *Broadcasting* 19 (July 7, 1941), 10.

48. "Institute News and Radio Notes," *Proceedings of the Institute of Radio Engineers* 30 (Jan. 1942), 52.

49. "Novel Commercials in Video Debut," p. 10.

Chapter 11: "World War II"

1. For RCA's magnificent war effort see, Sobel, *RCA*, pp. 137–39; Lyons, *David Sarnoff*, pp. 232–35, quotations p. 232; and Bilby, *The General*, pp. 157, 140–41.

2. VKZ MS, pp. 103–4; FOIA/FBI -3, memo from SAC (special agent in charge) to Director, FBI (J. Edgar Hoover), Aug. 19, 1944. On Oct. 27, 1993, I finally received some 272 pages from the FBI under the Freedom

of Information Act. This was out of 1294 pages reviewed by the FBI. As 100-292257 is the file number of every document and is redundant, I will give only the identifying number. I am also hoping to receive some twenty-four documents from the State Department, which so far they have been unwilling to give me access to.

3. "I.R.E. People: Vladimir K. Zworykin," *Proceedings of the Institute of Radio Engineers* 35 (June 1947), 590; "Television's Outstanding Pioneer: Dr. Zworykin of RCA Laboratories," publicity release, RCA Department of Information (Oct. 1950); "Contributors: Vladimir K. Zworykin," *Proceedings of the Institute of Radio Engineers* 39 (July 1951), 837; "Vladimir K. Zworykin," *Current Biography,* pp. 654–56. See G. A. Morton and L. E. Flory, "An Infra-Red Image Tube and Its Military Applications," *RCA Review* 7 (Sept. 1946), 385–413.

4. V. K. Zworykin, "Flying Torpedo with an Electric Eye," *RCA Review* 7 (Sept. 1946), 293–302; Bilby, *The General,* p. 157.

5. David Sarnoff, "Introduction to Technical Papers on Airborne Television," *RCA Review* 7 (Sept. 1946), 291–92; G. M. K. Baker, "Military Television," *Television* 4 (1942–46), 351–56 (published by *RCA Review,* RCA Laboratories Division, Princeton, N.J., Jan. 1947). See also Charles J. Marshall and Leonard Katz, "Television Equipment for Guided Missiles," *Proceedings of the Institute of Radio Engineers* 34 (June 1946), 375–401; Henry E. Rhea, "Airborne Television, " *Broadcast News* no. 43 (June 1946), 24–41; and Jordan McQuay, "Television Takes to the Air," *Radio News* 29 (Feb. 1947), 57–59, 101–2.

6. "2-Way Television on Planes Studied," *New York Times,* Dec. 24, 1939, p. 11:3; "Television in an Airplane," *Radio-Craft* 11 (Jan. 1940), 29; "Television Pickup from an Airplane," *Communications* 20 (Mar. 1940), cover and p. 2; "RCA and NBC Telecast Program from Plane in N.Y.C.," *New York Times,* Mar. 17, 1940, p. 25:5; "First Telecast from Plane Successful as RCA Demonstrates New Equipment," *Broadcasting* 18 (Mar. 15, 1940), 24.

7. Bilby, *The General,* p. 157.

8. M. A. Trainer and W. J. Poch, "Television Equipment for Aircraft," *RCA Review* 7 (Dec. 1946), 469–502; R. E. Shelby, F. J. Somers, and L. R. Moffett, "Naval Airborne Television Reconnaissance System," *RCA Review* 7 (Sept. 1946), 303–37; R. D. Kell and G. C. Sziklai, "Miniature Airborne Television Equipment," *RCA Review* 7 (Sept. 1946), 338–57; Bilby, *The General,* p. 157.

9. VKZ MS, p. 103.

10. "New Electron Microscope," *Radio-Craft* 14 (Jan. 1943), 206; "V. K. Zworykin Reports on Electron Microscope Possibilities," *New York Times,* Feb. 6, 1944, sec. 4, p. 9:5.

11. VKZ MS, p. 104.

12. VKZ MS, p. 104; FOIA/FBI -2, SAC (New York) to Hoover, July 11, 1944; FOIA/FBI -4, SAC (New York) to Hoover, Sept. 12, 1944. See Herbert Mitgang, *Dangerous Dossiers* (New York: Donald Fine, 1989), pp. 146–47; Athan G. Theoharis and John Stuart Cox, *The Boss: J. Edgar Hoover and*

the Great American Inquisition (Philadelphia: Temple University Press, 1988), pp. 14–15; and Don Whitehead, *The FBI Story: A Report to the People* (New York: Random House, 1956), pp. 270–71.

13. VKZ MS, p. 86; "Browder J. Thompson, 1903–1944," *Proceedings of the Institute of Radio Engineers* 33 (Feb. 1945), 71; "Tribute to Browder J. Thompson by the War Department," *Proceedings of the Institute of Radio Engineers* 33 (May 1945), 336. According to "Institute Notes and Radio News" (*Proceedings of the Institue of Radio Engineers* 19 [Apr. 1931], 510), Thompson did not come to RCA Victor until Mar.–Apr. 1931. This is probably when he took over as head of electrical research.

14. Alan Dower Blumlein (1903–42) was considered the genius who made the EMI/Shoenberg television system possible. He has been called the greatest circuit designer and one of the best electronics engineers Britain ever produced. His work on sound recording and television is legendary. See M. G. Scroggie, "The Genius of A. D. Blumlein," *Wireless World* 66 (Sept. 1960), 451–56; B. J. Benzimra, "A. D. Blumlein—an Electronics Genius," *Electronics and Power* 13 (June 1967), 218–24; and W. A. Atherton, "Pioneers, 14: Alan Dower Blumlein (1903–1942) the Edison of Electronics," *Electronics and Wireless World* 94 (Feb. 1988), 184–86.

15. VKZ MS, p. 84. For details of Holweck's life, see Georges Petitjean, "Fernand Holweck sa vie, ses travaux scientifiques," *Bulletin de l'association des Annis des P.T.T. d'Alsace* 30 (1984), 38–64; and Charles Fabry, "Fernand Holweck (1890–1941)" *Cahiers de Physique* 8 (June 1942), 1–16.

16. For details of "Magic City" during the Nazi Occupation, see Goebel, "Das Fernsehen in Deutschland," pp. 380–82. See also P. Hémardinquer, "Television in France," *Electronic Engineering* 17 (Sept. 1945), 692–93; "Television in France," *Journal of the Television Society* 4 (1944–46), 96, 224–25, 302. One result of this collaboration was that Barthélemy was able to demonstrate 1000-line pictures right after the war.

I visited the French Television Center in Paris in 1971 as I was quite interested in seeing how television was done in Europe. I had written them of my visit and on the appointed day was picked up at my hotel and transported to the center. It was not as modern as "Television City" in Hollywood, where I was working at the time, but I was impressed with the color pictures from the 625-line SECAM system that they were using. Sometime during my visit, I asked my guide why they had named this studio after René Barthélemy. The answer quite simply was that Barthélemy was one of France's pioneer television inventors and deserved that honor. I unwittingly reminded my hostess that Barthélemy had collaborated with the Nazis during World War II at "Magic City." She was quite offended and asked me who I would have chosen. I quickly replied that Fernand Holweck was a more important television pioneer and, with Belin, had made significant contributions to cathode-ray television. I also explained that he had been in the French Resistance and had been captured and shot by the Nazis. Surely, Holweck was a patriot who had earned the honor of having the center named after him.

With this, my hostess ushered my wife and me out of the center and coldly informed me that they had called for a taxi and that I would have to find my way home without their assistance.

17. Sobol, *RCA*, pp. 141–42; Lyons, *David Sarnoff,* pp. 247–68; Bilby, *The General,* pp. 142–52.

18. "V. K. Zworykin Gets Television Broadcasters Assn. Engineering Award," *New York Times,* Dec. 12, 1944, p. 25:6; "Post-War Competition Stressed at Television Conference Here," *New York Times,* Dec. 13, 1944, p. 25:8; "Radio-Electronic and Monthly Review," *Radio-Craft* 16 (Feb. 1945), 270; "Television Awards," *Proceedings of the Institute of Radio Engineers* 33 (Feb. 1945), 127. There were sixteen men honored in all; see Lyons, *David Sarnoff,* pp. 268–69.

19. Michael H. Gorn, *The Universal Man: Theodore von Kármán's Life in Aeronautics* (Washington, D.C.: Smithsonian Institution Press, 1992), pp. 98–104. This book sheds new light on Zworykin's role with von Kármán's Air Force committee.

20. VKZ MS, p. 103. This information about Zworykin's appointment and service with Theodore von Kármán in the office of the assistant chief of Air Staff Operations, Commitments and Requirements came from the National Archives and the Record Services of the Office of Personnel Management and is contained in about a dozen documents which were kindly furnished to me under the Freedom of Information Act by the St. Louis office of the Office of Personnel Management on Jan. 11, 1991. His title was Expert Consultant to the Commanding General, Headquarters, Army Air Forces on scientific matters. Zworykin's appointment is found in a letter from the War Dept. Headquarters of the Army Air Forces, Washington, D.C., to the Assistant Chief of Air Staff, Operations, Commitments and Requirements, Oct. 26, 1944. It was confirmed in a letter from Donald Wilson, Brigadier General, U.S. Army, Asst. Chief of Air Staff Operations, Commitments and Requirements to AC/AS Personnel, Civilian Division, Nov. 9, 1944. The appointment was to terminate on June 30, 1945. Zworykin's oath of office was dated Dec. 21, 1944. This material will be known as FOIA/Nat. Archives.

21. FOIA/FBI -4, incomplete report titled Dr. Vladimir Kosma Zworykin, no author given, Sept. 12, 1944.

22. FOIA/FBI -2, SAC (New York) to Hoover, July 11, 1944. After all of my research into Zworykin's war record, it appears that what he knew of the DSM project is the crux of the matter. While there is every indication that he knew of the project per se, there is no proof that he really knew what was being built there. Although there are several indications that he may have had the opportunity to find out about it, there is no hard evidence that he really had knowledge of the bomb project. This of course is verified in the many FBI and army intelligence reports that absolutely cleared him of this knowledge. See also FOIA/FBI -170, SAC, Newark, to Director, FBI, office memorandum, Aug. 20, 1945, about translations of part 41, entitled "Zworykin Series" (this is but one of the many Sound Scriber disks made surreptitiously in Zworykin's home). Amazingly enough, when Zworykin was asked

by Plevitskaya ("his paramour") about the atomic bomb after Truman announced it on the radio, he admitted that "RCA had contributed and manufactured a very tiny, tiny insignificant part connected with the microscope." If he knew more, he certainly kept it to himself.

At any rate, it now appears that Zworykin may have been aware of the enormous plants being erected and the need for, say, the RCA electron microscope to be used for analytical purposes, without actually knowing any details of the atomic bomb project. It is common knowledge that there were literally thousands of people employed on the project at the plants themselves who never had the foggiest idea of what was being produced there. Since the FBI was never sure of what Zworykin knew, this explains why they spent such an enormous effort to keep him in the United States. Every piece of hard information from the FBI and elsewhere indicates that (1) the Manhattan Engineer District which was running the atomic bomb project was not really interested in him and (2) the only real charge against him was his avowed membership in two pro-Soviet organizations during the war, which was not illegal.

23. FOIA/FBI -3, SAC, Newark, to Director, FBI, office memorandum, Aug 19, 1944.

24. FOIA/FBI -5, incomplete document entitled "Dr. Vladimir Kosma Zworykin," no author or recepient given, Oct. 27, 1944.

25. FOIA/FBI -5, "NK 100-28390," date blanked out.

26. FOIA/FBI -9, Hoover to Communications Section, date blanked out (but probably before Nov. 1944).

27. FOIA/FBI -14, D. M. Ladd to Director, office memorandum, Nov. 15, 1944.

28. FOIA/FBI -11, SAC, Newark, to Director, FBI, office memorandum, Nov. 18, 1944.

29. FOIA/War Dept. -18, Carl A. Holloman, CID agent, to Officer in Charge, Clinton Engineer Works, Oak Ridge, Tenn., memorandum, Nov. 21, 1944. I received this document from the Department of Army Intelligence on Nov. 18. 1993. Oak Ridge was the home of K25, Y12, and X10, where atomic bomb materials were being processed. The memorandum stated that Zworykin was in contact with a certain Lloyd P. Smith, who had worked on the DSM project since Sept. 1941 and later returned to the RCA laboratories and "is presently employed there. He [the informant] believed that Smith was a good friend of Dr. Zworykin and was working in the same laboratory with him. He advised that Smith does know the entire process of the project and could, for that reason, furnish any information." Another informant stated that Lloyd Smith, "formerly employed at Berkeley, California, on work on the project is now connected with the R.C.A. Laboratories." Another informant was quoted as saying that he believed that "Smith was a loyal American and did not believe that he had furnished any information about the subject to any unauthorized personnel." The report continued, "in Nov. 1943, 3 employees of RCA were brought to the Clinton Laboratories to install some equipment." Finally, the report said that someone whose

name has been blanked out stated that "he recalled while taking a course at the Metallurgical Laboratory, University of Chicago in the Summer of 1943, various lectures were made as to how television could be used in connection with work of the project. He believed that the talks on television were given by a representative of the RCA Laboratories." Finally Holloman concluded that "all the employees, interviewed individually in this office with possible contacts made with Dr. Zworykin or any representatives of RCA Laboratories with negative results." See also FOIA/FBI -18, SAC, Newark, to J. E. Hoover, Dec. 20, 1944, for an inquiry about Smith and Zworykin; FOIA/ War Dept. -72, Lt. Col. John Lansdale, Jr., to L. Whitson, FBI, Feb. 28, 1945. Lansdale was the head of the Manhattan Project security service.

30. FOIA/FBI -18, SAC, Newark, to Hoover, FBI, Dec. 20, 1944.

31. FOIA/FBI -11, Hoover to Attorney General, memorandum, Nov. 27, 1944; FOIA/FBI -23, Hoover to Communications Section (requesting that they transmit the following message to SAC, Newark), memorandum, date blanked out. The date was confirmed in FOIA/FBI -52, SAC, Newark, to all Bureaus, ten pages of an unsigned report (many pages were left out), Feb. 3, 1945.

32. FOIA/FBI -12, S. K. McKee (SAC) to Director, FBI, attention: Censorship Unit, Nov. 30, 1944. I didn't know this existed.

33. FOIA/FBI -12, SAC, Newark, to Hoover, Jan. 9, 1945.

34. FOIA/FBI -30, McKee (Communications Section AC), Newark, to Director, teletype, date blanked out.

35. FOIA/FBI -25, Hoover, Director, to Attorney General, memorandum, Dec. 11, 1944.

36. FOIA/FBI -23, Hoover to Communications Section (requesting that they transmit the following message to SAC, Newark), memorandum, date blanked out.

37. FOIA/State Dept., Guy Hottel, FBI Special Agent, to Mrs. Shipley, Chief of Passport Division, Dec. 19, 1944; FOIA/War Dept., Oath of Office of Vladimir K. Zworykin, Dec. 21, 1944.

38. FOIA/FBI -27, J. C. Strickland to D. M. Ladd, Assistant Director, FBI, memorandum, Dec. 27, 1944.

39. FOIA/FBI -64, J. C. Strickland to Ladd, memorandum, Jan. 3, 1945.

40. FOIA/FBI -45, report from [name blanked out] entitled "Vladimir Kosma Zworykin," Jan. 29, 1945; FOIA/FBI-52, SAC, Newark, to all Bureaus, unsigned ten-page report, Feb. 3, 1945.

41. The story of von Kármán's visit to Germany in May 1945 is told in Theodore von Kármán with Lee Edson, *The Wind and Beyond: Theodore von Kármán: Pioneer in Aviation and Pathfinder in Space* (Boston: Little, Brown and Co., 1967), pp. 272–83; Gorn, *Universal Man;* and the Theodore von Kármán archives at the California Institute of Technology, which I was able to inspect in Dec. 1991. Material from this archive will be referred to as the Kármán File. The list of men to be interviewed came from Kármán File, Doc. 91.1, entitled "Doctor von Kármán's Notes for Trip on Personnel," dated 1945.

42. Colonel Boris T. Pash, *The Alsos Mission* (New York: Charter Books, 1980); Gorn, *Universal Man,* p. 105.

43. FOIA/FBI -52, SAC, Newark, to all Bureaus, ten pages of an unsigned report, Feb. 3, 1945, p. 24.

44. FOIA/FBI -52, SAC, Newark, to all Bureaus, ten pages of an unsigned report, Feb. 3, 1945, p. 2.

45. FOIA/FBI -52, SAC, Newark, to all Bureaus, ten pages of an unsigned report, Feb. 3, 1945, p. 1.

46. FOIA/FBI -62, J. Edgar Hoover to Lt. Col. John Lansdale, Jr., U.S. Engineers Office, Manhattan District, Feb. 19, 1945.

47. FOIA/FBI -73, report of special agent [name blanked out], Mar. 6, 1945.

48. FOIA/FBI -84, report by [name blanked out], Mar. 22, 1945.

49. FOIA/FBI -74, SAC, Washington, to J. Edgar Hoover, Director, memorandum, Mar. 24, 1945.

50. FOIA/FBI -88, S. K. McKee (SAC) to Director, FBI, attention: technical section, Mar. 16, 1945.

51. FOIA/FBI -105, one page of a document entitled "IV. Miscellaneous Activities of Subject." One paragraph is left which mentions Mar. 26, 1945, but the document itself is undated.

52. FOIA/FBI -93, Hoover, FBI, to SAC, Newark, memorandum, Mar. 27, 1945.

53. The reports of these recordings are too numerous to mention. See, for example, FOIA/FBI -56, Feb. 6, 1945; -56, Feb. 22, 1945; -66, Jan. 29, 1945; -67, Feb. 8, 1945; -68, Feb. 22, 1945; -81, Mar. 2, 1945; -81, Mar. 12, 1945; -91, Mar. 8, 1945; -96, Mar. 20, 1945; etc.

54. FOIA/FBI -109, report from [name blanked out] but approved and forwarded by Guy Hottel (SAC), Apr. 13, 1945.

55. FOIA/FBI -111, S. K. McKee (SAC) to Director, FBI, Apr. 10, 1945.

56. FOIA/FBI -131, report by [name blanked out], May 17, 1945.

57. Kármán File, Captain Joseph T. Wirth, Air Corps, to Chief of Air Staff, Counter Intelligence Div., Station Intelligence Branch, routing sheet, subject: top secret clearance, Apr. 11, 1945.

58. FOIA/FBI -97, J. Edgar Hoover to Lt. Col. John Lansdale, Jr., U.S. Engineers Office, Manhattan District, Apr. 18, 1945.

59. FOIA/War Dept. PO52, War Dept. Foreign Duty Unit to Mrs. R. B. Shipley, Chief, Passport Division, request for a special passport, Apr. 19, 1945. FOIA/Nat. Archives, Maj. Frederick M. Roberts, Air Corps, Civilian Personnel Division, Office of Asst. Chief of Air Staff Personnel, to the Office of the Secretary of War, Civilian Personnel Division, classification review of request for extension of appointment of special consultant Vladimir K. Zworykin, May 1, 1945.

60. FOIA/FBI -120, FBI teletype from Newark to Director, Washington, no date.

61. FOIA/FBI -121, J. C. Strickland to D. M. Ladd, memorandum, Apr. 23, 1945.

62. Kármán File, Doc. 02.05, containing only the list of names. There is no other information on the page. For details of the trip, see Gorn, *Univer-*

sal Man, pp. 106–7. Schairer took Zworykin's place on the trip abroad. Theodore von Kármán went abroad in the uniform of a full major general. One of the mysteries in the Theodore von Kármán files is the name of Professor Melville Jones. I cannot find such a person elsewhere. I presume they mean Professor R. V. Jones, who played an enormous role in the British Scientific Intelligence Service between 1939 and 1945. See R. V. Jones, *The Wizard War: British Scientific Intelligence 1939–1945* (New York: Coward, McCann & Geoghegan, 1978).

63. Gorn, *Universal Man*, pp. 106–7.

64. FOIA/FBI -116, teletype from Newark to Director, Washington, Apr. 27, 1945.

65. FOIA/FBI -116, Hoover to SAC, Newark, teletype, Apr. 28, 1945.

66. FOIA/FBI -118, J. C. Strickland to Ladd, memorandum, Apr. 28, 1945, including a blind memorandum setting forth more information concerning Zworykin.

67. FOIA/State Dept. PO47, office memorandum by R. B. Shipley, May 16, 1945; FOIA/FBI -137, Strickland to Ladd, May 19, 1945; FOIA/FBI -155, part of file no. 100-28390 entitled "Proposed Trip to Russia," no author, no date.

68. FOIA/FBI -141, J. K. Mumford to D. M. Ladd, Apr. 28, 1945.

69. VKZ MS, pp. 103–4; Hillier to author, Dec. 18, 1991. FOIA/FBI -139, Strickland to Ladd, May 2, 1945.

70. FOIA/FBI -119, McKee to Director SAC, Washington, teletype, marked urgent, Apr. 30, 1945.

71. FOIA/FBI -129, S. K. McKee (SAC) to Director, FBI, May 14, 1945.

72. FOIA/FBI -122, S. S. Alden to Ladd, [title blanked out], May 1, 1945.

73. FOIA/Nat. Archives, Maj. Frederick M. Roberts, Air Corps, Civilian Personnel Division, Office of Asst. Chief or Air Staff Personnel, to Office of the Secretary of War, Personnel Division, classification review of request for extension of appointment of special consultant Vladimir K. Zworykin, May 1, 1945.

74. FOIA/State Dept. PO47, office memorandum by R. B. Shipley, May 16, 1945.

75. VKZ MS, pp. 103–4; Hillier to author, Dec. 18, 1991. FOIA/FBI -125, [name blanked out] to all Bureaus, report, May 9, 1945, p. 2.

76. FOIA/State Dept. PO47, office memorandum by R. B. Shipley, May 16, 1945.

77. FOIA/FBI -115, J. Edgar Hoover, FBI, to Lt. Col. John Lansdale, Jr., U.S. Engineers Office, Manhattan District, May 2, 1945.

78. FOIA/FBI -138, S. S. Alden to D. M. Ladd, May 12, 1945. See also von Kármán, *Wind and Beyond*, p. 269.

79. FOIA/FBI -137, J. C. Strickland to Ladd, [title blanked out], May 19, 1945.

80. FOIA/FBI -144, S. S. Alden to Ladd, memorandum, May 29, 1945.

81. FOIA/FBI -145, [name blanked out] to Director, FBI, June 1, 1945. See also FOIA/FBI -185, p. 2 of a report blanked out except for the information given above. Followed by a different paper dated Jan. 28, 1946.

82. FOIA/FBI -144, John Edgar Hoover, FBI, to SAC, Newark, June 4, 1945.

83. VKZ MS, p. 104.

84. Kármán File, Report of Panel No. 3 of the Scientific Advisory Board on Conclusions Reached during the Meeting at Wright Field—June 18th to 21st, 1946 (June 26, 1946). See group picture in Gorn, *Universal Man*, p. 82. On Sarnoff's relationship with J. Edgar Hoover, see Bilby, *The General*, pp. 158–59, 162; Dreher, *Sarnoff*, pp. 216–23; and Lyons, *David Sarnoff*, pp. 324–32.

85. FOIA/State Dept. PO26, passport application, Oct. 16, 1945.

86. FOIA/FBI -188, letter originating Newark, Nov. 30, 1945.

87. FOIA/FBI -194, S. K. McKee to Director, FBI, marked personal and confidential, Nov. 28, 1945.

88. FOIA/FBI -192, SAC, Newark, to Director, FBI, report, Apr. 11, 1946.

89. FOIA/FBI -206, SAC, Newark, to Director, FBI, memorandum, Sept. 17, 1946; FOIA/State Dept. PO53, Guy Hottel, FBI, to R. B. Shipley, Chief, Passport Division, Department of State, Nov. 1, 1946. See also FOIA/State Dept. PO12, sworn affidavit of Zworykin, Mar. 1, 1954.

Exactly what the Zworykin "matter" was has never been defined. It certainly was triggered by Zworykin's letter to Lawrence and it certainly concerned the atomic bomb project. See FOIA/FBI -145, to Director, FBI, June 1, 1945. But, as has been amply proven in documents received from the FBI, State Department, and War Department, there was no conclusive evidence that Zworykin really knew details about the project or that, if he did, he ever tried to transmit information of it to the Soviet Union. Zworykin was never formally charged and never lost his top secret clearance.

As stated earlier, the FBI has some 742 pages that may never see the light of day. The State Department also has twenty-four documents that cannot be released to me. My only question is that after forty-eight years, what can be in the files that deals with national security and cannot be disclosed? As other researchers have stated, it is perfectly clear that *personal and policy considerations,* not security, dictate the withholding of this information.

90. Sobol, *RCA*, p. 137; Lyons, *David Sarnoff*, p. 269; and Bilby, *The General*, p. 172.

91. A. Rose, P. K. Weimer, and H. B. Law, "The Image-Orthicon—A Sensitive Television Pickup Tube," *Proceedings of the Institute of Radio Engineers* 34 (July 1946), 424–32 (paper presented Jan. 24, 1946); Shelby, Somers and Moffett, "Naval Airborne Television Reconnaissance System," p. 336. The success of this tube led to much friction between Rose and Zworykin. Many articles claimed that Zworykin had invented the image orthicon; see, for instance, the reference in "TV Turns Off Its Father" to "Dr. Zworykin who invented the image orthicon tube which became the standard tube" (p. 16:3). Other articles showed pictures of Zworykin holding the image orthicon quite fondly. While he did not actually claim to have developed it, he did not deny it either. This did not sit well with Rose, as I learned during our correspondence.

92. Paul K. Weimer, Harold B. Law, and Stanley V. Forgue, "Mimo-Miniature Image Orthicon," *RCA Review* 7 (Sept. 1946), 358–66. Obviously this was an unsuccessful attempt to produce a much smaller tube for commercial use after the war.

93. "New Visions for Tomorrow's World," RCA advertisement, *Electronics* 17 (May 1944), 313.

94. Asa Briggs, *Sound and Vision*, vol. 4 of *History of Broadcasting in the United Kingdom* (London: Oxford University Press, 1979), p. 185.

95. "New Supersensitive Tube, Developed by RCA Demonstrated at NBC Studio," *New York Times*, Oct. 26, 1945, p. 21:1; "Dim Lights No Bar to New Video Tube," *Broadcasting* 29 (Oct. 29, 1945), 94; "Radio-Electronics and Monthly Review: New Television Camera Tube," *Radio-Craft* 17 (Dec. 1945), 128; "NBC 1st General Use of New Image *Orthocon* with Powerful Lens System to Pick-up Army-Navy Game and Broadcast to 3 Cities Described," *New York Times*, Dec. 2, 1945, sec. 5, p. 3:7; R. E. Shelby and H. P. See, "Field Television," *RCA Review* 7 (Mar. 1946), 72–93; "90% Cut in Studio Light Needs Claimed for RCA's New Image Orthicon Camera," *Broadcasting* 33 (June 23, 1947), 19.

96. R. B. Janes, R. E. Johnson, and R. S. Moore, "Development and Performance of Television Camera Pickup Tubes," *RCA Review* 10 (June 1949), 191–223; R. B. Janes, R. E. Johnson, and R. R. Handel, "A New Image Orthicon," *RCA Review* 10 (Dec. 1949), 586–92.

97. H. Delaby, "The Paris Television Transmitting Center," *Journal of the Television Society* 4 (1945), 307–13; "Television in France," *Journal of the Television Society* 4 (1945), 224–25; Michel Adam, "Recent Developments in Television in the USA and France," *Génie Civil* 12 (Mar. 1945), 33–35; Michel Adam, "Une nouvelle caméra pour télévision," *La Nature* 3025 (Apr. 1, 1945), 105–6; Hémardinquer, "Television in France," pp. 692–93; R. Barthélemy, "The Isoscope: A Slow Electron Image Analyser," *L'Onde Électrique* 25 (Nov. 1945), 103–15. The front cover of the Jan. 1946 issue of *L'Onde Électrique* shows a picture of the camera with an electronic viewfinder. P. Mandel, "1015 Line Television Apparatus of the Compagnie des Compteur Montrouge," *L'Onde Électrique* 26 (1946), 26–37.

98. V. K. Zworykin, G. A. Morton, E. G. Ramberg, J. Hillier, and A. W. Vance, *Electron Optics and the Electron Microscope* (New York: John Wiley & Sons, 1945).

Chapter 12: "Ad Astra per Video"

1. "Electronics to Aid Weather Figuring," *New York Times*, Jan. 11, 1946, p. 12:5; "Vladimir Zworykin, the Man Who Was Sure That TV Would Work," *Electronic Design* 18 (Sept. 1, 1977), 115; Irving Wolff, "Vladimir K. Zworykin," *Proceedings of the Institute of Radio Engineers* 45 (Apr. 1957), 445–47; "Radio-Electronics and Monthly Review, Electronic Computers," *Radio-Craft* 18 (Apr. 1947), 19. I tried to find more information about Zworykin's activities with von Neumann and his computer. Note 91 in "John von Neumann, 1903–

1957" (*Bulletin of the American Mathematical Society* 64, no. 645 [May 1958], 46) cites "Preliminary Discussion of the Logical Design of an Electronic Computing Instrument," part 1, vol. 1, W. Burks and H. H. Goldstine, report prepared for U.S. Army Ord. Dept. under Contract W-36-034-ORD-7481, June 28, 1946, 2d ed. (Sept. 2, 1947), 42 pp. There is no mention of Zworykin. On Zworykin's role, see Norman Macrae, *John von Neumann* (New York: Pantheon Books, 1992), pp. 314–15. Von Neumann was the youngest member of the Institute for Advance Study at Princeton University when it was founded in 1933. To my knowledge, Zworykin was never a member of this institute.

2. Details of this project are found in *The Legacy of John von Neumann*, ed. James Glimm, John Impagliazzo, and Isadore Singer (Providence, R.I.: American Mathematical Society, 1990), pp. 304–5.

3. *Radio-Craft* 18 (June 1947), 19; V. K. Zworykin and L. E. Flory, "Reading Aid for the Blind," *Electronics* 19 (Aug. 1946), 84–87; V. K. Zworykin, L. E. Flory, and W. S. Pike, "Research on Reading Aids for the Blind," *Journal of the Franklin Institute* 247 (May 1949), 483–96; Dobriner, "Vladimir Zworykin," p. 115.

4. Ross, *Television Jubilee*, pp. 66–68; "Getting Ready at A. P.," *Journal of the Television Society* 4 (Feb. 1946), 223; Briggs, *Sound and Vision*, pp. 196–202. Information on Baird comes from Margaret Baird, *Television Baird*, p. 156.

Of all the personnel who reopened the Alexandra Palace in 1946, I have had the pleasure of knowing Thornton (Tony) Howard Bridgewater, who has had an amazing career in television. Born on June 1, 1908, he trained at the London Telegraphic Training College and Regent Street Polytechnic. After school he was a seagoing wireless operator. In 1928 he went to work for the Baird Television Development Company in operations and general management at the Long Acre Studio. When the BBC took over the Baird transmissions in 1932, he joined their staff with Desmond R. Campbell. He was with the BBC when they moved to the Alexandra Palace in 1936, where he was senior studio and senior maintenance engineer. With the start of the war, he was responsible for the maintenance of radar stations and navigational systems, with the rank of squadron leader. After the war he rejoined the BBC as engineer in charge of television outside broadcasts. In 1962 he was chief engineer of television. He has been honored with the Coronation Medal and the O.B.E. (Order of the British Empire). In 1993 he was given an honorary doctorate by the University of Bradford, Bradford, England.

Bridgewater is the ranking British television historian, having written dozens of articles and the section on television for the fifteenth edition of the *Encyclopedia Brittanica* (1990). We have corresponded for many years and he is the source of my insights into the development of British television.

5. "Radio-Electronics and Monthly Review, Vladimir K. Zworykin," *Radio-Craft* 18 (May 1947), 181; "Zworykin Biography," *Proceedings of the Institute of Radio Engineers* 39 (July 1951), 837.

6. R. D. Kell, G. L. Fredendall, A. C. Schroeder, and R. C. Webb, "An

Experimental Color Television System," *RCA Review* 7 (June 1946), 141–54; "Radio-Electronics and Monthly Review, Color Television," *Radio-Craft* 18 (June 1947), 19. The system demonstrated by RCA was not compatible with the existing NTSC system then in use in the U.S. RCA tried for the next three years to fit it into the existing standards. This they did admirably.

7. FOIA/State Dept. PO19, "First Demonstration of American Television on Continent of Europe Planned by RCA," RCA press release, June 9, 1947; VKZ MS, p. 104. See also FOIA/FBI -208, K. G. Fitch to D. M. Ladd, Apr. 11, 1947.

8. Information on Zworykin's passport application came from FOIA/State Dept. PO36, Raymond L. Zwemer, National Academy of Sciences, to Ruth Shipley, Passport Division, May 23, 1947. Enclosed was a copy of a letter to Dr. Warren Kelchner, Division of International Conferences.

9. "Inventor of Television Aid Wins Poor Richard Prize," *New York Times,* Oct. 30, 1948, p. 32:4; on the presidential certificate and French Legion of Honor, see Zworykin's RCA Executive Bio., dated Sept. 1969.

10. "Lamme Medal Awarded to Expert in Television," *New York Times,* Mar. 26, 1949, p. 28:7; Nelson S. Hibshman, "Vladimir Kosma Zworykin: Lamme Medalist for 1948," *Electrical Engineering* 68 (Aug. 1949), 668–70; Vladimir K. Zworykin, "The Engineer's Role in the Program of Science," *Electrical Engineering* 68 (Aug. 1949), 670–72.

11. V. K. Zworykin and E. G. Ramberg, *Photoelectricity and Its Applications* (New York: John Wiley & Sons, 1949).

12. FOIA/FBI -233, SAC, Newark, to Director, FBI, report, July 13, 1949.

13. P. K. Weimer, U.S. Pat. no. 2,654,853 (filed Feb. 28, 1949; issued Oct. 6, 1953).

14. P. K. Weimer, S. V. Forgue, and R. R. Goodrich, "The Vidicon—A New Photo-conductive Television Pickup Tube," *Electronics* 23 (May 1950), 70–73. See also P. K. Weimer, S. V. Forgue, and R. R. Goodrich, "The Vidicon-Photoconductive Camera Tube," *RCA Review* 12 (1951), 306–13; and P. K. Weimer, "Photo-conductivity in Amorphous Selenium," *Physical Review* 79 (July 1, 1950), 697.

15. "Three Dimensions Achieved in Video," *New York Times,* Apr. 16, 1950, sec. 1, p. 82:3.

16. "I.R.E. Awards," *Proceedings of the Institute of Radio Engineering* 39 (May 1951), 570; V. K. Zworykin, "1951 IRE Medal of Honor—Speech of Acceptance," *Proceedings of the Institute of Radio Engineers* 39 (July 1951), 740.

17. Record of Superior Court of New Jersey Chancery Division, Nov. 13, 1951, final judgement (divorce).

18. Frank Lovece, "Zworykin v. Farnsworth, Part II," *Video* 9 (Sept. 1985), 98. The date of Zworykin's second marriage is given in FOIA/State Dept. PO06, Zworykin passport application, Mar. 15, 1962. Boris Polevitsky's death was noted in FOIA/FBI -259, SAC, WFO, to Director, FBI, memorandum, July 26, 1962.

19. At present no one seems to have written a book focused on the color television struggle between RCA and CBS. The subject is covered quite well

in Bilby, *The General*, pp. 171–205; for the RCA viewpoint, see Lyon, *David Sarnoff*, pp. 284–310, and for that of CBS, see Peter C. Goldmark, *Maverick Inventor: My Turbulent Years at CBS* (New York: Saturday Review Press, E. P. Dutton & Co., 1973), pp. 84–124.

20. "FCC, 5-2 Authorizes CBS to Begin Color Broadcasts November 20," *New York Times*, Oct. 12, 1950, p. 1:6; "Director Wilson Asks CBS to Suspend Plans for Mass Output of Color Sets because of Materials Shortage," *New York Times*, Oct. 20, 1951, p. 1:2.

21. The work of the second NTSC is best described in "Second Color Television Issue," *Proceedings of the Institute of Radio Engineers* 42 (Jan. 1954). For the technical details of the successful tricolor shadow mask tube, see "Petition of Radio Corporation of American and National Broadcasting Company, Inc. for Approval of Color Standards for the RCA Color System," before the Federal Communications System, June 25, 1952, pp. 261–303. It is also detailed in the *RCA Review* 12 (Sept. 1951), 443–645. The entire issue deals with the efforts of RCA to create a successful tricolor picture tube.

22. This is from FOIA/State Dept. PO12, Robert L. Werner, Vice President and General Attorney, Radio Corporation of America, to Ashley J. Nicholas, Assistant Director, Passport Office, Mar. 1, 1954. Zworykin's sworn affidavit is included in this document. See also FOIA/State Dept. PO11, Zworykin passport application, Mar. 1, 1954.

23. "Chinese Jet Expert Held as Alien Red," *New York Times*, Sept. 18, 1950, p. 12:4; "Peiping May Orbit Satellite by 1962," *New York Times*, Jan. 11, 1960, p. 12:3. See Gorn, *Universal Man*, p. 102.

24. Details of Zworykin's retirement party were given in a booklet entitled "Proceedings of the Seminar: Thirty Years of Progress in Science and Technology," convened at McCosh Hall, Princeton University, Sept. 18, 1954. About von Kármán's absence, see Kármán File, von Kármán to E. W. Engstrom, Aug. 12, 1954.

25. On Zworykin's "deception" of Sarnoff, see Dobriner, "Vladimir Zworykin," p. 114. Elmer Engstrom received an honorary degree of Doctor of Science from New York University in June 1949 (Engstrom's RCA Executive Bio., dated June 1981). "Engstrom Replaces Retiring Burns at RCA," *Broadcasting* 61 (Dec. 4, 1961), 76–77.

26. FOIA/FBI -233, A. H. Belmont to L. V. Boardman, Sept. 27, 1956; FOIA/FBI -232, Director, FBI, to SAC, Newark, Sept. 28, 1956; FOIA/FBI -235, administrative page, no date; FOIA/FBI -236, SAC, Newark, to Director, FBI, Dec. 6, 1956; FOIA/FBI -247, SAC, Newark, to Director, FBI, Jan. 23, 1959; FOIA/FBI -256, SAC, Newark, to ?, Mar. 24, 1961. The last FBI paper in my file is an answer (author unknown) to a request from White House security for information about Zworykin (FOIA/FBI -276, Sept. 19, 1975).

27. V. K. Zworykin, E. G. Ramberg, and L. E. Flory, *Television in Science and Industry* (New York: John Wiley & Sons, 1958).

28. "Election of Dr. V. K. Zworykin to Honorary Membership," *Journal of the British Institute of Radio Engineers* (Sept. 1959), 528–32. According to

FOIA/State Dept. PO08 (passport application) he arrived in England on June 9, 1959.

29. On Zworykin's visit to Moscow, see Joseph Roizen, "Vladimir Zworykin—A Personal Note," *Journal of the Society of Motion Picture and Television Engineers* 90 (July 1981), 578. I was also told about it in an interview with Roizen at his home in Palo Alto in 1986. According to FOIA/State Dept. PO08, Zworykin was in the USSR from July 6 through Aug. 4, 1959. Again, there is absolutely no mention of Zworykin's presence in Moscow in any of the American or British newspapers and magazines of the day. It was later mentioned in W. A. Atherton's "Vladimir Kosma Zworykin (1889–1982): Catalyst of Television," *Electronics & Wireless World* 93 (Oct. 1987), 1019–20; see also Novakosky, "100th Birthday of V. K. Zworykin," p. 68. Yet the obituary for Waldemar J. Poch (*Journal of the Society of Motion Picture and Television Engineers* 94 [Jan. 1985], 61–62) says that "Poch returned to Moscow in 1959 as a participant in the American Exhibition there." There is no mention of the fact that Zworykin was also there. For details of the "kitchen debate" see "Videotape Conquest," *Broadcasting* 57 (Aug. 3, 1959), 104; "U.S. Television Network Tapes Moscow Debate," *New York Times*, July 25, 1959, p. 3:3; "Nixon Opens American Exhibition in Moscow July 23, 1959," *Department of State Bulletin* 41:227 (Aug. 17, 1959).

30. "Montreux Television Symposium," *Wireless World* 66 (June 1962), 256–57. While Yagi's work is known worldwide, it is of interest that there was no mention of Kenjiro Takayanagi.

31. FOIA/FBI -262, SAC, Newark, to Director, FBI, Aug. 21, 1962.

32. V. K. Zworykin, "Magic Eyes for Medicine," *Saturday Evening Post* 230 (May 31, 1958), 26–27; "Vladimir K. Zworykin Honorary Vice President and Technical Consultant," Zworykin's RCA Executive Bio., dated Sept. 1969; Wolff, "Vladimir K. Zworykin," p. 446.

33. "11 Science Medals Given by Johnson," *New York Times*, Feb. 7, 1967, p. 18:3.

34. "TV Pioneer Gets Medal" *New York Times*, Feb. 22, 1968, p. 35:2.

35. See "80th Birthday Testimonial for Dr. Vladimir Kosma Zworykin," dinner and reception at the Princeton Inn Wednesday, Sept. 10, 1969. This came from a booklet entitled "Vladimir Kosma Zworykin, His 80th Year" (copy in author's collection); "Zworykin Award Established by RCA," *Broadcasting* 77 (Sept. 19, 1969).

36. "David Sarnoff Dies at 80," *New York Times*, Dec. 13, 1971, pp. 1:1, 43:3; "David Sarnoff," *New York Times*, Dec. 14, 1971, p. 44:2; "David Sarnoff," *Newsweek* 78 (Dec. 27, 1971), 47.

37. This came from the National Academy of Engineering of the United States of America. See James Hillier, "Memorial Tribute Vladimir K. Zworykin (1889–1982)," Oct. 6, 1982. For other personal references to Zworykin, see also Wolff, "Vladimir K. Zworykin," p. 447, and Roizen, "Vladimir Zworykin," p. 578.

I only met with Zworykin four times. I found him charming and friendly. He seemed anxious to please me. There were two problems—he was rather deaf and he spoke with a guttural Russian accent that made him hard to understand. I don't feel that I am qualified to judge him personally on the basis of our brief interviews. The only negative response I heard from his co-workers, excepting Arthur Vance, was to his habit of putting his name on every project that came out of his laboratory. This of course was a form of flattery. Adding his prestigious name insured attention to the project and gave his co-workers greater status by their association with him. As Hillier pointed out, in a way this inspired them to perform at levels far beyond their own expectations. And it was a two-way process—Zworykin got his best results in conjunction with a competent co-worker. As noted, he had a knack of picking people who were able to understand, execute, and finish the most difficult projects. In this he is not unlike Thomas Edison, who had a complete staff who did nothing but work on his ideas.

I did come across one negative observation of Zworykin from a former executive at RCA whom I presume never worked with him personally. He stated that Zworykin spent the remainder of his career in the glow of/and under Sarnoff's protection. According to Andrew F. Inglis, "He served as a 'gadfly'—not always welcome but always heeded because of his relationship with Sarnoff—to the company's technical community" (*Behind the Tube: A History of Broadcasting Technology and Business* [Boston: Focal Press, 1990], p. 172). According to the book jacket, Inglis was vice president and general manager of RCA's broadcast equipment division. Unfortunately, he seems to have done no research into the history of television and comes up with all the old war-horses that are still fouling up the subject. If there was one thing that Zworykin was not, it was a gadfly in the manner suggested by Inglis. He may have provoked and possibly annoyed his workers but it was always done in their best interests. According to Hillier, Zworykin was notorious for his daily visits to the laboratories. "His suggestions, his needling, his discussions, his proddings, while at times, made us uncomfortable, and at times we even rebelled, the fact is, we ended up being better engineers and scientists and, on occasion, surprised even ourselves with our accomplishments" ("Remarks Zworykin's Memorial Service").

38. "Dr. Vladimir K. Zworykin: 1889–1982," *Broadcast Management Engineering* 18 (Sept. 1982), 36–42; "Vladimir Zworykin, Father of Electronic Television Tells What He Likes Most about Television. The Turnoff Switch," *Television International* 24 (Feb. 1980), 14–17. Many details of Zworykin's visits to Florida were related to me in correspondence with Dr. Sidney Fox of the Institute for Molecular and Cellular Evolution, Miami, in Nov. 1988. I interviewed Fox in California in Dec. 1988. He told me that Zworykin was quite friendly and interested in many subjects that he [Fox] was involved in. For the "gelding" story, see Al Pinsky to author, Nov. 21, 1988; and Dobriner, "Vladimir Zworykin," p. 115.

39. R. Haitch, "TV Pioneer, 85, Finds Retirement Not to His Taste," *New York Times,* Nov. 17, 1974, p. 97:1.

40. "The National Inventors Hall of Fame," booklet, U.S. Dept of Commerce, Patent and Trademark Office, 1991, pp. 15–16.

41. "Dr. V. K. Zworykin, RCA TV Pioneer, Honored for Contributions to 'Fully Electronic' Television," RCA news release, Oct. 15, 1980.

42. "Dr. Vladimir K. Zworykin: 1889–1982," p. 42.

43. "TV Turns Off Its Father," p. 16:3.

44. See Lovece, "Zworykin v. Farnsworth, Part II," p. 138. While many of Lovece's details are correct, my research shows certain parts of his account are not quite accurate. For instance, Zworykin's ashes are not in either Katherine Zworykin's coffin or Tatiana's (she is still alive at the age of 102). See also Roizen, "Vladimir Zworykin," p. 578; and Robert McG. Thomas, Jr., "Vladimir Zworykin, Television Pioneer, Dies at 92," *New York Times*, Aug. 1, 1982, p. 32:1.

45. Lovece, "Zworykin v. Farnsworth, Part II," p. 138. Again contrary to Lovece, it appears that Zworykin's granddaughter was only carrying out his last wishes about scattering his ashes. This information comes from Zworykin's death certificate and a visit to the Princeton Cemetery. See also Hillier, "Remarks at Zworykin's Memorial Services," Aug. 3, 1982.

46. Interview with Donald and Iris Blackmore, in Dec. 1988, at Los Alamitos, Calif. Iris Blackmore is the granddaughter of Zworykin's second wife, Katherine. I learned from James Hillier that Zworykin's first wife, Tatiana, is still alive; Hillier to author, Sept. 6, 1993. Her birthdate is revealed in FOIA/FBI -180, [name blanked out, Philadelphia, to [name blanked out], (most of the report is also blanked out), Oct. 3, 1945. On the tribute to Zworykin in Russia, see N. D. Ustinov and V. P. Borisov, "Eminent Inventor and Scholar V. K. Zworykin (on the Occasion of His 100th Birthday)," *Photography BHET* (Moscow) (1990), 121–29. This article was given to Les Flory in 1990 when he was interviewed by Borisov. Flory gave it to me on my trip to Princeton in Aug. 1993.

47. "Revolutionary New TV Tape Process Unveiled by Ampex," *Broadcasting* 50 (Apr. 16, 1956), 72–74; H. F. Olson, W. D. Houghton, A. R. Morgan, J. Zenel, M. Artzt, J. G. Woodward, and J. T. Fischer, "A System for Recording and Reproducing Television Signals," *RCA Review* 15 (Mar. 1954), 3–17; "VERA: The BBC Vision Recording Apparatus," *Journal of the Television Society* 8 (Apr.–June 1958), 398–400; E. F. DeHaan and A. G. Van Doorn, "The Plumbicon, a Camera Tube with a Photo-conductive Lead Oxide Layer," *Journal of the Society of Motion Picture and Television Engineers* 73 (June 1964), 475. On the decline in sales of RCA color sets, see Sobel, *RCA*, p. 213. See also Margaret B. W. Graham, *RCA and the Video Disc: the Business of Research* (Cambridge: Cambridge University Press, 1966).

As this book is being written (July 1994) the American Ampex Corporation, which came out with the first successful video tape recorder in 1956, is barely defending itself against the larger, better-financed Japanese competition in video tape recording. It is down to one product, the Ampex DCT system. While a superb editing device, it has not found acceptance in the broadcast industry. Fortunately for Ampex, this machine is now finding extensive use in business, as it can store massive amounts of data, quickly and economically. Ampex and Zenith Electronics Corp. are the only major

American electronic companies left. While Ampex still produces all of its products in the United States, Zenith assembles most of its devices in Mexico. Even the giant N. V. Philips Industries of Eindhoven, Netherlands, is also in the throes of cutbacks, layoffs, and so forth.

48. The friendly (but deadly) takeover by General Electric is dealt with in Bilby, *The General,* pp. 295–316. See also Lessing's remark, "G. E. had been smarting against RCA ever since the anti-trust decree of 1932, when it felt that it had been maneuvered out of the radio business" (*Man of High Fidelity,* p. 193). Finally, according to Barnum, a senior vice president of GE stated that "as a GE executive, it has been gratifying to see the employees of Camden, adapt, respond to change, and to accept responsibility for their own future . . . recognizing that a profitable competitive business is the only real long run security for everyone" (*"His Master's Voice" in America,* p. 31). What security? Most of their jobs have long since vanished. How sad that GE's corporate policy has always put profits first, people last. This in contrast to RCA, where the individual came first (ask anyone who worked there under David Sarnoff) and the company profited through a work force that was second to none. See also, Grant Tinker and Bud Rukeyser, *Tinker in Television: From General Sarnoff to General Electric* (New York: Simon & Schuster, 1994). General Electric will never learn.

49. "Voyager 2's Amazing Journey to Outermost Planets Is Reviewed, as well as Plans for Future Unmanned Space Exploration," *New York Times,* Aug. 17, 1989, sec. 4, p. 1:3.

Index

France under Prof. Paul Langevin, 18; inducted into Russian army, 20; becomes an officer, 22; marries Tatiana Vasilieff, 22; goes to work for Russian Wireless and Telegraph Co., 31; escapes from Russia, 34–35; arrives in New York, 36; goes to work for Westinghouse, 42; goes to work for C and C, 43; returns to Westinghouse, 44; submits plans for all-electric television system, 44–45; becomes U.S. citizen, 49; starts work on television system, 49; demonstrates system to Westinghouse management, 50–51; first publicity, 53–54; gets Ph.D., 56; working on facsimile machine, 62; invents sensitive photocell, 62–63; working on Kerr cell for movies, 63; gets patent on CRT system, 70; leaves for Paris (1928), 70; visits Belin laboratories, 71; given demonstration of electrostatic tube, 72–74; arrives home from Paris, 75; meets with Sarnoff (Jan. 1929), 75–77; set up in his own laboratory at Westinghouse, 77; builds first successful picture tube, 80; demonstration to management, 80; builds first working camera tube with Ogloblinsky, 81–82; receives pictures at home, 83; reveals Kinescope to IRE, 84; applies for kinescope patent, 84; moves to Victor factory in Camden, 86–88; heads television group, 87; visits Farnsworth laboratories, 89–90; builds first planar camera tubes, 93–94; publishes first book, 94; sets up research television group, 96–97; loses patent rights to Kinescope, 101; develops Iconoscope, 105–7; joined by Engstrom, 115–16; loses patent interference 64,027 to Farnsworth, 118–20; reveals Iconoscope (June 1933), 124–26; visits London, Paris, Moscow, and Berlin (1933), 126–29; awarded Liebman Memorial Prize (1934), 133; meets Professor Takayanagi, 134; visits Soviet Union (1934), 137–41; visits Soviet Union (1935), 141–42; demonstrates new electron image tube, 146; visits Alexandra Palace (1936), 147; visits France, Berlin, and Hungary (1936), 147–48; describes projection kinescope, 153; hosts SMPE (1937), 154; granted patent on 1923 application, 157–58; awarded Doctor of Science degree, 158; describes work on electron microscope, 158; appears at RCA's television exhibit building prior to World's Fair opening, 160; celebrates fiftieth birthday at Taunton Lakes home (1939), 162; visits Palestine and Europe (1939), 162–64; publishes second book with George Morton (1940), 167; discloses commercial electron microscope, 167; named associate director of Princeton labs (1941), 169; awarded Mumford medals (1941), 170; moves to Princeton, 171; elected to National Academy of Science (1943), 173; becomes chairman of Russian War Relief (1943), 173–74; given award in engineering (1944), 176; appointed to von Kármán's staff by General Arnold (1944), 177; Zworykin sets off FBI investigation by writing Dr. Lawrence, 177; picked to go to Europe by von Kármán, 180; passport canceled, 184–85; trip finally canceled by

ALBERT ABRAMSON is a television historian and historical consultant, and the author of two books on the history of television, *Electronic Motion Pictures* and *The History of Television, 1880-1941,* as well as numerous articles. He is a life member of the Society of Motion Picture and Television Engineers, serving on their Board of Editors and Historical and Archival Committee. He is also a member of the British Kinematograph, Sound and Television Society, the Institute of Electrical and Electronics Engineers, the Comité d'Histoire de la Télévision, and the Academy of Television Arts and Sciences.

ALBERT ABRAMSON is a television historian and historical consultant, and the author of two books on the history of television, *Electronic Motion Pictures* and *The History of Television, 1880-1941*, as well as numerous articles. He is a life member of the Society of Motion Picture and Television Engineers, serving on their Board of Editors and Historical and Archival Committee. He is also a member of the British Kinematograph, Sound and Television Society, the Institute of Electrical and Electronics Engineers, the Comité d'Histoire de la Télévision, and the Academy of Television Arts and Sciences.